ROUTLEDGE LIBRARY EDITIONS:
POLLUTION, CLIMATE AND CHANGE

Volume 20

T0133944

THE CLIMATIC SCENE

THE CLIMATIC SCENE

Edited by
MICHAEL J. TOOLEY AND GILLIAN M. SHEAIL

Routledge
Taylor & Francis Group

LONDON AND NEW YORK

First published in 1985 by George Allen and Unwin (Publishers) Ltd

This edition first published in 2020
by Routledge
2 Park Square, Milton Park, Abingdon, Oxon OX14 4RN

and by Routledge
52 Vanderbilt Avenue, New York, NY 10017

Routledge is an imprint of the Taylor & Francis Group, an informa business

© 1985 M. J. Tooley and G. M. Sheail

British Library Cataloguing in Publication Data
A catalogue record for this book is available from the British Library

ISBN: 978-0-367-34494-8 (Set)
ISBN: 978-0-429-34741-2 (Set) (ebk)
ISBN: 978-0-367-36648-3 (Volume 20) (hbk)
ISBN: 978-0-367-36649-0 (Volume 20) (pbk)
ISBN: 978-0-429-34740-5 (Volume 20) (ebk)

Publisher's Note
The publisher has gone to great lengths to ensure the quality of this reprint but points out that some imperfections in the original copies may be apparent.

Disclaimer
The publisher has made every effort to trace copyright holders and would welcome correspondence from those they have been unable to trace.

——The——
Climatic Scene
Edited by
M.J. Tooley & G.M. Sheail

Department of Geography,   *formerly Department of Environmental*
University of Durham   *Sciences, University of Lancaster*

London
GEORGE ALLEN & UNWIN
Boston   Sydney

Frontispiece: Professor Gordon Manley: 3 January 1902–29 January 1980.

George Allen & Unwin (Publishers) Ltd,
40 Museum Street, London WC1A 1LU, UK

George Allen & Unwin (Publishers) Ltd,
Park Lane, Hemel Hempstead, Herts HP2 4TE, UK

Allen & Unwin Inc.,
Fifty Cross Street, Winchester, Mass. 01890, USA

George Allen & Unwin Australia Pty Ltd,
8 Napier Street, North Sydney, NSW 2060, Australia

First published in 1985

British Library Cataloguing in Publication Data

The Climatic scene: essays in honour of
Gordon Manley.
1. Climatic changes
I. Tooley, M. J.   II. Sheail, Gillian M.
III. Manley, Gordon
551.6   QC981.8.C5
ISBN 0-04-551089-X

Library of Congress Cataloging in Publication Data

Main entry under title:
The Climatic scene.
Includes bibliographies and index.
1. Climatology – Addresses, essays, lectures.
2. Climatic changes – Addresses, essays, lectures.
3. Manley, Gordon – Addresses, essays, lectures.
4. Meteorologists – Great Britain – Biography – Addresses,
essays, lectures. I. Manley, Gordon. II. Tooley, M. J.
III. Sheail, Gillian M.
QC981.4.C57   1985   551.5   84-21743
ISBN 0-04-551089-X (alk. paper)

Set in 10 on 12 point Bembo by Computape (Pickering) Ltd
and printed in Great Britain by Mackays of Chatham

Foreword

Professor Gordon Manley combined a prodigious memory for weather with an equally enquiring mind about evidence from Quaternary geology and biology for the interpretation of past climates. Any conversation with him on these matters invariably produced stimulating comments on past and present climates, with a welcome measure of sensible speculation. It was the same with his writings. There was no question of problems of interpretation outweighing the interest of the subject, and speaking as one on the Quaternary side of matters, I found this entirely beneficial and indeed proper for the pursuit of the subject.

In the course of Professor Manley's career a whole new range of data became available for the study and reconstruction of past climates, and he immediately made himself familiar with the facts and hypotheses associated with these novel areas. Thus his interest in late-glacial climates was strongly associated with the development of knowledge of late-glacial biota and environments in the early 1950s.

The scope of the essays in this book, written by his pupils and colleagues, reflects the width of Professor Manley's interests. His teaching stimulated his students at a time when studies of past climate were undergoing a great expansion, when new and varied kinds of evidence for climatic change and its associate phenomena were coming forward.

Here, then, in written form, are the consequences of his effectiveness as a teacher and colleague, combining to represent a proper recognition of his life's work.

R. G. West FRS
Cambridge, May 1983

Editors' preface

After Professor Gordon Manley's death and memorial service in 1980, the idea of a collection of essays in his honour was revived. It was felt appropriate that former students of Professor Manley, and other scholars who had been inspired by his life and work, might like to contribute to such a collection.

The essays collected here illustrate various aspects of Manley's work. His first permanent appointment was at the University of Durham, and it is fitting that the meteorological record at the Durham University Observatory should be treated both historically and statistically by Miss Joan M. Kenworthy and Dr Ray Harris. The instrumental record of rainfall from sites in northern England is considered by Dr Elizabeth M. Shaw. The following seven contributions treat proxy records of climatic change. Professor Hermann Flohn presents a critical assessment of proxy data for climatic reconstruction. Professor Hubert Lamb and Dr Jean Grove treat of the 'Little Ice Age' using historical documentary evidence. Dr Christian Pfister develops the theme on snow cover and snowlines in Central Europe, upon which Professor Manley had contributed several papers. Natural records of climatic change inferred from peat stratigraphy (Dr Keith Barber), the magnetic intensity of sediments (Professor Frank Oldfield and Mr Simon G. Robinson), and coastal sedimentation (Dr Michael J. Tooley) comprise the themes of the last three essays on proxy evidence. The last two essays are applications of studies of climate on plant distributions by Mr Richard M. Carter and Dr Steven D. Prince, and on the diseases and pests of agriculture by Dr Austin Bourke.

All the contributors have agreed to forego their royalties, which will be used to endow a prize, to be called the 'Gordon Manley Prize', administered by the University of Durham. The 'Gordon Manley Prize' may be awarded annually on the recommendation of the Chairman of the Board of Studies in Geography at the University of Durham for a postgraduate student working on an aspect of climatic change in *one* of the departments of geography at the Universities of Durham, Cambridge, and London (Bedford College), or the Department of Environmental Sciences at the University of Lancaster. In this way, it is our hope that work on climate in general, and climatic change in particular, will be noted and encouraged, and that knowledge of Professor Gordon Manley's scholarship and teachings will be perpetuated through the 'Gordon Manley Prize'.

We present this collection of essays to him posthumously as a mark of our continuing respect.

Durham and Hilton   Michael J. Tooley
                                                                 Gillian M. Sheail

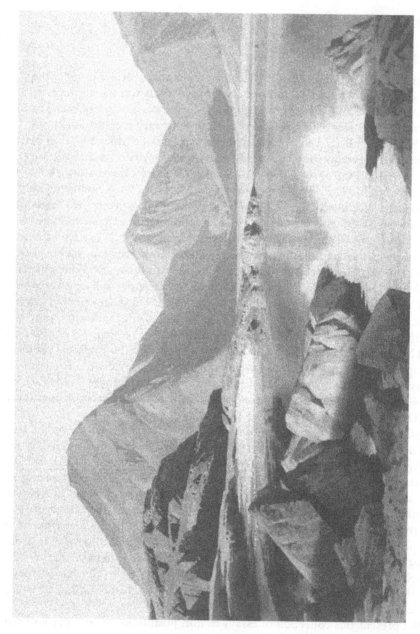

Plate 1 *Clear evening, Wastwater,* May 1967, W. Heaton Cooper.

Wastwater, May 1967

MICHAEL J. TOOLEY

The painting of Wastwater by W. Heaton Cooper (Plate 1) was chosen as the subject for the dust jacket of this memorial volume for three reasons.

1. Professor Manley was particulary fond of paintings of the Lake District by W. Heaton Cooper, and this painting was chosen by Professor Manley and presented to him on his retirement from the Department of Environmental Sciences in the University of Lancaster by members of the department in 1967.
2. It shows atmospheric conditions over the Lake District during a burst of maritime arctic air in late spring 1967.
3. Mosedale and Piers Gill, flanking Wasdale Head, were areas from which Professor Manley collected data on Late Glacial snowlines in the Lake District (Manley 1961).

Clear evening, Wastwater was painted by W. Heaton Cooper towards the end of May 1967, during a camping excursion in Wasdale. The northeastern end of Wastwater is shown beyond Long Crag. Great Gable is flanked by Kirkfell and Lingmell. Piers Gill, a tributary of Lingmell Beck, swings behind the shoulder of Lingmell, and Mosedale lies to the west of Kirkfell.

May 1967 was exceptionally wet, with thunderstorms, and was in fact the wettest May since 1729 (Anon. 1967a). Cyclonic weather was characteristic, and in the central Lake District at Ambleside rainfall receipts of 211 mm were 216% of the average (Anon. 1967b). The incidence of bright sunshine was well below the expectation for May. The weather pattern relented on two occasions in the month – on 18 May and 30–31 May, when a weak ridge of high pressure extended across the country. As the ridge was replaced, a burst of maritime arctic air occurred, bringing clear, sunny weather to the Lake District and cloudy, dull weather to north-east England. On 31 May 1967, the maximum temperature at Ambleside was 18.9°C, whereas at Durham University Observatory it was 13.3°C (Anon. 1967c). In an early paper, Manley (1935) drew attention to the noteworthy differences in weather conditions between the coastlands of north-west Britain between the Ribble and Ayrshire, and north-east England between the Humber and the Tweed, particularly in May and June when anticyclonic conditions obtained. Under these conditions, with a northerly or northeasterly airflow, a 'sea-fret' with formless cloud and damp, chilly air is characteristic of the coasts of Northumberland and Durham, but pushing occasionally as far west as Stainmoor and Crossfell, whereas the 'Lake

District enjoys long spells of superbly clear sunshine' (Manley 1952). Such were the conditions that were experienced briefly during the painting of *Clear evening, Wastwater.*

References

Anon. 1967a. Weather log. May 1967. The wettest May since 1729. *Weather* **22**(5).

Anon. 1967b. May 1967. *Monthly weather report of the meteorological office* 84(5), 65–80.

Anon. 1967c. *Daily meteorological observations, 1967.* Durham: Durham University Observatory, Department of Geography.

Manley, G. 1935. Some notes on the climate of north-east England. *Q.J. R. Met.Soc.* 61(262), 405–10.

Manley, G. 1952. *Climate and the British scene.* London: The New Naturalist; Collins.

Manley, G. 1961. The Late-Glacial climate of north-west England. *Lpool Manchr Geol. J.* 2, 188–215.

Acknowledgements

The editors are pleased to acknowledge the help of Mrs Audrey Manley, and are grateful to her and to Cambridge University Library for permission to reproduce extracts from letters and unpublished documents in the collection of Manley Papers in Cambridge University Library. We thank Mr W. Heaton Cooper for allowing us to reproduce his painting of Wastwater.

Acknowledgement is also made to the typing and technical staff of the Department of Geography, University of Durham, for re-typing several chapters and re-drawing and photographing many of the figures. Mr D. Hudspeth photographed Durham University Observatory. Gratitude is expressed to the Chairman of the Board of Studies in Geography, Professor J. I. Clarke, for making these facilities available. We are pleased to acknowledge financial assistance from the University of Durham Publications Board to assist in the publication of this book.

The editors would like to thank Dr P. Anne GreatRex and Mrs Rosanna M. Tooley for help with the index.

Contents

List of tables

List of contributors

Dr K. E. Barber, Department of Geography, The University, Southampton SO9 5NH, UK

Dr A. Bourke, 143 Ballmyn Road, Dublin 9, Ireland

Mr R. N. Carter, Department of Plant Biology and Microbiology, Queen Mary College, University of London, Mile End Road, London E1 4NS, UK

Professor H. Flohn, Meteorologisches Institut, Auf dem Hügel 20, D–5300 Bonn, West Germany

Dr J. M. Grove, Girton College, Cambridge CB3 0JG, UK

Dr R. Harris, Department of Geography, University of Durham, Science Laboratories, South Road, Durham DH1 3LE, UK

Miss J. M. Kenworthy, St Mary's College, University of Durham, Elvet Hill Road, Durham DH1 3LR, UK

Professor H. H. Lamb, Climatic Research Unit, School of Environmental Sciences, University of East Anglia, Norwich NR4 7TJ, UK

Professor F. Oldfield, Department of Geography, The University of Liverpool, Roxby Building, PO Box 147, Liverpool L69 3BX, UK

Dr C. Pfister, Historisches und Geographisches Institut, Universität Bern, Engehaldenstrasse 4, 3012 Bern, Switzerland

Dr S. D. Prince, Department of Plant Biology and Microbiology, Queen Mary College, University of London, Mile End Road, London E1 4NS, UK

Mr S. G. Robinson, Department of Geology, The University of Liverpool, Brownlow Street, Liverpool L69 3BX, UK

Dr E. M. Shaw, Department of Civil Engineering, Imperial College of Science and Technology, Imperial College Road, London SW7 2BU, UK

Mrs G. M. Sheail, 3 Mill Hill End, Hilton, Huntingdon, Cambridgeshire PE18 9NX, UK

Dr M. J. Tooley, Department of Geography, University of Durham Science Laboratories, South Road, Durham DH1 3LE, UK

1

The life and work of Gordon Manley

MICHAEL J. TOOLEY AND GILLIAN M. SHEAIL

This account of the career and academic achievements of Gordon Manley sets out some of his aspirations for the future of environmental studies and climatic research, and indicates the way in which he hoped his later work might be presented. An article by Manley on British weather originally published in *Anglia* (1963, in Russian) is reproduced.

1.1 Introduction

Gordon Manley had a very rich academic life, as evidenced in the papers by Lamb (1980, 1981) and others (Anon. 1980, Aitken 1981, Vollans 1979–80), which have served as a basis for the present description and evaluation. The present authors experienced his enthusiasm briefly at Lancaster, where they were, respectively, a research student and a lecturer in the Department of Environmental Sciences in the 1960s. Discussions with colleagues and friends have disclosed the range of Manley's interests and abilities, as well as his compassion, intellectual humility and quixotic sentiments. He pursued his research doggedly and cheerfully, and received international accolades as a scientist for his work on the range of variation of British climate. All this was done frugally and thriftily, without munificent research grants or the aid of continuous research assistance. He worked alone, but his output of scientifically sound papers was enormous: reference to his bibliography shows that he wrote 182 papers in the period spanning more than 40 years from 1927 until 1981. His book *Climate and the British Scene* still stands as a worthy monument to his ability as a writer of scientific prose, but at the same time it is an eminently readable volume. Manley's wit, his apposite turns of phrase, his joy of learning and of music, and his infectious enthusiasm for his subject were characteristics from which all his students benefited. Disappointments were seldom shared, but the disenchantment with the subject he had chosen at Cambridge and the lack of enthusiastic, strongly motivated students to continue his research occasionally emerged in conversation.

An outline of his career is given in the following sections, and the developments and directions of his research are described. One section is devoted to the reproduction of an article by Manley on British weather that was published in *Anglia* in 1963, because it epitomises in style and content Manley's prowess as a writer of scientific prose. It is redolent of the writings that emanated from the gatherings of great scientists in the provinces in Manchester, Birmingham and Edinburgh, to which Manley liked to refer as 'the "outlying foci" of early scientific progress'.

1.2   Early life and work, 1902–47

Gordon Manley was born on 3 January 1902, at Douglas in the Isle of Man, the son of Valentine Manley, a chartered accountant. He was brought up in Blackburn, and educated at Queen Elizabeth's Grammar School. Professor Fisher (1980) has recounted that his origins were 'the cold north-western shoulder of the Rossendale Uplands of Lancashire, where annual rainfall can and often does exceed 60 inches and peaty streams coalesce to form a black burn and a black pool'. From the age of 12, Manley took meteorological readings on the hills around his home, and showed the same precocious interest in natural phenomena that characterised the Elton brothers (William and Sir Charles) in their record of the plants, animals and weather on the Ainsdale Dunes of the Lancashire coast. In October 1918, he went up to Manchester University, where he read for an honours degree in engineering under the Beyer Professor of Engineering and Director of the Whitworth Laboratory, a post first held by Sir Joseph Ernest Petavel (who became Director of the National Physical Laboratory) and then by Professor Arnold Hartley Gibson (who specialised in hydro-electric engineering). Manley graduated in July 1921.

From Manchester, Manley went on to Gonville and Caius College, Cambridge, where he became an affiliated student and took geography shortly after the tripos had been established. He showed considerable aptitude for geography, and gained a double first in 1923. He was an exhibitioner of the college in 1922–3. At Cambridge, he met Mr (later Professor) Frank Debenham, who was a Fellow of Caius College from 1919, a lecturer in surveying, and, from 1923, tutor. Subsequently Debenham became Reader in Geography in 1928, and Professor of Geography from 1930 to 1949. Debenham had been the geologist on the British Antarctic Expedition from 1910–13 under the command of Captain Robert Falcon Scott, and he founded the Scott Polar Institute at Cambridge, being a Founder Director from 1925 to 1946. Debenham's inspiring teaching had a profound effect on Manley, and his interest in polar and high-altitude environments stems from this influential period at Cambridge. Many years later, (in 1947), when he was awarded the Murchison Grant by the council of the Royal Geographical Society, Manley acknowledged his debt to Debenham 'for putting me on to the polar track'.

The early training that Manley received at Manchester in a practical and applied science, and the introduction to polar and high-altitude environments at Cambridge provided the firm foundation upon which he built his academic career. Measurement, primary data and field experience, particularly of high altitudes and latitudes, together constituted his lodestone.

In 1925, he entered the Meteorological Office, and was stationed at Kew Observatory, where there was a long meteorological record which had been taken irregularly from 1773, and where he had training in the use and maintenance of meteorological instruments.

Manley quit the Meteorological Office in 1926 to take up an assistant lectureship in geography at the University of Birmingham. Between these two posts, in the summer of 1926, he joined the Cambridge Expedition to East Greenland under Mr (later Sir) James M. Wordie, who was to become the Honorary Secretary and President of the Royal Geographical Society. Manley worked with Sir Gerald Lenox-Conyngham (lately retired from the Indian Survey) on Sabine Island, where they undertook survey work, gravitational measurements and meteorological readings. The results were written up by Manley as appendices to the expedition report published in the *Geographical Journal*. The gravitational measurements that Manley made were, at that time, the northernmost record, and these were published in 1932 in *Meddelelser om Grønland*.

In 1927, Manley became a Fellow of the Royal Geographical Society. He was appointed Lecturer in Geography at the University of Durham from October 1928. The establishment of geography at Durham had been mooted two years earlier by Canon Braley, the Principal of Bede College. Manley was required to organise and conduct the teaching of geography at Durham in the Faculties of Arts and Science, and to co-operate with Professor Arthur Holmes in the Department of Geology. Notwithstanding a small operating budget, and small staff and student numbers, Manley pursued an active teaching and research programme. He taught classes in physical geography, including climatology and biogeography, and the regional geography of North America, in which he was no doubt helped by visits to Pittsburgh as a boy, when he had accompanied his father on business trips. Important aspects of the honours and pass degree courses were practical mapping and fieldwork.

Manley was appointed Curator of Durham University Observatory in 1931, and he began his work on standardising the long temperature record by adjusting the values for exposure, changes in the position of the instruments and inadequate recorders, as Miss J. M. Kenworthy recounts in Chapter 2. In addition, from 1932, he began to collect data at Moor House on the climate of the northern Pennines, particularly the helm wind of Crossfell. Manley's empirical study led to theoretical work on standing waves in the atmosphere in the lee of mountain ranges by such meteorologists as R. S. Scorer. In 1937, he received a grant for meteorological research from the Leverhulme Trust that enabled him to establish a meteorological station adjacent to the summit of Dun Fell at 2780 ft (847.3 m), and maintain with others a record at three-hourly intervals of weather conditions for three years from 1938 to 1940. This was the first series of mountain observations to become available in England; it remains the longest unbroken mountain record to this day. These investigations led to more than a dozen papers on the weather of the northern Pennines and adjacent areas, published between 1932 and 1947.

At the same time, Manley's eclectic interests carried him into other areas, and he published papers on antiquarian maps, atlases, the geography of Durham City and on snowfall and transport problems. In 1938, he was successful as an external candidate for the degree of MSc in the Faculty of Science at the University of Manchester; and in the following year he gave up his post as Senior Lecturer in Geography and Head of Department at Durham to take up a demonstratorship in Geography at the University of Cambridge. He retained a great affection for Durham throughout his life, and it was during his time there that he met and married (in 1930) Audrey Fairfax Robinson, with whom he shared almost 50 years of his life. Audrey Manley was the daughter of Dr Arthur Robinson, who was Master of Hatfield College until 1940 and who previously had been a Lecturer in Classics from 1899, Professor of Logic and Psychology from 1910, and Vice-Chancellor of the University of Durham.

At Cambridge, Manley renewed his contact with Professor Debenham, and found a department in which Steers, Lewis and later Peel were members. At this time, physical geography played a leading rôle at Cambridge, and its exponents were practical field scientists concerned with measurement in the field as a sound basis for interpretation and explanation. Nearby, Dr (later Professor Sir) Harry Godwin had begun his seminal and stimulating work on the postglacial history of British vegetation, and Debenham continued as Director of the Scott Polar Institute. Manley was greatly interested in all these topics, which he enhanced by his scholarship, and was in turn enhanced by them.

From 1942 to 1945, he was a Flight-Lieutenant in the Cambridge University Air Squadron, but he still carried on his research and maintained a miscellaneous teaching programme not only for Cambridge undergraduates, but also for undergraduates from Bedford College, University of London, who had been evacuated there.

While at Durham, Manley had been elected a Fellow of the Royal Meteorological Society, on 16 November 1932. In 1943, he was awarded the Buchan Prize, jointly with Dr T. E. W. Schuman, for three papers which he had published in the Quarterly Journal: the first was on the analysis of meteorological observations taken during the British East Greenland Expedition at Kangerdlugssuak; the second was on the occurrence of snow cover in Great Britain; and the third was on the Durham meteorological record, 1847–1940. The following year, on 19 April 1944, he gave the G. J. Symons Memorial Lecture, and he chose as his topic recent contributions to the study of climatic change. This was undoubtedly inspired by the work of H. W. Alhmann of Stockholm on glacier variations. In 1945, he became President of the Royal Meteorological Society, a post he held for two years; this was a signal honour for a geographer, and was in recognition of his outstanding research in meteorology. During the period of his presidency, Manley helped to organise (and gave a paper at) a joint meeting with the British Ecological Society, chaired by Sir Harry Godwin, on 'Ecology and the study of climate'. It is no coincidence that Weather was founded in May 1946 during Manley's presidency. It was subtitled, 'a monthly magazine for all interested in meteorology', and its objective was to make contemporary developments in meteorology available to a wider public.

1.3 Later life and work, 1948–80

In 1948, Manley became Professor of Geography in the University of London at Bedford College for Women, and he stayed there for 16 years. Lamb (1981) described this period in Manley's career as probably one of his happiest. He was no longer in one of the 'outlying foci' of scientific progress, and had access to the richest resources (in terms of people and material) that were available in the capital. He was a committed, enthusiastic Head of Department, cherishing his students and giving entertaining lectures and tutorials. The department was pervaded with his genial personality and friendly manner.

In the pre-Robbins period, there was little opportunity for university expansion, and, while Manley preferred a slow growth of student numbers and the maintenance of high standards, it effectively prevented the enlargement and diversification of the research interests of his staff. Nevertheless, Manley was not slow to grasp the opportunity provided by the creation of a small number of tutorial fellowships by the college. During Manley's tenure the tutorial fellows were the late Michael Holland, Jane Soons, Joan Kenworthy and Brenda Turner, and Elizabeth Shaw became his research assistant. Manley also appointed to the staff of the geography department Michael Chisholm, Clifford Embleton, Jean Grove and Eleanor Vollans (Eleanor Vollans, personal communication 1984).

Manley maintained his links with Cambridge while at Bedford College, one of the results of which was the joint participation of Bedford and Cambridge undergraduates in expeditions to Norway and Iceland (Vollans 1979–80).

In 1949, he gave his inaugural lecture, 'Degrees of freedom' (Manley 1950), which is replete with aphorisms for the intending geography undergraduates and professional colleagues. Some of the definitions are perhaps worthy of recall: 'If British geographers are to be productive and not merely clerkly, they must gain experience in the field. The country is our laboratory'; 'I feel strongly that we could fittingly pay regard to the work of such men as Ahlmann of Stockholm, Werenskiold of Oslo and Nielson of Copenhagen, animated as it is by the inspiration of an exploratory tradition, painstaking and energetic field work with a massive power of integration of material, by an immense fertility in by-products and by no heedless drawing of frontiers between subjects'; 'the geographer can and should carry his capacity for integrating observed material to that of integrating results derived from the peripheral sciences'.

In 1952, Collins published *Climate and the British scene*, in the New Naturalist series. It drew together all the strands of Manley's research interests and presented them in a clear, concise way. Lamb (1981) described it as 'a feast of his most attractive writing'. This flair for writing equipped Manley to write a long series of articles for *The Manchester Guardian* between 1952 and 1961. Meanwhile, his output of research papers and monographs continued unabated as he consolidated and extended his earlier work on variations of British climate, upland climates, climatic change, snow cover and early meteorological observers.

While at Bedford College, he became a member of the Council of the Royal Geographical Society (1952–4), and served on four national committees of the Royal Society. He was the UK national correspondent for glaciology, appointed by the British National Committee for the International Geophysical Year, from 1955 to 1961. He was a member of the Air Ministry subcommittee for meteorological research from 1958 to 1962, and he served on the Commission for Snow and Ice of *L'Association d'Hydrologie Scientifique* of the UGGI whose biennial assemblies he attended in 1936 and from 1948 onwards, publishing reports on their proceedings.

In 1953, Manley became a member of the 'Exploration Group at Brathay' at Windermere. The establishment of the group in 1947 arose from W. Vaughan Lewis' interest in glaciology and the depths of mountain tarns in the Lake District; Lewis had been with Manley in the Department of Geography at Cambridge. In 1966, a committee was set up to advise on the establishment and running of the Field Study Centre at Brathay; together with Professors Steers, Clapham, and Newbould and Dr Winifred Tutin, Manley served on this committee (Campbell & Campbell 1972).

His university commitments continued, and he played his part in establishing and maintaining examination standards in the greatly enlarged departments of geography in the UK by serving as an external examiner at the universities of Bristol (1958–61) and St Andrews (1962–5). At the same time, he gained a DSc degree at the University of Manchester, which was conferred in July 1958.

In 1964, at the age of 62, Manley could have retired from university life and continued his eclectic and stimulating scientific research from his home, supported by a personal research award from Shell. However, it was entirely characteristic of him that instead he rose to the challenge of founding a new department in the equally new University of Lancaster, at the behest of the Vice-Chancellor, Professor (later Sir) Charles Carter, who had been a lecturer at Cambridge and a professor at Manchester.

Manley drew to him a group of like-minded, energetic colleagues and research students, as well as lively groups of undergraduates, and from an elegant 18th-century house in St Leonard's Gate, and later the converted furniture workshop of Gillows of Lancaster, he presided over the new interdisciplinary subject that he called environmental studies. It is well worth recalling the hopes and expectations of this new subject, and, why, by implication, Manley sought to establish environmental studies rather than a department of geography, which, as a subject, had nurtured him for 43 years. In his inaugural lecture, Manley (1967) makes the following points:

It is appropriate, indeed inevitable, that I should try to comment on some aspect of this pursuit that I have called 'Environmental Studies', in order to illustrate what I hope to do. I have to explain an attitude of mind, to justify a departure, to suggest a philosophical framework. In my view, and in that of many younger workers today, we are no longer satisfied with the description, arrangement and assemblage of the knowledge of that which exists at the Earth's surface in a gleaming present, and in the depiction of such things whether with the aid of maps, or with the style and elegance of well written regional geography; for in the past 40 years such depiction has become a very proper part of the education provided by a good school. I am far from being alone in the belief that from such descriptive accomplishment we must go forward. Just as the sixteenth century was the age of assertion of existence, of exploratory discovery, of expansion of spatial horizons, of mapping, and of those exercises in mathematics that were appropriate to the establishment of position – so there followed that seventeenth century that became the age of the philosopher, of questioning of existence, of prepossession with the lapse of time; the century in which the observational sciences began to grow.

Suffice it accordingly to declare that if universities exist to advance knowledge, rather than merely to convey it, then from the description of what happens on this Earth's surface, it is appropriate to go forward to the discussion of the evolution of the physical landscape and the rate at which that complexity of variables that we see is capable of change, especially having regard to the fact that we men can now interfere with the land, the air and the sea with ever-increasing power. We live, indeed, in a mobile world; we judge, all day and every day, the changing location of fast-moving objects relative to each other; in fifty years our whole personal concepts of space and time have changed; we live in a world of trends. The study of movement and of process and of change in the physical world, whether on the geological time-scale or on that of the recent storms, seems to me a logical necessity.

Research of both a pure and applied nature was pursued energetically in the new department (the name of which soon changed to Environmental Sciences), but in 1967 Manley retired and moved back to Cambridge. Some of those who had formed the nexus of the new department moved back into geography, and later a department of geography was established at the University of Lancaster.

Manley remained a Research Associate in the Department of Environmental Sciences at the University of Lancaster, and during this period the research on Manchester rainfall and central England temperatures (that had taken more than 40 years to assemble) was published. It gave Manley much pleasure to know that the temperature record he had assembled from overlapping series at different sites constituted the longest instrumental run and was a world record: it vindicated, for him, the years of patient recording on the part of many observers, whose work was for the first time used and noted. Futhermore, the temperature record stands as a test for all the proxy records of climatic change for this period, and an example of scientific scholarship. In 1969–70, Manley was a Visiting Professor of Meteorology at the A and M University of Texas, a position he enjoyed immensely and one which enabled him to pursue his interests in applied meteorology and environmental problems.

Substantial articles continued to be published on snowfall, temperature records and observers, and extreme events, such as the summer of 1976, were noted in short communications. How he would have rejoiced in noting the extreme winter of 1981/2 and the warmest July on record in 1983 at the Durham University Observatory. Many substantial papers remained unpublished, such as *Snow in London, 1699–1958, Monthly means 1786–1833 kept by Cary in The Strand,* and *Corrected temperature data 1771–1798 at Lyndon in Rutland.* In 1979, he was honoured by the University of Durham, which conferred on him the degree of Doctor of Science *honoris causa.*

At the time of his death, Manley was assembling instrumental data for the North of England and Scotland back to the 18th century, and he had been particularly excited by the discovery of a long weather record, hitherto unknown, at Rothesay Castle, to which he had been given access by the Marquis of Bute.

1.4   The weather in Britain, by Gordon Manley

In 1963, the editor of *Anglia*, Mr Wright Miller, from the Central Office of Information, asked Manley to write an article on British weather for Russian readers. It was published in August 1963 in Russian (Manley 1963). The style and prose are examples of Manley at his best, and the article is reproduced here for an enlarged readership.

Europeans are supposed to cherish a traditionally dim view of our British climate. For in every country in Europe we know that if we move to the north and west we come to a region with more cloud and wind, with damper air, with more frequent showers. The Leningrad winter with its bitter wind off the Gulf of Finland, and the summer with its cloudy and damp August among the marshes were proverbially disliked by the men of Moscow and Kiev. That dry glowing summer that goes with the *chernoziom*, that Russian term for the black earth that your famous soil scientists have made known to the whole world, that produces the wheat, that lies as a beautiful fertile dream in the minds of those who sit outside the Kievskaya station in Moscow, only penetrates to the shores of the Baltic for a few days at a time.

Equally the Germans comment on the fogs and the raw winter cold of Hamburg, while to the Frenchman the sad grey skies of Lille are something to avoid. To the Italian from Florence or Rome, the chilly winter mists of Milan are unpleasant; for Milan in January and February is colder than London, and Zurich is just as sunless although five degrees of latitude further south. And most Europeans are ready to tell us that the Dutchman's placid temperament and liking for good food and schnapps is because of the damp, the grey skies and the cold wind off the sea that he is supposed to endure.

So it is no matter for wonder that Continental Europe is ready to think that the British climate must be horrible, just as it is ready to think that the Russian climate is horrible for other reasons. Much of this is fashion deriving from the Germanic–Romantic ideas of the earlier 19th century. Travel was becoming common, and the rich were always ready to excite envy in their poorer neighbours by telling them how much better the weather was in countries elsewhere. Tourist agencies now try to tell us the same story.

Because we in England so well understand love for the country in which we have grown up, those of use who read the great Russian writers appreciate their pleasure in the natural environment in which their characters move. The brilliance and the uncompromising hardness of February, the pale birches coming into leaf in the Moscow countryside, the glowing contentment of the last warm days in September, the grey skies and the harsh breath of oncoming winter when the last shooting party is held in November, have gone to mould a people. The great, indeed solemn annual strokes of Nature, so well described by your novelists, ring in the mind like the chiming of the great bells of Moscow. The magnificent description of the Russian winter, by the first Englishmen to reach Moscow in 1556, still stands.

By contrast, Britain is a place where Nature operates in a more subtle way; by the repeated small but insistent blows characteristic of a variable climate under the dominance of the vast and ever-open seas, she tempers those who dwell in it. For us, not the great bells, but the steady underlying throbbing of the plucked strings and the muted drums. We live on the battleground between the surges of different types of maritime air, now from the Arctic, now from the subtropical anticyclones to the south west; sometimes the dry Continental air reaches us, but rarely for more than a few days at a time; and even that is modified a little by the narrow seas that divide us from France and Holland and Denmark.

We have many enjoyable summer days with a fresh west wind when that Atlantic anticyclone spreads towards us, giving an afternoon temperature of 23° in southern England, and a sky dappled with little cumulus clouds that disappear towards sunset. Further north the west wind blows more strongly, and the cloudy, damp windswept western coasts and islands off Scotland lie too near the normal track of the depressions; rain, in the Hebrides, falls on nearly three days out of four, and it is almost as cloudy as it is along the coasts east of Murmansk. But in central and eastern Scotland the west wind is coming down again off the mountains; skies are much clearer, and there is the agreeably bright summer climate of Edinburgh and Aberdeen. Edinburgh nevertheless is breezy, and cooler than London by about the same amount as Leningrad is cooler than Moscow, and for the same fundamental reasons.

If such an anticyclone moves east into Europe we can get short spells of great heat; the air comes across the warm land from Bulgaria or Hungary and if the anticyclone is big enough it may come all the way round to us. Temperatures can go up to 37° in southern England, and are capable of exceeding 30° from May to the end of September. But such conditions can very rarely develop for more than about four days in succession; then comes a thunderstorm, and the fresh Atlantic air brings the welcome relief from the heat.

East and north of the mountains of Wales and Scotland our lowland climate is just as sunny and free from rain as that of our neighbours. We can and do grow a magnificent wheat crop as far north as the latitude of Moscow, a fact that always surprises our French visitors who have like so many been ready to believe that we were a country of eternal rain and fog. London has very little more rain in an average year than Paris or Berlin or Moscow. Edinburgh, and even Manchester, have less rain than Brussels or Milan. As regards the number of days on which it rains, London is very much like Paris; but the number of days with small amounts of rain does definitely increase north-ward and westward. This helps to explain the greater sensation of dampness, as with the lower summer temperatures the rate of evaporation also decreases, so that the ground remains moist. But this is an advantage in another way; our pastures remain green and the grass grows throughout most of the year. Indeed, down in Cornwall you almost always need to mow your garden lawn in January.

Summer temperatures at London are a little warmer than at Leningrad, but are not quite warm enough for us to be able to produce wine from grapes on a regular commercial scale. For our summers are variable; the weather goes wrong in those seasons when the track of depressions lies south of its normal North Atlantic route past Iceland. When depressions go up the Baltic we are likely to get a cool and rather dry summer, as in 1962; whereas Northern Russia then gets a cool and wet summer. With us 1962 was distinctly cool, and it gave rise to a lot of grumbling because the favourite summer holiday period early in August was wet in the south, and the Londoners did not like it at all. To the Continental visitor the chief feature he notices is that there is more wind, and the skies change more quickly; many days begin badly, but quickly clear up to give sunshine in the evening.

But if he comes in the winter the story is different, because the effects of the sea become so much more evident. Both London and Edinburgh are less cold than Paris in the midwinter months; indeed London's temperature in January is just above that of Yalta. Most of the winter weather is the result of the passage of Atlantic depressions; strong and stormy south winds on the coast, then a few hours' rain, a clearance, a short spell of sun with a brisk west wind and a shower or two, then may come a clear night when the air falls calm, the ground is wet, and so in all the valleys there is a thick mist and a touch of frost the next morning. When the sun is low in November and December you must often wait for a wind to get up before the mist and fog formed on a quiet night will disappear; the sun is not strong enough.

Now most of our big cities lie in the broad river valleys where the mist can gather, and nearly twenty million – about two-fifths – of the British people live in the seven great conurbations. Of these, London, with its ten million people spread over a built-up area nearly 50 by 40 kilometres wide, is the biggest; and is still the habit of the vast majority to burn coal in an open fire. It is the domestic fire that makes most of the smoke; industry accounts only for about one-third of it. Hence the bigger the city population, the bigger the accumulation of dirty smoke that makes the city fogs much thicker and more unpleasant than in the country; and much more persistent, although less than 200 metres deep. In December 1962 London had a spell of anticyclonic weather with five days of miserable fog, while a few miles away five cloudless days gave a total of thirty-five hours of bright sunshine.

If we were to stop burning coal in open fires, city fogs would give much less trouble; hence, we now have our smokeless zones where open fires are prohibited. Manchester has pioneered this reform, and already there is a quite noticeable improvement. For the effects of smoke are not only seen with regard to fog; the haze, even on fine autumn and winter days, cuts off some of the sunshine. Fifty years ago the centre of London was considered to lose about 15%, and at Manchester 30% of the sunshine that would have been recorded if the city had not been built. But we are now seeing an improvement; and while from November to mid-January there are always a few foggy mornings inland, London has fewer nasty fogs than there were in the days of Sherlock Holmes.

Still, nothing can be more hopelessly depressing to a visitor than a cold, damp and foggy evening with a temperature of 3° in, say, Glasgow at the end of November. I should certainly regard it as quite as bad as a blizzard in the streets of New York, or the awful warm rain that fell on me for twelve hours one summer day in Boston, or the unpleasant local fog, polluted by the exhausts of thousands of cars, that they get in Los Angeles.

It is because our winter visitors have generally found their way to the cities that we get the reputation for fog. But Prague and Hamburg and Milan are also foggy; and a visitor who spent his winter in the English country would have a very different impression. Dickens, after all, was a Londoner who wrote in that Romantic Age when it was the fashion to exaggerate. Those who read his novels, with their accounts of snow at Christmas, will find out if they look at the statistics that London has had only four 'white' Christmas Days in this century.

Snow we do have; when depressions take the track across France and on their northern side the cold air reaches us from Russia, snowfall can be very heavy. This year, 1963, has seen the greatest fall in southern England since 1881. But the cold air coming from the Continent becomes not only moister, but also appreciably milder as it crosses the North Sea, so that quite frequently the snow that falls melts at once on the low ground. Yet it only wants two hundred metres of altitude to make sure that the snow lies, and, if the hills are close to the sea, they are treeless and windswept; the snow begins to drift, and within a very few kilometres of the green fields you may find that the upland roads are completely blocked. Then, far on into the spring months, cold air from the Arctic can come down on us. Almost always we have the snow showers here and there in April, and in the north the ground is sometimes covered for a few hours in May. It is in the springtime, too, that very sharp frosts damage the early potatoes and the fruit blossom.

In an average winter at London there are only about five mornings when the ground is covered by snow, and only ten at Edinburgh; yet at 700 metres there are upwards of a hundred days. Then in any given winter you may have a snowy December, a mild January so that the first crocuses are to be seen before the end of the month; then will come a biting cold March with the north-easter bringing the snow showers off the North Sea, and daffodils will not appear before the middle of April, a month after the normal date. I have seen the shade thermometer at 15° in December, at 23° in March, at 28° in October. But last June at Cambridge we had three nights when it was down to 0° at midnight and we ran round the garden putting newspapers over our early potatoes. Rather rarely, some places in southern England may record − 20°; yet in many winters the thermometer will never go below − 4°, and in 1938 the coldest night of 'winter' came on the 8th of May.

That the changeable British climate has its advantages, we firmly believe. It is a national characteristic to be adaptable; never to have so rigid a plan that it cannot be adjusted to a change in the weather. Our farming is characteristically 'mixed'; rootcrops and potatoes do well if it is wet, wheat and barley if it is dry. So wide are the local variations in the intensity of frost, the character of the soil, the exposure to wind, the amount of rainfall and evaporation, that we have evolved an astonishing variety of stock and of crops; for example Hereford cattle, Lincoln or Cheviot sheep, Clydesdale horses, our endless breeds of dogs; then the Scottish seed potatoes, the marvellous apples such as the Cox and the Ribston pippin, the many varieties of plums, even the hardier peaches – all these we grow. We have exported them too, with our poets, our physicists, our politicians – all over the world. Even in my own village, my neighbour sells his pedigree pigs to Russian breeders. Down in the mild south-west we can grow our Mediterranean evergreens, even the eucalyptus and small palms; yet our tree line on the mountains stands at only 600 metres, far below that in the Urals or the Altai.

So we can claim that this island of frequent changes, of the terrible Atlantic gales whose endless roar besets our coasts in winter, of the exquisite long June days celebrated by our poets throughout the centuries, of the harsh biting north-easter in April, the wind-driven rain day after day if there comes a wet autumn, the occasional spell of three weeks of snow and frost, the persistent dryness that quite frequently leads to shortage of water in early summer – all these give us much cause to grumble, but even more cause to enjoy the march of the seasons and the opportunities for such a variety of flowers that the poorest man can still grow them in his garden. Then there are the vegetables and fruits and the trees from all over the world so that you will find the Russian fir, the cedar of Lebanon, the araucaria from Chile and the larch from Japan beside our native oak, beech, ash and elm that testify to the fact that our climate has many, many advantages. It may lack the grand drama of the bursting of the ice in spring, the fierce and implacable dry heat when the south-easter comes up the Volga, the crackling cold under the northern stars; but it has the stimulus of constant variety and protection against it has always lain within the scope of the single man working by himself. Yet equally with his brother the Russian he must work, and be ready for all the blows that his variable, if rarely lethal, climate can deal him.

1.5   Collectanea meteorologica Britannica

By 1972, Manley's book *Climate and the British scene* had been reprinted five times, and had appeared as a Fontana paperback. It had, however, never been revised, and after his retirement Manley turned his attention to a new book. By this time, he had accumulated a considerable amount of material from archives on past meteorological records relevant to the British Isles and to Western Europe, and was in the process of integrating these data. It was also apparent that he wished to use these data to demonstrate fluctuations of climate in general, within the context of a book, the scope of which would be climate as an environmental factor and as an element in British history.

Manley had been moved by an invitation to contribute a foreword to the translated edition of E. Le Roy Ladurie's book *Histoire du climat depuis l'an mil*, and felt that the time was opportune for a British equivalent, using the long British sequence of daily meteorological observations that he had established and extending it back to Elizabethan times. This would make up a book of 300 standard octavo pages of text, and 30 pages of tables, grouped into eight chapters.

It was proposed that the introductory chapter should contain material on the British climate in general, and on the range of variation of temperature, rainfall, and the rainfall–evaporation ratio in particular.

The second chapter should be devoted to the sources of the instrumental record, beginning with the official meteorological tabulations and covering the profusion of earlier amateur observations in the 16th, 17th and 18th centuries. The final part of the chapter would treat the proxy record. Manley was anxious to take the sources in this order, as a scientific principle reasoning from the known to the unknown.

The third chapter would be concerned with the makers of observations. Manley's research had shown how rich Britain was in observers, and through their recorded observations he had gained an insight into their personalities, which he wished to record as a contribution to the history of science.

The fourth chapter would be the core of the book, containing pressure, temperature, rainfall and snowfall data. For central England, he proposed tables of mean temperatures for each month from 1660 to 1972, with estimates for the seasons from 1580. The rainfall record from 1727 would be considered, with a table to show the ratio of the monthly rainfall to the average. Regional variations in rainfall pattern would be stressed, especially with reference to north-west England and Scotland. The section on pressure would show a table of monthly mean pressures at London since 1690. Manley felt that the best available long-term assemblage for snowfall was likely to come from London since 1668, and from the Pennines since 1798. The final weather variate treated would be bright sunshine, records for which could be found at a few places since 1881.

The fifth chapter would be devoted to extremes of behaviour. It was not Manley's intention to consider in detail frequency of frost or thunder accompanied by lightning, but rather selected seasons or incidents. He had in mind descriptions of the severe winter of 1740, the phenomenal gale of January 1839, the record snowstorms of January 1940 and February 1941, the very severe cold of June 1749, and Defoe's great gale of 1703, the very great heat of certain days in 1521, 1602, 1676, 1707, 1808, 1911 and 1932, and some phenomenally cold days in 1739 and 1838 when major rivers and lakes froze.

The sixth chapter would be devoted to upland and mountain climates. In general, the more rapid decline of temperature and increase of snowfall with altitude northwards would be considered as an introduction to a discussion of the frequencies of gales and increased exposure to wind. Here, Manley would describe the helm wind, and the frequency of upland snow upon which he had collected data in the field.

The seventh chapter would be devoted to the effects of man. Manley wished to draw together data on extremes of behaviour relevant to the work of architects, engineers and ecologists. He wished to note the remarkable improvement in sunshine duration and intensity as a result of smoke abatement policies. Further, he wished to add comments on the effects of climate and extreme weather on man, notably the effects on fuel consumption and the demand for energy in severe winters and springs, and the impact of frost, catastrophic thunderstorms and flooding. Manley acknowledged the controversial nature of the effects of climate on economic history, and was anxious to show how recognisable climatic fluctuations within history had considerably different effects from one part of the country to another.

In the eighth chapter, Manley wanted to consider the longer-term climatic fluctuations during the past 11 000 years, drawing on his own work and that of others.

If the book had been published, it would have represented the culmination of a lifetime's work, in which the main strands of Manley's research interests would have been drawn together in a volume of far-reaching implications of both a pure and applied scientific nature. Sadly, this book was never written. Instead, after 1972, the contents of some of the chapters were published in a series of papers in 1974, 1975, 1977, 1978, 1980 and 1981.

1.6 Conclusion

Manley was happiest in northern Britain, amongst the mountains that he knew intimately and loved in all their seasons and atmospheric moods. This love of the north found expression in his cherished election in 1966 to the Manchester

Literary and Philosophical Society, in his association with the Brathay Exploration Group and Field Study Centre and, much later, (in 1974), in membership of the Cumberland and Westmorland Antiquarian and Archaeological Society. After the Royal Geographical Society and the Royal Meteorological Society, Manley's membership of the Scottish Mountaineering Club since 1937 was the longest, and his annual pilgrimage to the summit of Ben Nevis to record the position of the semi-permanent snow beds at the foot of Zero and Five Point Gullies in the *Ben Nevis Snow Book* is an expression of his tenacity as a scientist.

Manley left an awesome and inspiring collection of published and unpublished papers. He set the highest standards in his scientific investigations, begun when he was still at school, and maintained to the year of his death. Few approach Manley in his scientific erudition and integrity in an area of learning that will always be associated with him.

Szafer (1968) classified scientists into five distinct types, and of the fifth type, which he believed contributed most to the development of science and was the most creative scientifically and socially, he wrote:

> Outstanding self-dependent scientists, enthusiasts of science, endowed with imagination and intuition, searching for new ways, unrelentingly facing difficult problems, sociable, usually excellent teachers, sympathetic supervisors of both individual and collective works by their disciples, good organisers, devoted social workers and good popularizers.

This was Gordon Manley.

Acknowledgements

We are gratful to Mrs Audrey Manley and the Cambridge University Library for permission to consult and publish material from Professor Gordon Manley's personal papers, which are now deposited in the Manuscripts Room of the Cambridge University Library. We are also grateful to Mrs Audrey Manley for reading and commenting upon a draft of this paper, and to Miss Eleanor Vollans for notes on the work of Professor Gordon Manley at Bedford College.

References

Aitken, R. 1981. In memoriam. Professor Gordon Manley. *Scott. Mountaineering Club J.* **32**, 195–6.

Anon. 1980. Gordon Manley MSc, MA 1902–1980. *Geog. J.* **146**, 475–6.

Campbell, B. and M. Campbell 1972. *Brathay. The first twenty-five years 1947–1972.* Ambleside: The Brathay Hall Trust.

Fisher, W. B. 1980. Professor Gordon Manley: DSc. *Univ. Durham Gaz.* **25**, 38.

Lamb, H. H. 1980. Professor Gordon Manley. *Q. J. R. Met. Soc.* **106**, 656–7.

Lamb, H. H. 1981. The life and work of Professor Gordon Manley (1902–1980). *Weather, Lond.* **36**, 220–31.

Manley, G. 1950. *Degrees of freedom. An inaugural lecture given at Bedford College for Women, January 27th 1949*. London: Christopher Johnson.

Manley, G. 1963. The weather in Britain. *Anglia* **8**, 34–43 (in Russian).

Manley, G. 1967. This North-Western environment. In *Inaugural lectures 1965–1967, University of Lancaster*, 1–11. Lancaster: University of Lancaster.

Szafer, W. 1968. Creativity in a scientist's life: an attempt of analysis from the standpoint of the science of science. *Organon* **5**, 3–39.

Vollans, E. 1979–80. Professor G. Manley. *J. Bedford College Association*, 22–3.

2

The Durham University Observatory record and Gordon Manley's work on a longer temperature series for north-east England

JOAN M. KENWORTHY

Manley had resolved to place the temperature record of Durham University Observatory on a similar basis to that of the Radcliffe Observatory. He then hoped to use it as the basis for a still longer temperature series representative of north-east England. The founding of the observatory at Durham is described and the basis of meteorological observations in the period 1847–1979 is discussed. By using records obtained from Losh and other observers in north-east England, Manley was able to construct a tentative series of mean monthly temperatures for the region from 1801.

2.1 Introduction

It is a fitting tribute to Gordon Manley to examine a topic in which he had resumed an active interest and almost completed material for writing up in the years immediately before he died, and on which he began work when he was the first head of the Department of Geography at Durham from 1928 to 1939, and a curator of the observatory from 1931 to 1937. It is appropriate, immediately following the sesquicentenary of the University of Durham in 1982, to pay tribute to the contribution made to meteorological science by the university, in the establishment of a meteorological station at the observatory in the 1840s and in the maintenance of continuous records from that time to the present day.

Manley had resolved to place the record of the observatory on a similar basis to that of the Radcliffe Observatory at Oxford, for which published tables of the record from 1815 were available. His resulting account of a series of 'adopted means' of temperature for Durham University Observatory for the period 1847–1940 (Manley 1941a) established the record as the second longest for a university in Britain and, in the year of its publication, happened to celebrate the centenary of the foundation of the observatory in 1841. Adopted mean monthly temperatures for 1941–50 were added by E. F. Baxter (Baxter 1956, Manley 1980) and, in his last, posthumous publication on the temperature observations started by himself in 1932 at Moor House in the northern Pennines, Manley

(1980) included a table to bring the series of monthly means of temperatures at Durham from 1941 up to January 1979.

In the later years of his retirement, Manley undertook the somewhat frustrating but challenging work of constructing on a much broader basis a still longer temperature series for north-east England, using other available records to extend the Durham series back to 1794. This work was comparable in method to that for 'central England' (Manley 1953, 1974) and for 'Lancashire' (Manley 1946a, 1946–7). On the occasion in 1979 when he was made an honorary Doctor of Science by the University of Durham, he commented privately that the series for north-east England would be his return gift to the university. Sadly, he did not live to complete it. It is left therefore to one whose life was enhanced by his visits to Durham and by his reports in letters (dated in the style 30.i.78) on the progress of his investigations to attempt to set out the context of the work – and thereby it is hoped some basis for future research – and to illustrate something of the spirit of adventure implicit in Manley's approach to the task.

2.2   The founding of the observatory at Durham

Manley (1946b) indicates how, following the establishment of daily observations at Greenwich under James Glaisher, several observatories in the north of Britain began their meteorological record in the 1840s; 'that at Durham Observatory (1843) being one of the first, followed by Liverpool and Glasgow (1845) and Stonyhurst (1847)'. Some uncertainty as to the exact time when meteorological observations began at Durham will be discussed later.

In the 18th century there was an increase in the concern of men of science to establish observatories supplied with the necessary instruments for the observation of astronomic and magnetic phenomena. In the first half of the 19th century, this trend matured, and associated with it was an increase in meteorological observations, made initially as an aid to astronomical measurement (knowledge of air temperature was required, for example, in the determination of refraction). Changes over time in the instruments and the methods used in making meteorological observations affected the usefulness of the earliest records, but their length makes many of the series important and some have been 'reduced' from several stations in a neighbourhood. Meteorological observations may pre-date the founding of an observatory, as at Edinburgh (1825), where temperature records are available for 1731–6 – mean temperature at 9 a.m. – and for 1764–1896 – mean temperatures 'reduced' to a standardised series (Mossman 1897). At Kew Observatory (built for George III in 1768–9), meteorological observations were started as far back as 1773 (Whipple 1937, Drummond 1947, Jacobs 1969). At the Radcliffe Observatory at Oxford (built between 1772 and 1794), a continuous series of daily records of temperature is available from January 1815, and daily readings at the same site are available for approximately half the period 1767–1805 (C. G. Smith 1968, 1980).

The 1840s were a critical period in the development of meteorology. At

Greenwich Royal Observatory (erected in 1675 by Charles II for John Flamsteed, the first Astronomer Royal), a meteorological department was created in conjunction with a department for the study of terrestrial magnetism in 1840. James Glaisher was appointed as the first Superintendent of Meteorological Observations (Witchell 1947), and from 1847 Glaisher prepared meteorological tables for the *Registrar-General's quarterly returns of births, marriages and deaths*. In 1842, the observatory at Kew was taken over by the British Association for geophysical and meteorological studies, and much pioneering work on instrumentation took place there in the mid-19th century. Interest in meteorology was blossoming. The Meteorological Society of London (founded in 1823) published a volume of *Transactions* in 1839 and two volumes of *The Quarterly Journal of Meteorology and Physical Science*, published under the immediate sanction and direction of the Meteorological Society of Great Britain in 1842 and 1843. In 1848, an earnest effort was made to revive the then declining society, but it was overtaken in size and in the soundness of its scientific deliberations by the British Meteorological Society, formed in 1850 and later to become the Royal Meteorological Society (Cockrell 1968, Ratcliffe 1978).

It is in this overall context that the development at Durham should be viewed. Though founded by clerics in 1832, the University of Durham did not neglect the sciences in its early years. As well as being the first university in the country to offer courses in engineering, Durham invested at an early stage in astronomical equipment towards the founding of an observatory (Rochester 1980, Heesom 1982). Bishop Maltby wrote to Lord John Russell on 19 February 1839 referring to steps for the establishment of an observatory, and on 23 February 1839 he expressed a desire to see a chair of Astronomy, and saw the establishment of an observatory in Durham as highly desirable in the interest of science (Whiting 1932). Astronomical instruments were purchased from the Reverend T. J. Hussey of Hayes Court in Kent, who had offered them to the new university in 1838 following an accident that left him unable to use his own observatory. The money involved was raised by private subscription, and a site for the observatory was made available by the Dean and Chapter (Whiting 1932).

A prime mover in the negotiations was the Reverend Professor Temple Chevallier (1794–1873), who became the first director of the observatory. Professor of Mathematics and Astronomy, Chevallier took a great interest in the development of an engineering school, and later in the college of science (Armstrong College) founded at Newcastle in 1871. He taught Hebrew to theological students, served as registrar to the university for 30 years, and still found time to preach in his parish at Esh. One of the first group of Fellows of the British (later Royal) Meteorological Society, his contribution to the meteorological record is indicated in the obituary in the society's journal:

' . . . it was Mr Chevallier's care alone that has prevented those interruptions in meteorological work which have so injurious an effect upon the deduced results. His interest in meteorology is shown by his early enrolment among

the Fellows of this Society, having been elected May 7th, 1850' (Meteorological Society 1873–5).

Clearly, Temple Chevallier can be included in that band of 'keen, enthusiastic
and purposeful early Victorians who were responsible for so much scientific
progress', to whom Gordon Manley refers in a foreword written when he was
President of the Royal Meteorological Society for the first issue of *Weather* in
1946. Moreover, it seems that Temple Chevallier possessed some of those
qualities that were much appreciated in Gordon Manley himself, for, as one of
his pupils wrote:

'Mr Chevallier . . . was animated and interesting in his way of teaching, and
would do anything in his power to help one either in or out of lecture.
Sometimes he would go off into some digression which might have little or
nothing to do with the matter in hand, but which was sure to be inspiring and
instructive' (Fowler 1904).

2.3   The beginning of meteorological observations at Durham

As elsewhere, the prime purpose of Durham University Observatory and that
of the appointed observers was astronomical. Fowler (1904) gives a list of
observers from 16 June 1840, and the building was finished in 1841 (Plate 2),
though it was not until 1843 that observations were made to determine its
latitude (Whiting 1932). Baxter (1937, 1956) comments that the first mention of
meteorological work in the observatory reports is that for 1849–52 by the
observer R. C. Carrington. For some time no manuscript of meteorological
records had been found earlier than that for 1850–52 by Carrington, 'very finely
written up, with a useful preface' (Manley 1941a), and Manley had suggested
that there was no evidence of meteorological observations being kept until
sometime in the latter part of 1846. It was established from Glaisher's early
summaries (1848) that regular observations were sent in throughout 1847.
Glaisher's published yearly mean for 1847 and quarterly means and extremes for
the last quarter of 1847 and for 1848 were the earliest records used by Manley in
compiling the temperature series to 1941. Baxter (1956) reports that Manley
'assessed approximate mean annual temperatures for each of the years 1843 to
1846', but it seems likely that he refers to approximations arrived at by reference
to another station or stations (perhaps York).

Both Sargent (1923) and Whiting (1932) state that 'the meteorological record
is an unbroken one from the opening of the Observatory in 1841', though it is
not clear whether their comments are based on knowledge of existing records.
Manley's notes from 1938 suggest ' . . . (observations began in 1843) records
lost for 1843–1847', but do not indicate whether reference is to meteorological
or astronomical observations. Later (27.ix.78), Manley comments that 'observations at Durham were certainly being made in 1843', and he refers (1.iii.79) to

Plate 2 Durham University Observatory, designed by Anthony Salvin in 1840

odd mid-Victorian summaries of observations where he originally found reference to 'the beginning of Durham about 1843' and to 'mentions of odd readings for individual days in 1841'. The challenge of locating earlier records remained.

An incomplete series of daily observations was located in copies of the *Durham Advertiser* in weekly summaries from 15 May 1842 to 27 December 1843, signed by A. Beanlands, Observer. Further investigation showed that daily observations had been published in the *Durham Advertiser* by the observer, John Stewart Browne, from 13 June 1841, though again the series is incomplete.

Manley had been increasingly convinced that observations were carefully recorded during the early years of the observatory. In a letter (3.vi.79) he reported:

'What I *have* found recently while re-examining a very obscure little book by the redoubtable amateur *E. J. LOWE* of Nottingham is (1) that he dedicated his little book on 'Atmospheric Phenomena' (1846) to the *Rev. Temple Chevallier*, Prof. of Maths and Astronomy at DURHAM, and (2) in a discussion of the prolonged cold winter of 1845 he quotes the monthly means for Durham for Oct. 1844 to March 1845 inclusive.'

Manley had 'searched the Observatory collection forty years ago . . . ' for any fragments of MS. before Carrington began his very neatly kept MS., but without success' (1.iii.79), and his later enquiries in the Durham University Library seem to have been related to other sources, such as diaries and weather notes in various special collections. However, his concern to turn up early observatory records was well known, and it is therefore sad to report that, after 50 years of interest in the Durham meteorological record, he was to die before the cataloguing of the collection of records at the observatory in 1982 led indirectly to the discovery in the University Library of a manuscript record of observations from 23 July 1843 to 31 December 1847 and one for 1 January 1848 to 30 April 1850 (monthly means from January 1850 had been available previously from Carrington's 1850–52 manuscript mentioned above). It may be noted that there is some ambiguity in Manley's account, and it is not clear what source he used for data for 1849 ('alas, no record has been found for April to June, 1849' – Manley 1941a).

2.4   The Durham temperature record, 1847–1979

The standardisation of a long temperature series, particularly one that goes back to the days before the Stevenson screen, is a complex and essentially meticulous task. As Manley (1941a) states: 'The fundamental difficulties are of course well known; differences in exposure, in hours of observation, in instrumental corrections and even in the observers' predilictions must all be accounted for.' Manley had aimed to produce 'a "University" record to accompany the senior one (Oxford, Radcliffe obsy. since 1815), and to be in between Oxford and Edinburgh' (27.ix.78). 'Durham ranks after Oxford as a continuous *University* record in one place. (Cambridge, Glasgow and Aberdeen all broke down somewhere!)' (Manley 5.xi.79).

The qualities of the site and its representativeness are described in his paper (Manley 1941a). The local exposure of the observatory is good, although the basin-like character of the Wear valley causes extreme minima, even on the observatory ridge (336 feet, 102 m), to be 'somewhat lower than the exposure would at first suggest'. The fact that very little building had taken place within the vicinity of the observatory added to the value of the record and, as it seemed likely that this state of affairs would continue, it was 'very desirable that the meteorological observations on this site should be carefully maintained'. This point holds good today (Harris 1982, and Ch. 3 of this volume).

The complexity of the problems met with in the standardisation of the Durham record through a period of 90 years makes the account by Manley (1941a) of an 'almost over-elaborate job' (Manley 27.ix.78) a classic illustration of the fact that it is not enough simply to accumulate instrumental measurements of climatic conditions. Manley's concern to make a long run of data usable in the interest of studies of climatic variation over time also reflects his great respect for those who devoted time and effort to collecting the data in the first place. He expresses the hope that his work will:

'afford some belated recognition of the faithful and painstaking maintenance of the meteorological record over many years by the fourteen successive University Observers, a task which has generally been unassisted and carried out in isolation from the rest of the University' (Manley 1941a).

For a full account of the adjustment of the monthly mean temperatures for the period from 1847 to 1979, the reader can refer to Plummer (1873), Manley (1941a, 1980) and Baxter (1956). In brief, 'adopted means' (A), are derived from a combination of the mean of the daily extremes (M) with the means of the 9.00 h and 21.00 h fixed-hour readings (F), so that:

$$A = \tfrac{1}{2}M + \tfrac{1}{2}(F + K)$$

When monthly values of K are as given by Manley (1941a) and allow the effects of known weaknesses in the observation of extreme values, in particular of the maxima, to be lessened by reinforcement from fixed-hour observations. A complicated procedure of correction is also described that allows for variations in the times of the fixed-hour observations (Manley 1941a).

2.5   The Durham temperature record, 1842–7

The discovery of weekly summaries of daily data from 13 June 1841 and, since Gordon Manley died, of the manuscript volume for 23 July 1843 to 31 December 1847, confirms previous indications that a continuous series of meteorological observations was made from 1842 at least, and probably from 1841. As will be seen, data for those years can help bridge a gap in the construction of a longer temperature series for north-east England, but the data have importance in any case in extending the record for the observatory back in time and allowing it to reflect the outstanding weather events expected to be evident in those years: 'cool weather in May 1843; excessive cold mid-Feb. 1845 and all through a very snowy March; a tremendously hot June of 1846, probably mitigated at Durham by the sea-breeze; pretty cold in Jan. 1847; dry Dec. 1844' (Manley 1.iii.79). Elsewhere, Manley (8.viii.79) refers to 'about five extremely hot, or cold, months that would be worthwhile–MARCH *1845*, JUNE–JULY– AUGUST *1846* (especially the phenomenally hot June) and perhaps December 1844, and December 1846.' An entry in the *Durham Chronicle*, Friday, 19 June 1846, confirms that conditions in that month were exceptional:

> It is so melting hot this week,
>   That, though our readers flout us,
> The truth we must in conscience speak,
>   We've scarce our wits about us.

There were three observers during the early period: John Stewart Browne in 1841, Arthur Beanlands, 18 February 1842 to 3 February 1846, and Robert

Anchor Thompson, 3 February 1846 to 25 June 1849; while two others stood in: Le Jeune in June, 1849 and Robert Healey Blakey from June to October 1849, before Richard Carrington took over (Fowler 1904). Observations were probably made initially on the north wall of the observatory, 4 feet (1.2 m) above a small, flat, jutting roof, 17 feet (5.2 m) from the ground (Manley 1941a):

> This was a fashion set by the renowned Radcliffe obs^y. at Oxford, to read the thermometer outside the 'Transit Room'; They built the Cambridge obs^y. (1823, begins) in the same fashion and we have fragments here indicating that the thermometer was outside the window (above a little jutting wing) 15'3" above the ground. Gradually they changed towards the 'Glaisher stand' type of exposure, after Glaisher began to publish his 'Quarterly R' in 1847. This Victorian rivalry as regards 'type and routine of exposure' went on for a long time, and bedevils everything before 1881 or even 1900, in some cases: (and *1968* for *Kew Observatory*!) (Manley 1.iii.79).

The thermometers at Durham were apparently not screened (by the 'north-shed') until October 1851 (Manley 1941a), and Manley assumed:

> that as the north wall readings were well protected from radiation, the fixed-hour observations at 9 h and 21 h could be treated in similar fashion to those in a Stevenson screen. The evidence of the apparent mean daily range indicated that otherwise the corrections for the maximum and minimum were considerably less than those in the 'north-shed', and again a combination of the four readings has been used to give a mean temperature comparable with those which would be recorded under present-day conditions. Both the 'north-wall' and 'north-shed' readings, until at least the beginning of 1854, were kept with great care' (Manley 1941a).

His working notes suggest corrections of $+1°$ on the maxima and $-1°$ in winter on the minima ($-2°$ May–September) for 'estimates for north wall extremes'.

With regard to the data from 1842, Manley (1.iii.79) suggests that 'one can continue backwards with confidence, I feel sure, given a summary of the observations', and, from his use of data extracted from the *Durham Advertiser* (Manley 8.viii.79), it is clear that he has reasonable confidence in a mean derived conventionally from the maximum and minimum readings at least as a working base, and that he now considers it probable that they were already screened by a 'north-shed'. He takes the mean values which he found published for October 1844 to March 1845 to be a combination of 'M/m, 9.00 h/21.00 h' and makes some adjustment, although these means have since proved to be from the fixed-hour observations alone.

Much more remains to be done in the careful examination of the manuscript records for 23 July 1843 to 31 December 1847, and for 1 January 1848 to 30 April 1850, but for the readers' interest the mean monthly temperatures derived by the

Table 2.1 Durham University Observatory: mean monthly temperatures (°F) derived by the contemporary Observers from readings at '9 a.m.' and '9 p.m.'.

	Jan.	Feb.	Mar.	Apr.	May	June	July	Aug.	Sept.	Oct.	Nov.	Dec.	Mean
1843	—	—	—	—	—	—	58.5	56.9	56.6	44.2	40.5	44.1	—
1844	37.9	33.1	38.7	48.0	48.5	53.0	58.6	55.3	54.3	46.6	41.7	34.0	45.8
1845	35.9	32.1	35.4	42.8	45.9	57.8	57.5	52.8	50.2	47.2	42.2	37.6	44.8
1846	42.2	42.3	40.1	42.6	51.9	65.0	60.6	60.1	57.0	46.9	42.6	32.9	48.7
1847	34.2	34.8	39.8	42.3	51.5	55.3	61.8	57.4	50.5	47.7	44.4	39.8	46.6
1848	33.1	39.7	39.5	43.1	56.4	55.4	59.0	54.5	53.1	46.6	40.1	38.9	46.6
1849	37.2	40.7	40.8	41.1	(49.8)	(53.9)	57.7	57.7	53.1	44.7	41.2	(38.1)★	46.3

Figures in brackets are calculated from incomplete daily data, the contemporary Observer having chosen not to calculate a mean.
★The contemporary Observer used data for 1–23 December only.

Table 2.2 Durham University Observatory: mean monthly temperatures (°F) derived from the mean maximum and mean minimum values entered for each month by contemporary Observers (*max. + min./2*).

	Jan.	Feb.	Mar.	Apr.	May	June	July	Aug.	Sept.	Oct.	Nov.	Dec.	Mean
1843	—	—	—	—	—	—	57.7	57.5	58.0	45.7	41.9	45.3	—
1844	39.1	35.0	40.3	49.1	48.5	54.5	53.5	55.5	55.1	47.9	44.3	35.3	46.5
1845	37.7	34.1	36.5	44.5	46.9	57.0	55.9	55.3	51.4	48.7	44.6	38.6	45.9
1846	41.1	42.9	40.4	43.2	51.5	63.3	60.7	59.7	56.8	48.2	43.9	33.1	48.7
1847	34.7	35.4	40.7	41.5	51.5	55.7	61.5	56.5	50.7	48.1	45.3	39.9	46.8
1848	33.2	40.5	40.6	43.7	55.5	55.3	58.3	53.9	53.3	47.6	40.9	40.1	46.9
1849	37.7	41.9	40.7	41.3	50.2	(53.3)	57.1	57.9	53.4	45.5	41.1	(38.1)★	46.5

Figures in brackets are calculated from incomplete daily data, the contemporary Observer having chosen not to calculate the mean maximum and minimum values for the month.

★ The contemporary Observer used data for 1–23 December only.

contemporary observers from the fixed-hour readings at '9 a.m.' and '9 p.m.' are given in Table 2.1. These are the values likely to have been quoted in other contemporary sources. For comparison, the mean monthly temperatures derived from the mean maxima and the mean minima are given in Table 2.2. The contemporary observers did not calculate these means.

2.6 A temperature series for north-east England

For many years Manley had known of the manuscript volumes in the library of the Newcastle Literary and Philosophical Society, which comprise the daily meteorological readings and notes in ledger form of James Losh (1762–1833), a prominent Newcastle lawyer, and cover the period from 1802 to 1833 (September) for Jesmond, Newcastle. He believed that the record might be adjusted and linked to the Durham University Observatory record, thus extending the temperature series.

Material would be needed to bridge the gap between the Jesmond and the Durham series and, as Manley (27.ix.78) suggests:

> It is always possible that there was a 'keen local amateur' in Durham. Quite often it was a subtle means by which the local medical man (or apothecary perhaps) advertised himself. If so there *might* possibly be records in the 1830s. It would be pleasant if there were!

No such records have been found, but a number of stations in north-east England for which data are available for the first part of the century could be used to derive values representative of Durham. Moreover, Manley had worked in Hull on a record of daily observations for South Cave, west of Hull, for the period 1794–1814, and this could be adjusted and overlapped with the data from Losh. So 'there is sufficient material to extend "Durham" back to 1795' (Manley 1980). The chief reason for considering and attempting the latter extension was to include 'the exceedingly severe winter of 1795 and the very mild 1796' (Manley 27.ix.78). The proposal as it seemed in 1978 is shown in Table 2.3, taken directly from one of Manley's letters, and it illustrates his approach.

Manley had long tried to find local sources:

> I've tried repeatedly in the past to find if the (reported) diary and weather notes of Timothy Whithingstall about 1636–1670 (mentioned in Arch. Aeliana, about 1924), who seems to have farmed towards Lanchester, have ever been found in MS. in the Diocesan, or other Cathedral collection (?Mickleton MSS). (Manley 30.i.78).

The search for a local source for the gap in the 19th century was similarly frustrating:

> There is YORK (1831 onward) but I'm having great difficulty in finding any

Table 2.3 Composition of a temperature series for north-east England: proposals (Manley 27.ix.78).

1794–1814	Daily obs. *SOUTH CAVE*, west of Hull: could be overlapped and used as a tentative extension for Durham.
1798–1808	*Braithwaite above Keighley*; indoor obs. only, providing pretty rough monthly means.
1802–1833	NEWCASTLE (JESMOND), Daily obs. 9^h, 14^h, 23^h. (*Means available, 1812–1818 only.*) Appear to be a *very good series*, closely representative of Durham.
1801–1824	*York*: moderate quality. (1817–1824 Malton)
1841–1890	*York*: good series for most part. Possibly back to 1831, not yet found.
1824–1852	ACKWORTH Luke Howard's record. Appears very reliable.
1841–1845	KELSO. Other Berwickshire fragments, 1832 onward: possibly more Kelso available.
1809–1857	BRAITHWAITE above Keighley. Apparently v. good; but 750 ft. above sea.
More Distant	*EDINBURGH* 1764–1896, carefully reduced. 'Lancashire' 1753–1978 available.
DURHAM	University Observatory 1847–1940. 'Adopted means' based on instruments in Q.J Roy. Met. S. 1941 a variety of exposures, carefully overlapped: using max. min. *and* the fixed-hour obs. 9h and 21h.
	1941–1950 adopted means continued
	1951–1958 might be continued but after 1958 evening obs. abandoned
	1959–1978 mean based on ½max. and min. with small adjustments.

York before 1841. There's an earlier *YORK* set, 1801–1824 that inadequately overlaps *ACKWORTH* (Manley 27.ix.78).

The Royal Hist. MS. Commission mentions a Hindmarsh MS. Diary near ALNMOUTH for 1833–39. I don't know what it holds or if there are instruments (Manley 27.ix.78).

I've found fragments for *North Shields* 1841 and 1842; *Middlesbrough* 1842; but no real 'bridge' *locally* to link with *Losh's 1802/1833* set (Manley 1.iii.79).

Darlington, altho' its a Quaker town, is no good. I've never tried the *Teesdale Mercury* at Barnard Castle! There is another fragment in Berwickshire that might 'support a pattern' in 1835–39, and I might have another try at Alnwick Castle. Without doubt one can make a very reasonable 'bridge' for 1833 to 1843, but if one could find something nearer Durham it would be preferable! (Manley 1.iii.79).

We don't seem to have had many 'science-minded gentry (or doctors) round Durham, but Bishop Barrington's secretary *Mr. EMM*, at Auckland Castle was keeping observations of some kind in 1807 and he's a possible source (again, it would probably be a 'yearly summary'). Of course if any other Durham county contributor [to the *Durham Advertiser*] covered 1833–1842 it would save a great deal of rather 'uncertain estimation' from the several records at a distance that I've used (Manley 3.vi.79).

There's a 'Tyneside Natural History Society' – with volumes beginning in 1846, but they are rather disappointing until later in the 1850s when their secretary collected quite a number of rainfall totals – including Durham from the observer, A. Marth (Manley 3.vi.79).

I have found, at *York*, the complete York series (on Ford's earlier site) that run from 1832–1846 and I know they run tolerably well with Durham; and this last week I have collected Mackerstoun (Berwickshire) 1842–1855 (a *very* superior 'observatory'). South of Durham we have also Luke Howard's *ACKWORTH* – the Quaker School – that runs from 1824–1850 but alas, its figures give reason for suspicion over 1848–50 and also about 1835–6 (Manley 8.viii.79).

2.7   The Losh record, 1802–33

The daily meteorological readings of James Losh cover a period of almost 32 years (1802–33), and are entered in columns for three fixed-hour observations each day. None seems to have been previously converted to mean values, except for the seven-year period 1812–18, for which means were presented by Losh in a paper to the Newcastle Literary and Philosophical Society and printed in their *Transactions*. These means were quoted by Heinrich Dove (1838), and they were also used by a local botanist (Winch 1838).

On the evidence of his extremes, which are not exceptional, Manley concludes that Losh may well have had his thermometer on the north wall of the house, possibly outside his study window (Manley 27.ix.78), but later suggests the east wall (Manley 5.xi.79). The house (Jesmond Grove), and park were well up on the gentle slope that declines towards Jesmond Dene, and Manley (27.ix.78) suggests that Durham at 336 feet (102 m) above sea level is probably about 0.6°F cooler than Jesmond, at about 150 feet (46 m). For the period 1802–18, Losh used a variety of hours and left some days out. Observations were mainly at 8.00 h, 16.00 h and 23.00 h for 1802–03; 9.00 h, 15.00 h and 23.00 h for 1806–11; and 9.00 h, 14.00 h and 23.00 h for 1812–18 – being at 9.00 h, 14.00 h and 22.00 h thereafter.

Having calculated fixed-hour means, Manley adjusted them to the mean of the extremes (maximum and minimum values), using the Greenwich hourly correction, 'which appears to give more satisfactory results than those for either Kew or Rothesay, Aberdeen or Eskdalemuir' (Manley 5.xi.79). Scrutiny of the means for 1812–18 (quoted by Dove and Winch) led Manley to think that large

Table 2.4 Durham: Monthly Means (°F) reduced from Losh and other observations on to Durham* University Observatory (Manley 5.xi.79, 27.xi.79).

	Jan.	Feb.	Mar.	Apr.	May	June	July	Aug.	Sept.	Oct.	Nov.	Dec.	Year
1801	(39.0)	(40.0)	(43.0)	(45.5)	(51.0)	(56.0)	(58.5)	(61.0)	(56.5)	(49.5)	(40.0)	(33.5)	(47.8)‡
1802	35.7	38.0	41.2	46.2	46.5	55.0	54.8	60.5	55.5	49.5	41.5	37.7	46.9
1803	34.0	36.5	41.2	47.0	47.6	53.8	60.7	58.5	51.5	47.6	39.2	36.5	46.2
1804	39.8	34.5	37.7	41.0	52.6	57.5	58.0	57.0	57.8	51.2	42.3	35.6	47.1
1805	36.3	37.0	41.4	44.0	46.0	52.5	60.5	60.5	57.0	45.5	39.5	37.1	46.4
1806	35.3	36.6	(38.5)	(42.0)	(50.0)	(56.0)	(59.0)	(59.8)	55.0	49.3	44.1	41.1	47.2
1807	37.4	36.5	35.7	43.4	50.1	53.4	60.2	60.8	48.3	50.8	35.6	37.1	45.8
1808	36.1	36.0	38.0	40.0	52.2	(55.9)	(61.2)	(61.4)	52.6	43.5	42.4	36.8	46.4
1809	32.7	42.1	42.3	41.1	53.3	54.5	57.2	58.1	52.1	51.6	41.5	39.1	47.1
1810	37.0	37.5	38.8	45.0	47.6	57.1	58.1	58.5	54.7	49.8	41.2	37.0	46.9
1811	33.8	38.1	44.9	46.7	51.3	54.4	58.7	57.8	55.5	53.8	45.2	36.9	48.1
1812	35.4	39.6	36.0	39.9	49.1	54.6	56.0	56.2	54.7	47.8	40.6	36.0	45.5
1813	35.6	42.2	43.5	45.0	50.8	54.4	59.0	56.2	55.3	45.8	38.6	38.8	47.1
1814	27.4	34.6	37.0	48.2	45.8	50.9	58.8	58.6	55.3	46.8	40.3	37.6	45.1
1815	32.4	41.9	42.0	44.4	51.5	55.2	56.4	57.3	55.5	49.2	41.6	32.2	46.6
1816	35.8	34.8	(36.8)	(40.3)	46.7	52.5	54.7	55.0	51.0	(47.8)	38.0	35.8	44.1
1817	39.4	41.9	40.0	44.4	47.1	55.5	55.6	54.0	54.3	42.7	44.5	34.1	46.1
1818	36.8	34.8	36.9	40.8	48.7	59.8	61.0	56.9	55.2	52.0	47.4	39.7	47.5
1819	37.4	36.6	42.1	45.3	49.1	54.4	59.3	60.4	54.6	46.9	38.1	32.0	46.3
1820	31.1	38.8	38.5	47.2	50.3	53.8	56.7	56.4	53.0	46.6	40.4	39.8	46.1
1821	37.1	36.9	41.0	46.6	46.4	50.7	56.0	58.0	56.3	49.1	42.7	40.1	46.7
1822	39.0	41.4	42.9	44.2	51.6	59.4	57.6	57.4	50.0	48.2	44.4	35.3	47.6
1823	32.3	34.0	38.6	41.3	51.2	51.3	56.0	53.3	52.5	46.3	44.8	37.9	45.0
1824	38.9	38.1	37.9	43.6	48.6	54.0	59.2	56.6	54.2	46.6	41.7	38.7	46.5

(1801 Estimated from Lancs. Edin. and York by extrapolation)

DURHAM
Decade   Lancs   Edin.
46.8   47.7   46.8

Revised from 1812 to Jan. 1818 for errors in assumed hours of observation

DURHAM
Decade   Lancs   Edin.
46.3–   47.1   46.1

Year													Annual
1825	38.3	39.1	38.9	46.4	49.7	53.9	60.3	60.6	57.9	50.4	38.7	39.0	47.8
1826	32.3	41.5	40.4	45.8	50.1	62.2	64.2	61.1	54.5	50.4	38.8	42.0	48.5
1827	34.8	34.0	39.2	44.7	50.1	56.4	60.2	56.0	54.6	51.3	42.1	42.9	47.2
1828	39.9	39.8	42.3	44.1	51.3	57.4	58.7	57.7	55.1	48.4	44.6	44.0	48.6
1829	32.4	37.6	38.6	41.1	53.5	56.1	57.0	54.8	50.5	46.1	40.0	35.3	45.3
1830	32.6	35.1	45.9	45.5	50.3	51.6	59.0	54.5	52.2	49.3	42.9	35.0	46.2

DURHAM Decade *46.9*   Lancs *48.2*   Edin. *47.2*

Year													Annual
1831	35.1	38.9	42.7	46.6	49.4	56.7	60.4	59.8	54.5	55.2	40.4	41.9	48.5
1832	38.6	39.1	41.9	44.7	49.0	55.7	57.2	56.8	55.1	52.7	42.1	39.5	47.7
1833	34.2	39.7	38.1	43.9	56.5	55.8	59.0	55.0	53.0	48.5	42.3	41.6	47.3
1834	42.0	40.6	43.1	44.0	52.1	56.0	60.3	58.8	55.0	49.1	43.1	41.7	48.8
1835	35.9	39.9	40.9	44.3	48.7	55.1	58.7	59.6	54.0	46.3	41.9	38.5	47.0
1836	37.7	37.3	39.7	42.7	50.3	56.5	57.3	55.3	51.7	46.2	39.8	38.9	46.1
1837	35.2	38.9	35.0	38.3	47.9	56.5	59.9	56.5	52.9	49.3	40.3	40.5	45.9
1838	29.7	30.5	39.4	40.7	47.2	55.1	59.1	56.8	53.9	48.2	38.6	39.4	44.9
1839	35.9	37.6	38.1	42.6	48.7	55.6	57.6	56.7	53.7	47.8	43.3	38.2	46.3
1840	37.2	36.2	38.0	48.5	50.0	57.2	56.5	60.8	51.3	44.5	39.5	34.0	46.1

Revised amended a little Nov. 1979

DURHAM Decade *46.9*   Lancs *48.1*   Edin. *46.7*

Year													Annual
1841	32.5	36.0	45.0	45.5	54.2	54.0	57.0	56.5	54.0	45.5	38.0	38.5	46.3
1842	32.0	37.5	42.0	44.0	51.3	57.0	58.0	61.5	54.2	42.2	40.0	45.5	47.2
1843	38.2	33.0	40.9	46.0	47.0	51.3	57.7	57.9	57.8	45.8	40.4	45.3	46.8
1844	38.5	33.5	39.5	47.5	49.5	55.5	57.7	55.0	54.5	47.2	42.3	34.2	46.2
1845	36.3	32.3	35.8	44.8	47.5	57.0	57.0	55.3	52.0	48.2	41.8	38.5	45.5
1846	41.7	43.0	42.0	44.2	52.5	62.6	61.0	61.0	57.7	47.6	44.0	33.8	49.2
1847	35.2	35.4	41.2	42.3	51.9	56.3	62.5	58.3	51.0	48.2	44.7	39.6	47.2
1848	32.7	40.8	40.7	44.2	55.3	55.2	59.3	54.3	53.4	47.2	41.2	40.3	47.1
1849	37.7	42.5	42.0	41.7	50.7	53.8	57.6	57.9	53.5	45.9	41.5	38.3	46.9
1850	32.3	42.8	39.7	45.1	47.4	57.5	57.4	55.9	51.1	44.8	43.0	38.5	46.3

DURHAM Decade *46.9*   Lancs *47.9*   Edin. *47.3*

For 1851–1940, see Manley (1941a); for 1941–79 see Manley (1980); for onwards see monthly *Summaries of the meteorological readings at Durham University Observatory*, Department of Geography, University of Durham.

* Wording of Manley's title unclear here.   ‡ Brackets are not explained.

errors had crept in, whether through misprints, bad arithmetic or gaps in the record, but further checks against the original manuscript revealed that through 1812 to January 1818, observations by Losh were at 9.00 h, 14.00 h and 23.00 h (as indicated above), and not, as he had previously supposed, at 9.00 h, 14.00 h and 22.00 h. An adjustment was therefore needed to his first calculations for Table 2.4.

Losh took his 9.00 h observation as usual on the day when he collapsed and died. Scientifically minded, 'intelligent all round' (Manley 27.ix.78), and a backer of George Stephenson, James Losh would surely have approved with pleasure the use now being made of his careful observations, though he may have had some regrets that they serve to enhance the record of the observatory set up by the University of Durham. He was a supporter of the 1831 proposal to establish a university in Newcastle, a proposal temporarily foiled by the foundation of the University of Durham in 1832. In 1833 'James Losh reluctantly wrote to his friend Henry (now Lord) Brougham, that though "some progress" had been made, "as Durham University seems likely to be carried into effect, our plan will be at least suspended for the present" ' (Heesom 1982).

2.8 Bridging the gap: the period 1833–47

By calculating the departures for each month of the series on which it was eventually decided to depend for an overlap with the Losh series for Jesmond, Newcastle, to September 1833 (Table 2.5), Manley arrived at 'provisional monthly means' for Durham. The deduced values for each of the series quite often departed from the others, and sometimes instrumental error was suspected. Clearly Manley had no great confidence in his conclusions at this stage, and he still considered the usefulness of more local 'fragments':

> N. Shields 1842, Middlesbrough 1842–43 (Yarm 1842), Allenheads 1842 and perhaps 1844, Newcastle 1846 to give an idea of 'pattern' – they're all

Table 2.5 'The stations on which one must depend in order to get an overlap with the Losh series from Newcastle, 1802 to September 1833' (Manley 3.vi.79).

(1) Ackworth 80 miles SSE: Luke Howard's series running from *1824–1850* (perhaps '52).
(2) York 60 miles SE by S: John Ford 1841–1852 (and later) but with a break in 1845 (change of location) and 1848 (instruments).
      [York: There are also *averages* for each of the months for the period 1832–44]
(3) Kendal 60 miles SW: Samuel Marshall, a full series 1823–60, almost homogeneous.
(4) KEIGHLEY (BRAITHWAITE) (Abraham Shackleton MS) 1798–1857 65 miles SW.
(5) EDINBURGH 100 m NW [a very carefully reduced table (1764–1896), but pretty distant].

There are also some shorter records in the Border Counties – none right through the gap: and one in Yorkshire for 1831–40 that I haven't yet examined (*if* I can find it). *Also I MIGHT run Allenheads to earth on my next visit to Newcastle: it covers 1836–1876. Ackworth* is a less perfect record than I hoped; *York* is troublesome; *Kendal* is across the Pennines. *EDINBURGH* is not only a long way off but I want to avoid using it if I can, and keep 'Durham' independent.

rather uncertain alas. Really its a most troublesome period, fortunately the general agreement between *inland* N. Yorkshire–Durham–Northumberland–South Berwickshire is good (Manley 8.viii.79).

On 9 September 1979, he was still dissatisfied:

The general pattern displayed by Losh's observations against 'Central England' is good. But getting a satisfactory 'bridge' between Losh at Jesmond and Durham Observatory is going to be quite tricky. I can see how to do it in general terms, but life is complicated by John Ford the Quaker of York, who moved his station; by Sir J. Brisbane (ex-Australia), who at Makerstoun in Berwickshire had a superbly equipped observatory, but DID NOT OBSERVE ON THE SABBATH.

2.9 A tentative series of mean monthly temperatures for north-east England from 1801

Having arrived at the reduction to Durham of the data for 1802–32 from Losh for Jesmond, Newcastle, and having looked at the various possibilities for use of the data from stations further afield, Manley arrived at a tentative series of mean monthly temperatures for north-east England (see Table 2.4). Clearly he felt that it would be quite some time before it was completed and written up, and stated explicitly, 'this isn't yet for publication, merely retention' (Manley 5.xi.79), and, on sending a slightly revised version (incorporating the modification referred to above on the Losh data for 1812–18): 'This is really an insurance, in case of loss!' (Manley 27.xi.79).

Manley's own words best summarise what had been done after the reduction to Durham of the Losh data:

There is an overlapping series kept by a Yorkshire squire at Brandsby 13 m north of York from 1811–1830 (*and afterwards*; but we cannot find his MS. at the York Philosophical Society after 1830). For 1811–1830 there is a summary in the County Archives at Northallerton. Two short gaps can be filled from a Malton record (1817–1825). The Brandsby man observed at 8½ h., 14 h., and 23 h., but he doesn't tell us *how*; but his exposure was more 'open' than Losh, who I think used the *East* wall of his house, possibly outside his study window with a board to screen it, and several feet above the ground.

Much more distant, there is the long 'Edinburgh' table (by Mossman) since 1764 of which I have some little doubts; and there is the long 'Lancashire' table since 1753 compiled by Manley, of which other people may have some little doubts, or even bigger ones.

I think they must be used to provide an 'overall control', that is the fluctuations of the *decadal* means, and perhaps the *annual means* at Durham, after all reductions have been made, should 'fit' with Lancashire–Edinburgh.

Table 2.6   Data available 'to link all before 1833, on to Durham which so far has been all after 1846' (Manley 5.xi.79).

	(1) *Durham Advertiser*: (a) close estimates based on bits of about 5 months on 1842, (b) nearly all of 1843.
	(2) Published figures for the *five-year mean for Jan. Apr. July Oct.* for 1843–47 (found in Phillips' 'Yorkshire', 1853).
	(3) YARM for 1840 and 1842: obs. at 8h. 12h. 16h. and 20h. daily, capable of reduction.
	(4) North Shields 1842: Middlesbrough 1841 and 1842: Allenheads 1841 and 42 (dubious).
(Quaker record)	(5) *YORK* 1832 onward; careful, but in sheltered garden; change to *more open site* end 1846, at Yorks Philosophical Society.
(Quaker LUKE HOWARD)	(6) *ACKWORTH* 1824–1850. Occasionally suspect.
(Quaker J. Gray)	(7) *YORK 1800–1824*: 8 a.m. *only* and very sheltered, *Wykeham*, inland from Scarborough, 1831–1836.
(Quaker A. Shackleton)	(8) *KEIGHLEY 1800–1857*: 10 a.m. only, and 'indoors' until 1809.

All these are N.E. England

SCOTTISH BORDERS	(9) *KELSO* said to begin 1832 but not found until 1842.
	(10) *MAKERSTON* nearby, 1841–1845, but local change in 1849.
	(11) *Abbey St. Bathams 1835–1839.* (Hawick and Creswell-Twizell in Northumberland in 1840s rejected).

WEST OF PENNINES   *KENDAL 1832–1869, Carlisle 1802–1824, Applegarth (Dumfries) 1827–1851.*

All these have been incorporated in my 'Lancashire' reduction.

Having attempted a compilation of sorts of Losh and the Brandsby set the big problem has been to link *all before 1833*, on to Durham which so far has been *all after 1846* How? [Table 2.6].

Well: off all those I worked out a series of the *most probable 'monthly anomalies'* applicable in NE England, a few miles inland from the coast. I extended one set *forward*, from the earlier *Losh–Brandsby* series 1801–1832: and another set *backwards*, from the later *Durham* series 1847–1856. The agreement wasn't too bad and I've taken the mean.

So you can see; quite a fierce job, of 'trimming' between records that just do *not overlap*, with the exception of rather doubtful Keighley, Kendal beyond the Pennines, and the more distant Lancashire and Edinburgh that I've wanted to avoid (Manley 5.xi.79).

Manley was still in pursuit of data: 'I've been trying Raby Castle (*they* had a record in 1860!) but haven't got any further' (5.xi.79); 'I went back via *Hull*. They're going to provide me with microfilm to photocopy 1794–1803. After that another visit to Northallerton County Record Office will enable me to 'top off' to 1784 (*sic*)' (Manley 27.xi.79).

2.10 Conclusions

The maintenance of regular meteorological observations at one site for a period of more than 140 years is no mean achievement. Not least of the problems has been two world wars. Sargent (1923) makes a special point of the fortunate fact that volunteers were forthcoming to help bridge the war period of 1914–18. The Second World War was not without its difficulties for the observatory, and Manley (27.ix.78) wonders, 'when we had *two* hours daylight saving . . . whether the readings were *really* taken at 11 h by the "civil Double Daylight Saving time" and at 23 h. Another period of difficulty was in the 1850s:

> It's surprising what a lot of tiresome little troubles have come up with regard to the Durham observations, quite apart from the 'Drunken observer' of 1854–55. It certainly reflects on the period of torpor that seems to have supervened when the initial urge of astronomy and mathematics in the hands of ordained churchmen began to be replaced by a depressing lack of energy (Manley 27.xi.79)

Indeed it is not surprising that for one reason or another, over such a long period, the record should have flaws. What Gordon Manley's work demonstrates is that patient analysis can lead towards the means of standardisation, making such a record both acceptable and useful.

Not to make use of the careful weather observations of James Losh of Newcastle would not only be to neglect the opportunity to allow the series to reflect conditions throughout the 19th century (bar two years), but also be to disregard the splendid effort made by an early amateur of meteorology in the north-east. One senses not only Manley's admiration for the man, but also his surprise that such a man as Losh apparently never himself 'slogged out his monthly means after 1818!' (Manley 9.ix.79).

Manley is characteristically bemused by the fact that:

> Curiously enough Durham and Northumberland are rather lacking in earlier weather records; nothing 'daily' of any length before 1802. This is probably, in part, accidental; it would be hard to sustain the view that counting and measuring (i.e. 'Science') were better developed in Lancashire and Westmorland and Yorkshire; possibly there was more imagination over the west side, and its easy to run off into speculation about the Anglians of Northumbria and the greater Danish – Norse infiltration to the south and west that makes such a dialect difference. Still, Robert Stephenson was a more delicate imaginative chap than his father, and his mother was a Hindmarsh: a very good Tyneside name (Manley 30.i.78)

What is the value of a temperature series for north-east England based on Durham? Certainly, as Manley himself makes clear, 'The Durham Observatory record, valuable as it is, by no means summarizes the characteristics of the county' (Manley 1936b). Those concerned to make use of the temperature series in other local research, whether biological or historical, should make a careful study of comments made by Manley and others on the effects of topography on local variations in climate. (Manley 1932, 1933, 1935, 1936a & b, 1939, 1941a & b, 1943, 1946b are but a selection, see also K. Smith 1970.)

Of value in the long and standardised series is the indication of the likely range of variation of temperature experienced in the region under sequences of weather conditions that may in themselves reflect longer term fluctuations in climate over the decades. Such a temperature series can be compared with those available for other parts of the country; when combined with knowledge of the effects of topography, it can give at least an indication of the likely mean and extreme values at other exposures within the region, for example on high ground. When combined with information on other factors of climate, such as rainfall and hours of sunshine – and there is still work to be done on these aspects of the longer Durham record – a temperature series can give some indication of variation in the conditions for plant growth.

From Table 2.4, some extremes can be derived to compare with those extracted from the Durham record for 1847–1981 by Harris (1982), regarding with caution the earlier extremes, of course, in keeping with Manley's own reservations about the completion of his work on the series. Months when higher mean temperatures are indicated than the extremes given for the 1847–1981 series are: November 1818, October 1831, December 1842 and June 1846 (closely rivalled by June 1826). Months when lower mean temperatures are indicated than the extremes given for the 1847–1981 series are: September and November 1807, January 1814, August 1823, April 1837 and October 1842. To the lowest mean annual temperature given for the 1847–1981 series (i.e. 1897) can be added to the identical value for 1816. The warmest year remains that of the original series (i.e. 1949), as do the warmest decade (1940s) and the coldest decade (1880s), although the latter is closely rivalled by the decade 1811–1820.

We must not forget that the temperature series produced in Table 2.4 is a working document, which Manley did not consider ready for publication. Much remains to be done to integrate the additional data now available for 1843–7 and to complete Manley's proposed extension of the series back to 1794.

Acknowledgements

The author wishes to acknowledge the help of Mrs Gillian M. Sheail in extracting data from a notebook entitled *The meteorology of Durham* (dated May 1938) in the collection of Manley papers in Cambridge University Library.

References

Baxter, E. F. 1937. The Observatory. In *The University of Durham, 1937*, C. E. Whiting (ed.), 59–64. Durham: Centenary Committee of the University of Durham.

Baxter, E. F. 1956. Durham University Observatory. *Weather* **11**, 218–22.

Cockrell, P. R. 1968. The Meteorological Society of London 1823–1873. *Weather* **23**, 357–61.

Dove, H. 1838. Über die geographische Verbreitung gleichartiger Witterungserscheidungen. Über die nicht periodischen Änderungen der Temperaturvertheilung. Zweite Abhandlung. *Abhandl. K. Akad. Wissensch.* Berlin, 380. (In National Meteorological Library, Bracknell.)

Drummond, A. J. 1947. Kew Observatory. *Weather* **2**, 69–76.

Fowler, J. T. 1904. *Durham University. Earlier foundations and present colleges.* London: F. E. Robinson.

Glaisher, J. 1848. Remarks on the weather during the quarter ending December 31st 1847. *Phil Mag.* **32**, 130–7.

Harris, R. 1982. Extremes in the Durham University Observatory meteorological record. *Weather* **37**, 287–92.

Heesom, A. 1982. The founding of the University of Durham. *Durham Cathedral Lecture.* Durham: Dean and Chapter of Durham.

Jacobs, L. 1969. The 200-years' story of Kew Observatory. *Met. Mag.* **98**, 162–71.

Losh, J. 1802–33. *Meteorological observations, made by James Losh, at Jesmond Grove, in the years 1802 to 1833,* vols. 1–6, manuscripts. Newcastle: Newcastle Literary and Philosophical Society.

Manley, G. 1932. The weather of the high Pennines. *Durham Univ. J.* **28**, 31–2.

Manley, G. 1933. Notes on the weather in Upper Teesdale, 1933. *Durham Univ. J.* **28**, 304–07.

Manley, G. 1935. Some notes on the climate of north-east England. *Q. J. R. Met. Soc.* **61**, 405–10.

Manley, G. 1936a. The climate of the northern Pennines: the coldest part of England. *Q. J. R. Met. Soc.* **62**, 103–13. Discussion 114–15.

Manley, G. 1936b. On the instrumental climatology of the Durham district. *Durham Univ. J.* **30**, 43–8.

Manley, G. 1939. High altitude records from the northern Pennines, 1938–39. *Met. Mag.* **74**, 114–17.

Manley, G. 1941a. The Durham meteorological record, 1847–1940. *Q. J. R. Met. Soc.* **67**, 363–80.

Manley, G. 1941b. Climate and agriculture in County Durham. In *The land of Britain. The report of the land utilization survey of Britain,* Part 47: *County Durham,* Ada Temple (ed.), 193–200. London: Geographical Publications.

Manley, G. 1943. Further climatological averages for the northern Pennines, with a note on topographical effects. *Q. J. R. Met. Soc.* **69**, 251–61.

Manley, G. 1946a. Temperature trend in Lancashire, 1753–1945. *Q. J. R. Met. Soc.* **72**, 1–31.

Manley, G. 1946b. The centenary of rainfall observations at Seathwaite. *Weather* **1**, 163–8.

Manley, G. 1946/7. The climate of Lancashire. *Mem. Proc. Manchr. Lit. phil. Soc.* **87**, 73–95.

Manley, G. 1953. The mean temperature of central England, 1698–1952. *Q. J. R. Met. Soc.* **79**, 242–61.

Manley G. 1974. Central England temperature: monthly means 1659–1973. *Q. J. R. Met. Soc.* **100**, 389–405.

Manley, G. 1980. The northern Pennines revisited: Moor House, 1932–78. *Met. Mag.* **109**, 281–92.

Meteorological Society 1873–5. Obituary. The Rev. Temple Chevallier, BD, FRAS. Q. J. Met Soc. 2, New Series, 80–81.

Mossman, R. C. 1897. The meteorology of Edinburgh, Part II. Trans R. Soc. Edin. 39, 63–207.

Plummer, J. I. 1873. On some results of the temperature observations at Durham, Q. J. Met. Soc. H.S. 1, 241–6.

Ratcliffe, R. A. S. 1978. The story of the Royal Meteorological Society. Weather 33, 261–8.

Rochester, G. D. 1980. The history of astronomy in the University of Durham from 1835 to 1939. Q. J. R. Astron. Soc. 21, 369–78.

Sargent, F. 1923. The University Observatory. Durham Univ. J. 23, 567–70.

Smith, C. G. 1968. The Radcliffe Meteorological Station. One hundred and fifty years of Oxford weather records. Weather 23, 362–7.

Smith, C. G. 1980. Two hundred years of Oxford weather. In The Oxford region, T. Rowley (ed.), 23–46. Oxford: Oxford University Department of External Studies.

Smith, K. 1970. Climate and weather. In Durham County and city with Teesside, J. C. Dewdney (ed.), 58–74. Durham: Local Executive Committee of the British Association.

Whipple, F. J. W. 1937. Some aspects of the early history of Kew Observatory. Q. J. R. Met. Soc. 63, 127–35.

Whiting, C. E. 1932. The University of Durham, 1832–1932. London: Sheldon Press.

Winch, N. J. 1838. A footnote. Trans. Nat. Hist. Soc. Northumberland, Durham & Newcastle upon Tyne 2, 137.

Witchell, W. M. 1947. Greenwich Observatory – a sketch of its history and functions. Weather 2, 23–9.

3

Variations in the Durham rainfall and temperature record, 1847–1981

RAY HARRIS

The annual and monthly rainfall and temperature records for Durham are analysed. The data are: rainfall totals for the period 1886–1981; temperatures recorded in a Stevenson screen for the period 1900–81; mean temperatures, including 'adopted means', for the period 1847–1981. The data are smoothed and analysed for dependencies and periodicities.

3.1 Introduction

Durham University Observatory has provided a long record of observations of weather conditions in the north-east of England. Since 1843, weather observations had been made on a daily basis, giving Durham the second longest continuous record in Britain after the Radcliffe Observatory in Oxford (see discussion on p. 17). Durham has the advantage that its site has been relatively unchanged since the building of the observatory in the middle of the last century, and is still essentially a rural one. This is a significant factor in the analysis of trends and changes in the meteorological record at Durham, because the observed changes are the result of broad atmospheric variations rather than those brought about by local, microclimatic modifications such as increasing urbanisation.

While the site of the observatory has remained relatively unaltered during the last 140 years, the exact location and type of the instruments have not (see Manley 1941). Before the year 1900, temperature was recorded at different times on the north-facing balcony of the observatory in a louvred screen and in a Glaisher stand on the south lawn. On 1 March 1900, the latter was replaced by the present arrangement of a Stevenson screen. The temperature data used in this discussion are principally those from the post-1900 record, and include maximum and minimum temperatures recorded in a Stevenson screen. The data are those given by Manley (1941) and in recent published records of the observatory. The annual mean temperature data for the longer period 1847–1981 are taken from Manley (1941, Table I) and recent published data. For the period 1847–1957, the yearly means are what Manley termed 'adopted means', and after 1957, the mean values are the mean of the maximum and minimum for the 24-hour period 09.00–09.00 GMT. The adopted mean was used by Manley to account for variations in the times of temperature observations at the observatory, but recently he noted that 'the means as now derived from the 9 h extremes at Durham are so close to the equivalent of any older "adopted means" that they can be treated as a continuous series' (Manley 1980). The rainfall data discussed here are taken from the records for the rain gauge located on the south lawn since 1886.

Several authors have examined the record of Durham Observatory, giving both the history of the observatory and discussion of the meteorological record. The seminal paper by Manley (1941) gives a standardised data set for Durham of temperature and sunshine for the period 1847–1940, and the data have been updated (Anon. 1948, 1958, Manley 1980). Other authors who have discussed the observatory and its record include Plummer (1873), Baxter (1956), Smith (1970), Harris (1982) and Kenworthy (this volume).

The objective of this chapter is to examine the trends and changes in the temperature and rainfall record of the observatory. This meteorological series now offers a sufficiently long data set to be able to make generalisations about changes in climate at Durham since the last century. The tools employed are statistical in character, using the routines available in the computer packages MIDAS (Fox & Guire 1976), and SPSS (Nie & Hull 1981).

3.2 Precipitation 1886–1981

3.2.1 Pattern of change

Fig 3.1 shows the annual precipitation for each year for the period 1886–1981. The mean annual precipitation for the series is 645.2 mm, with a standard

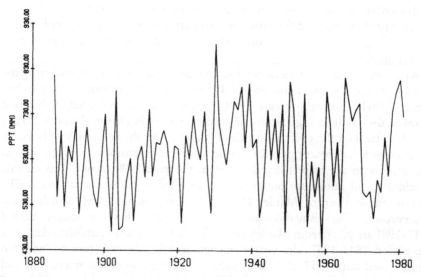

Figure 3.1 Annual precipitation (mm) 1886–1981. Mean annual precipitation is 645.2 mm.

deviation of 99.2 mm. The erratic nature of annual precipitation, with dry years often followed by wet years (e.g. the lowest annual total in the series, 439.7 mm was recorded in 1959, and followed by 782.3 mm in 1960), is also shown. The frequency distribution of annual precipitation shows a near normal shape, although it is multimodal, but the distribution of monthly precipitation totals shows a positive skew (skewness = 1.27), and a unimodal distribution. Monthly rainfall totals vary from 1.3 mm to 206.4 mm, with a standard deviation of 31.9 mm about a mean of 53.7 mm.

In order to investigate the overall trends in the rainfall series, a number of smoothing operations were performed on the annual rainfall data by using running means with unweighted windows of 5, 10, 20 and 30 years. The series filtered with a 5-year window (Fig. 3.2) shows annual rainfall of *c.* 600 mm in the last years of the 19th century, falling to a minimum of *c.* 540 mm in 1906, before beginning an upward trend to reach a peak near 750 mm in 1937. Since the late 1930s, the rainfall pattern has been one of decrease to the late 1950s, a peak in the mid-1960s, followed by a further trough. This pattern is also evident in the series filtered with 10- and 20-year windows (not shown), with the highest rainfall of the 1930s decade emphasised. Figure 3.3 shows a very smoothed, running mean series transformed with a uniform 30-year filter. The graph shows annual totals of *c.* 615 mm at the turn of the century, rising steeply to *c.* 670 mm during the 1930s, followed by a fall to *c.* 640 mm in the 1950s and 1960s.

The changes shown in Figures 3.2 and 3.3 reveal the most recent dry period occurring at the turn of the century, and the wettest period around 1930–40. The range of the smoothed annual precipitation shown in Fig 3.3 (*c.* 60 mm) is approximately 10% of the mean of the series, and less than the standard deviation of 99.2 mm.

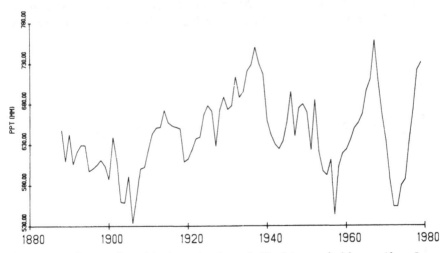

Figure 3.2 The annual precipitation series shown in Fig. 3.1 smoothed by a uniform 5-year window filter.

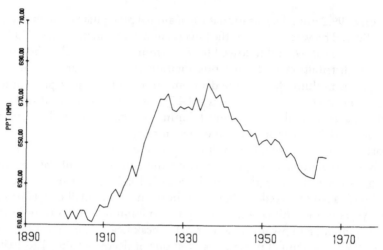

Figure 3.3 The annual precipitation series shown in Fig. 3.1 smoothed by a uniform 30-year window filter.

3.2.2   Autocorrelation

The temporal dependencies of precipitation at Durham were analysed by examining autocorrelation functions (correlograms) of annual and monthly rainfall. The autocorrelation function of annual precipitation shown in Figure 3.4 reveals autocorrelation coefficients close to zero for most lags. On Figure 3.4 (and all other correlogram plots) the lines of two standard errors are shown in dashed lines. With the proviso that 1 in 20 autocorrelation coefficients from a random series would be expected to fall outside two standard errors (Chatfield 1975), three lags appear to have significant or near-significant autocorrelation coefficients: 15, 24 and 25 years. At 15 and 24 years, the correlation is positive (0.188 and 0.159 respectively), and at lag 25, the correlation is negative

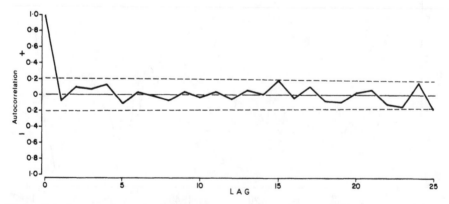

Figure 3.4 Autocorrelation function for the annual precipitation data shown in Fig. 3.1. The dashed lines are two standard error limits.

(−0.184). At lag 3, 4 and 5 years, the correlation is consistently positive (near 0.1), but for the remainder of the correlogram the values oscillate near zero. The correlogram resembles white noise, and suggests that the process is stationary, so that the precipitation data may be used in their original form for later analysis. The correlograms for the first and second halves of the series reveal similar patterns to the overall series, and the mean precipitation for each half is within 10 mm of the series mean. The correlogram of the differenced series has a correlation coefficient of − 0.564 at lag 1 year, and other significant correlations at longer lags. This high lag 1 correlation in the differenced series reflects the highly varied and erratic nature of the rainfall by illustrating the persistence of *change* of rainfall totals from one year to the next.

In order to explore further the correlations within the rainfall data, the autocorrelation function of rainfall for each month was examined, and the plots of these are shown in Figure 3.5. For each month, the lag 1 correlation is not significant, and only for May is the lag 2 coefficient significant (= − 0.314). Some grouping of months with similar correlogram characteristics can be undertaken. Significant correlations (both positive and negative) at lags between 5 and 10 years are present for the months of January, May, June, September, October and November. These display some seasonality in that early summer and autumn show dependencies in the data at these lags, and these may be associated with transition periods from one season to another. For longer lags beyond 15 years, significant correlations are present in January, February, April, June and August, suggesting a possible winter and summer dependency, although at such long lags interpretation of the results is less certain. May stands out markedly from the other months, with significant correlations at lags of 2, 5 and 7 years. The correlation at lag 2 suggests that the *trend* of precipitation occurrence for successive Mays might be predictable, particularly if this were supported by evidence from 5- and 7-year lags.

3.2.3 Spectral analysis

Meteorological series often show cyclical or quasi-cyclical change. In order to investigate the presence of any cyclical change in the Durham rainfall record, a spectral analysis was performed on the data, and the spectral density function is shown in Figure 3.6. The two principal peaks present are at periods of 15 and 2.1 years, although neither is significant at $\alpha = 0.1$. The spectral density function echoes the autocorrelation functions in resembling a white noise spectrum, and the median spectrum statistic test for a flat spectrum confirms this, as the test statistic is not significant at $\alpha = 0.1$. The rainfall record for Durham shows few dependencies and little cyclical change. It appears to be a stationary series with a wide range of annual totals about the mean. The smoothed series does show some changes, with a wetter period in the 1930s, but beyond that, there is little predictable change.

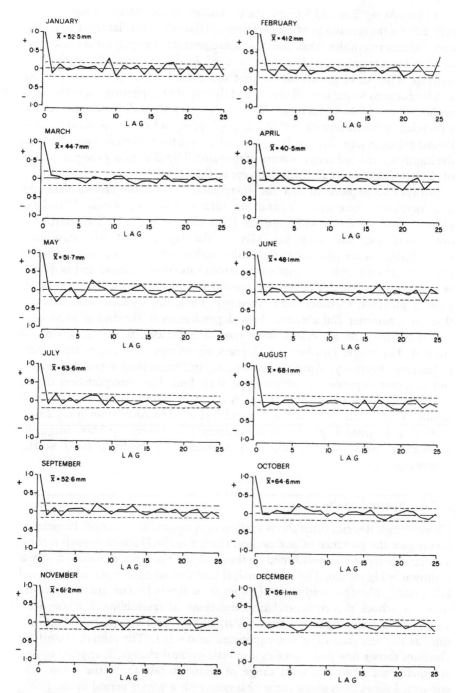

Figure 3.5 Autocorrelation functions for monthly precipitation for the period 1886–1981. Mean precipitation for each month ($\bar{x}$) is also given.

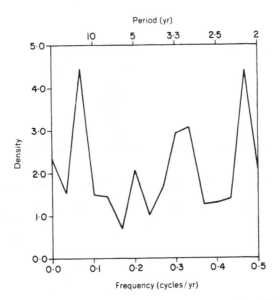

Figure 3.6 Spectral density function of the annual precipitation series. A Hanning window and a maximum lag of 40 were used.

3.3   Temperature 1900–81

The data used for this section are taken from the daily observations of maximum and minimum temperature recorded in a Stevenson screen on the south lawn of the observatory. The mean temperature is calculated as (maximum + minimum)/2. The data are those published in Manley (1941) and in recent published records. Table 3.1 gives the first two moment statistics for the three variables. Maximum, minimum, and mean temperatures are closely related when examined in aggregate, and this is shown in the correlation matrix in Table 3.2, so that the later comments in this discussion should be seen in the light of these strong linear associations.

3.3.1   Changes since 1900

The Durham *mean temperature* pattern for this century shows changes found at many meteorological stations in the Northern Hemisphere, i.e. a warming to the 1940s followed by a cooling trend since then (Mason 1976, Lamb 1982). Figure 3.7 shows the graph of the annual mean temperature for Durham for the period 1900–81, and Figure 3.8 shows the same data smoothed with a 20-year uniform window filter. The mean for the series is 8.4°C, with a standard deviation of 0.49°C, and the frequency distribution shows a small positive skew. The smoothed series in Figure 3.8 indicates yearly mean temperatures near 8.2°C at the beginning of the century, rising to near 8.7°C in the 1940s, followed by a cooling to around 8.4°C in the 1960s. The 1970s saw a slight

Table 3.1 Moment statistics for average annual mean, maximum and minimum temperatures (°C), 1900–81.

	Mean	Standard deviation	Maximum	Minimum
mean temperature	8.4	0.49	9.9	7.4
maximum temperature	12.2	0.58	13.9	11.2
minimum temperature	4.7	0.47	5.9	3.4

Table 3.2 Correlation matrix of annual mean, maximum and minimum temperatures (°C), 1900–81.

	Mean	Maximum	Minimum
mean temperature	1.0	0.707	0.937
maximum temperature	0.707	1.0	0.904
minimum temperature	0.937	0.904	1.0

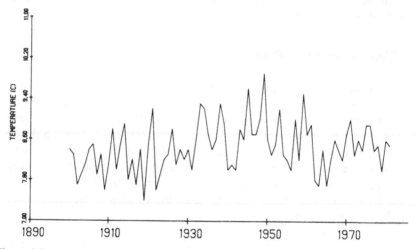

Figure 3.7 Annual mean temperatures (°C) 1900–81. Mean annual temperature is 8.4°C.

increase in Durham temperature by about 0.1°C, and this is more clearly revealed in the data smoothed with a 5-year window (not shown).

As the mean temperature for the period is constructed from the recorded maximum and minimum temperature, the latter two reveal similar patterns to the mean temperature graph. The peak in the *maximum temperature* series comes in the late 1940s, after the peak for the Northern Hemisphere as a whole (Mason 1976). Figure 3.9 shows the maximum temperature series smoothed with a 10-year filter. Maxima increase from near 11.9°C at the beginning of the century to near 12.6°C in the late 1940s, followed by a decline after 1950 and an upturn in

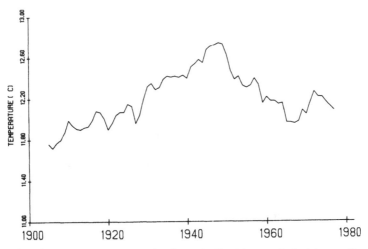

Figure 3.8 The mean temperature series shown in Fig. 3.7 smoothed with a uniform 20-year window filter.

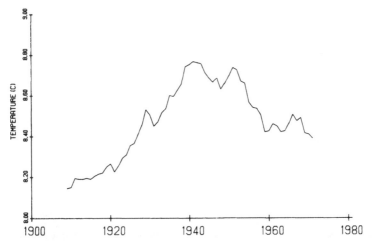

Figure 3.9 The maximum temperature series (°C) 1900–81 smoothed with a uniform 10-year window filter.

the 1970s. The bump on the maximum temperature curve in the 1970s was largely caused by high mean maxima in 1971, 1973, 1975 and 1976, the last two associated with the drought and the hot summers of those years. However, this was then followed by a mean maximum for 1979 (11.3°C) almost as low as any in the series.

The plot of mean annual *minimum temperature* shows a similar pattern, except for the 1940s. Figure 3.10 shows the low minima recorded for the years 1940–43, and these cooler conditions are still evident on the series smoothed by a 10-year uniform window and shown in Figure 3.11. The low minima for this

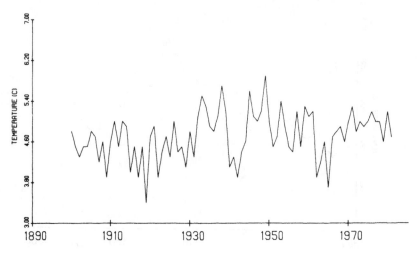

Figure 3.10   The annual minimum temperature series (°C) 1900–81.

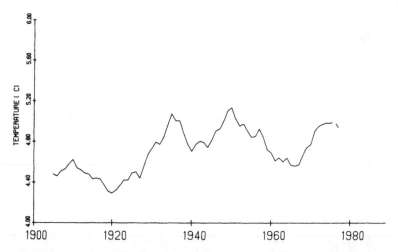

Figure 3.11   The minimum temperature series shown in Fig. 3.10 smoothed with a uniform 10-year window filter.

period were caused by relatively low cloud cover allowing an increase in incoming solar radiation, and consequently the upward trend in maximum temperatures, but also allowing an extensive loss of terrestrial radiation at night bringing cooler minima. Other than this trough in the minimum temperature curve, the progression of temperature change through the century is similar to that observed for maximum temperatures.

3.3.2   Autocorrelation

Figure 3.12 shows the autocorrelation function for each of the annual mean,

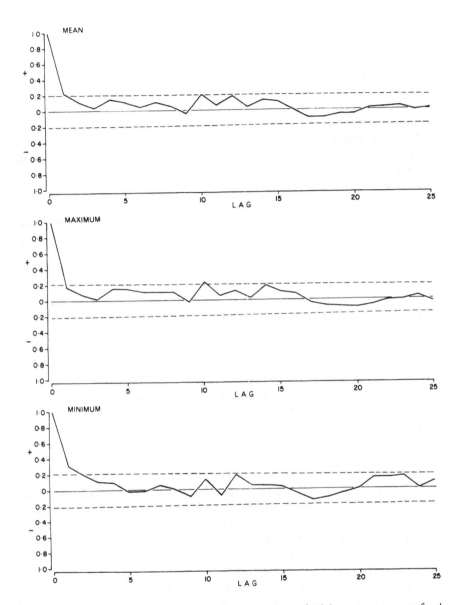

Figure 3.12 Autocorrelation functions of mean, maximum and minimum temperatures for the period 1900–81.

maximum, and minimum temperature series for the period 1900–81. Four principal features can be identified from these functions.

(a) Dependency at a lag of 1 year is shown for mean and for minimum temperatures, with significant positive correlations (0.22 and 0.31 respectively) at $\alpha = 0.05$. This suggests a weak persistence in temperature trends from one year to the next. For maximum temperatures, the lag 1 autocorrelation (0.169) is not significant, but it does have the same sign as the mean and maximum temperature figures. The minimum temperature correlogram also shows a lag 2 coefficient (0.201) which is also nearly significant at $\alpha = 0.05$.

(b) For lags 1 to 15, all three graphs show generally positive correlations, with the common exception of lag 9 which is negative in all three cases.

(c) From lags 16 to 20, correlations are generally negative, although this feature is more pronounced for the case of maximum temperature. This pattern reflects the trends shown in Figure 3.9.

(d) Lag 10 in the mean and maximum temperature correlograms shows significant correlations. This may be weakly associated with the 11–year sunspot cycle, but as the correlations are barely significant at $\alpha = 0.05$, this feature is indicated rather than firmly supported by the data.

Overall, only weak autocorrelation is shown in the mean, maximum, and minimum temperature series for 1900–81. The significant autocorrelations at lag 1 for mean and minimum temperatures indicate some dependency and the possibility of predicting tendencies, but even these correlations are weak, and barely significant at $\alpha = 0.05$.

The autocorrelation functions of the temperature data for January and July were chosen to examine high and low season patterns of dependency in more detail, and these are shown in Figure 3.13.

January   One consistent feature of the three graphs for January is the significant negative autocorrelation at a lag of 8 years. This suggests an 8-year alternation of January temperature tendencies, although the precise explanation of this is not clear. A further significant correlation is found in the minimum and mean temperature correlograms at a lag of 23 years, which may be related to the double sunspot cycle and thereby weakly linked to the 10-year lag noted in the annual correlograms (see Figure 3.12). The minimum temperature correlogram is the only one to show a near-significant lag 1 dependency ($= 0.197$), and thereby some evidence of persistence in the series, while the other two correlograms have lag 1 autocorrelations near zero. With the exceptions noted here, the autocorrelation functions for January resemble white noise, with oscillations about zero, and little consistent pattern.

July   The three correlograms for July each show only one significant autocorrelation: for maximum and mean temperature this occurs at lag 19 (-0.256 and

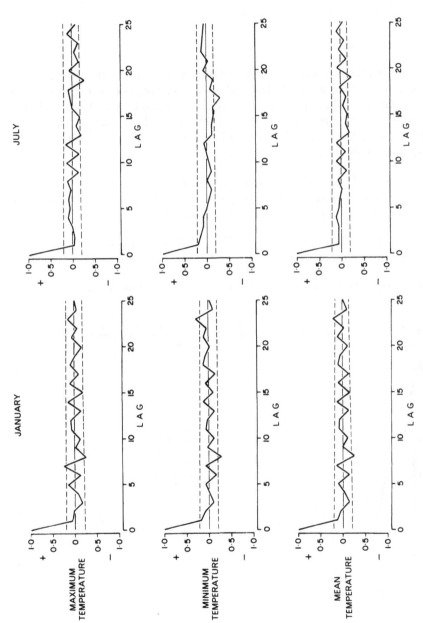

Figure 3.13 Autocorrelation functions of maximum, minimum and mean temperatures for the months of January and July, for the period 1900–81.

− 0.224 respectively), and for minimum temperature it occurs at lag 16 (− 0.286). The minimum temperature correlogram shows the only lag 1 autocorrelation of note (0.2), a feature also found in the January correlogram. Autocorrelations for maximum and mean temperatures are mainly positive up to lag 8, but after this become highly variable.

3.3.3 Spectral analysis

As with the precipitation data discussed in section 2, the 1900–81 temperature series were analysed using spectral analysis to identify periodic variations. The plots of the spectral density functions are shown in Figures 3.14 and 3.15. The smoothed temperature series in Figures 3.8, 3.9 and 3.11 suggest that some cyclical change had occurred in the Durham temperature record. The results of the spectral analysis are therefore given for:

(a) the original maximum and minimum temperature series;
(b) the residuals from a four-term polynomial fitted to the original series. These explain the following proportions of the variance of each data set:

maximum temperature   17.7% variance explained.
minimum temperature   15.4% variance explained.

In both cases, the relatively high density at frequency 0 in the spectral density

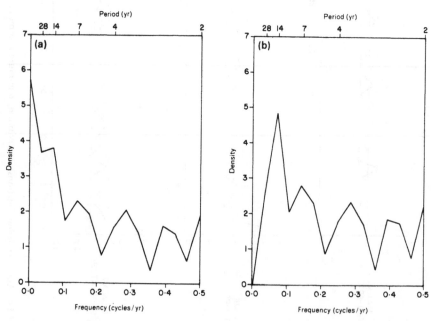

Figure 3.14 Spectral density functions of the 1900–81 minimum temperature series. (a) Original data. (b) Residuals from a 4-term polynomial explaining 15.4% of the variance. A Hanning window and a maximum lag of 35 were used.

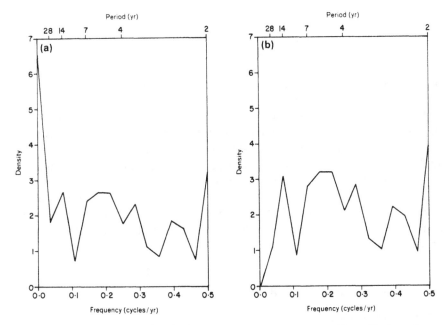

Figure 3.15 As for Fig. 3.14, but for the maximum temperature series. The residuals are from a 4-term polynomial explaining 17.7% of the variance.

function for the original data was replaced in the spectral density graph for the residuals by a value close to zero, indicating the effective removal of the longer term trend by the polynomial. However, beyond this change at frequency zero the paired graphs for original data and residuals closely resemble each other, and this shows that the polynomial term is extracting the underlying variation of the data for the longer term, and that the indications given by the peaks in the spectral density curves are real (although relatively small) variations about a long-term trend. As no spectral density peak (other than frequency zero in the original data graph) is significant at $\alpha = 0.05$, comments here are restricted to discussion of the relative peaks in the spectral density functions.

The graphs for minimum temperature (Fig. 3.14) reveal the most pronounced relative variations, with peaks at 14, 7, 3.5, 2.5 and 2 years. This suggests:

(a) a controlling frequency of 14 years; and
(b) the presence of the quasi-biennial oscillations, noted in other temperature series, with a cycle of c. 2 years.

The 14-year cycle may be a disguised form of a sunspot-related cycle: the periodogram for the minimum temperature series reveals a period of 11.7 years (significant at $\alpha = 0.05$) in the graphs for both the original data and for the residuals.

This pattern is reflected in a more subdued way in the maximum temperature spectral density curves (Fig. 3.15), with peaks at 14, 5.6, 3.5, 2.5 and 2 years. For maximum temperature functions, the 2-year cycle is relatively more pronounced than with the minimum temperature functions, although the smaller frequencies are not so pronounced. The spectral density functions for mean temperature and its residuals (not shown) reveal similar patterns to those observed for maximum and minimum temperature data because of the close linear association (see Table 3.2) of the variables.

The spectral density functions give indications of tendencies rather than significant cycles, and this echoes the results from the autocorrelation functions where few significant dependencies are found in the data. The periodograms for all three temperature series reveal a relative peak at about 11 years, although it is only barely significant for the minimum temperature case. One reason for this evident lack of periodicity may be the fairly short time period of the series. In the next section, the time period is lengthened to examine the mean temperature data since the start of the published meteorological recordings at the observatory.

3.4   Mean temperature 1847–1981

3.4.1   Commentary

Although mean temperatures calculated from the recorded Stevenson screen maximum and minimum temperatures are only available for the period since 1900, Manley (1941) published 'adopted mean' temperatures for the period from 1847, where the mean for each month and each year was standardised to take account of differing observational procedures. It is these annual mean temperatures which are used in this section, taken up to 1981 by using the annual means drawn from the observatory temperature records (see section 3.1).

The graph of annual mean temperature for the period 1847–1981 is shown in

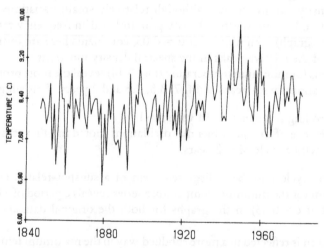

Figure 3.16   Annual mean temperature (°C) 1847–1981. Average of the series is 8.3°C.

Figure 3.16. The pattern shown is a rather erratic one, although the frequency distribution shows a near-normal distribution, with a mean annual temperature of 8.3°C, and a standard deviation of 0.54°C. The lowest mean annual temperature (6.7°C) occurred in 1879, and the highest (9.9°C) in 1949.

In order to observe the extent of underlying changes and trends in the series, uniform window filters of various lengths were used to smooth the data, and of these the 10-year and 30-year smoothed series are shown in Figures 3.17 and 3.18. For the period 1900–81, the series is identical to that discussed in section 3.3, but the addition of the 19th-century data reveals the trough in the

Figure 3.17 The mean temperature series shown in Fig. 3.16 smoothed with a 10-year uniform window filter.

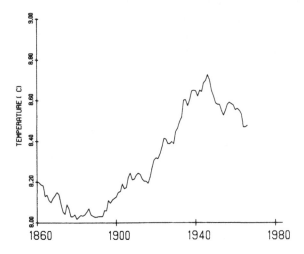

Figure 3.18 The mean temperature series shown in Fig. 3.16 smoothed with a 30-year uniform window filter.

temperature curve at the end of the last century preceded by milder temperatures similar to those of the earlier part of this century. Figure 3.17 shows the 10-year smoothed temperatures of around 8.2°C in the 1860s and 1870s, falling to 7.8°C around 1890, before beginning the improvement which lasted until the late 1940s. This pattern is repeated in a smoother way with the 30-year filtered data (Fig. 3.18). In this graph, temperatures were around 8.0°C at the end of the last century, rising to 8.7°C at a maximum in 1947, before falling by about 0.2°C since then. This pattern agrees well with conditions in the Northern Hemisphere during the period, although the timing of the main turning points on the curve appears to be later at Durham.

3.4.2   Autocorrelation

The autocorrelation function for the mean temperature series (Fig. 3.19) shows eight significant correlations (at $\alpha = 0.05$) out of 25 lags. Significant correlations are found at lags 1, 2, 4, 5, 7, 10, 15 and 21 years, and all of these are positive, indicating a persistence of trends of mean annual temperature through time. The pattern shown by the correlogram is one of positive dependency at all lags except one (-0.005 at lag 19). This exception is part of a section of correlations near zero for lags 17 to 20 inclusive, which may be associated with turning points on the temperature curve.

Because the correlogram (and later analyses) may be affected by non-stationarity of the temperature series, a four-term polynomial accounting for 15.7% of the variance was fitted to the data, and a correlogram of the residuals produced. This correlogram is shown in Figure 3.20. The shape of the curve is similar to that for the raw data, but the values of the correlations are approximately 0.1 lower, with a variation about zero. None of the coefficients is significant after the removal of the polynomial function from the data. Figures 3.19 and 3.20 demonstrate the importance of the long-term trend in the temperature series: beyond this there appears to be no significant dependency in the data, and the correlogram resembles white noise.

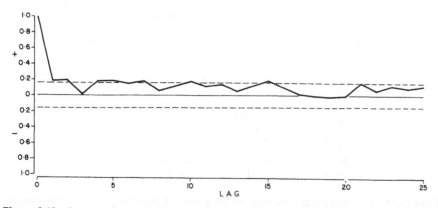

Figure 3.19   Autocorrelation of mean annual temperature for the period 1847–1981.

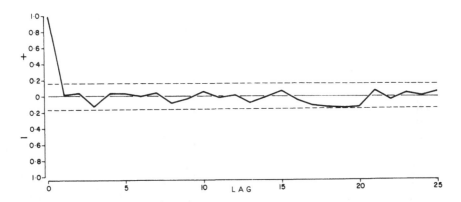

Figure 3.20 Autocorrelation function of the residuals from a 4-term polynomial fitted to the 1847–1981 mean annual temperature series which accounted for 15.7% of the variance.

3.4.3 Spectral analysis

As in Sections 3.2 and 3.3, the 1847–1981 mean temperature series was analysed using a spectral analysis technique. Both the original data and the residuals from the four-term polynomial referred to in the last section were analysed, and the spectral density functions are shown in Figure 3.21. The principal peak in Figure 3.21a is at frequency zero, and is significant at $\alpha = 0.01$. This frequency is removed by the polynomial model, and Figure 3.21b shows the zero frequency with a density of 0.093. After the removal of this frequency, the patterns of the two curves are very similar as was the case with the autocorrelation functions in Figures 3.19 and 3.20. On the residuals spectral density function (Fig.3.21b), peaks occur at periods of 46, 11.5, 7.7, 5.1, 3.5, 3, 2.3 and 2.1 years, although none is significant at $\alpha = 0.05$, and all are simply of relative importance. This pattern is also present in the curve for the raw data. Of these peaks, the two with direct interpretations are the 2.1/2.3-year and 11.5-year peaks: the former connected with the quasi-biennial oscillations, the latter with the 11-year sunspot cycle. The peaks at 3.5 and 7.7 years repeat similar peaks found with the 1900–81 maximum and minimum temperature series (Section 3.3.3)

3.5 Discussion and conclusion

The data and analyses in this chapter indicate the highly variable nature of the meteorological record at Durham. Even though Durham has one of the longest meteorological records in Britain, in many ways it is still too short to allow reliable statistical trends to be extracted. The correlogram and spectral density analyses show that no statistically significant cycles emerge from the rainfall and temperature data, although the relative peaks are useful indications of trends.

The smoothing procedures employed in earlier sections are useful ways of describing the changes that have occurred in the record, but their use for

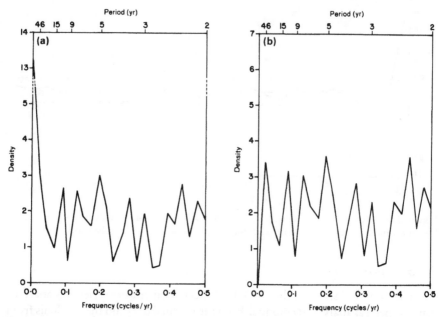

Figure 3.21 Spectral density functions of the 1847–1981 mean temperature series. (a) Original data. (b) Residuals from the 4–term polynomial. A Hanning window and a maximum lag of 60 were used.

prediction is limited. The smoothed series support other records for the Northern Hemisphere in indicating cooler and drier conditions at the turn of the century, followed by a warmer and wetter climate towards the middle of the 20th century, and a cooling and lower rainfall trend since then. The polynomials fitted to the meteorological series illustrate these trends, but their use for prediction of climate must be limited. Simply because there have been turning points on the smoothed curves for temperature and rainfall for the last century does not encourage extrapolating repetitive trends of the same character in the future given that:

(a) the polynomials account for relatively small proportions of the variance of each series examined (less than 20%); and
(b) the data all have high variability from one year to the next.

Extrapolation of future turning points on the temperature and rainfall curve could be performed, but this would be a dangerous and misleading prediction outside the range of reliable data.

In summary, the following main points emerge from the discussion in this chapter:

(a) Annual rainfall and temperature data are highly variable.
(b) Significant autocorrelation and cyclical change are limited.

(c) Of the relative regularities in the data, there is some evidence of a dependency at lag 1 in the temperature series (see Section 3.2.2), and some suggestion of cycles at 14, 11 and 2 years.

The Durham record, while being a relatively lengthy one, is still too short for significant trends to emerge. What is needed is a meteorological record for northern England extending back not just to the middle of the 19th century, but to the middle of the 17th century, as Manley (1974) has compiled for central England. Such a longer record would allow better statistical summaries to be prepared, and a firmer data base for the analysis of predictable change.

References

Anon. 1948. *Temperature, rainfall, sunshine and wind readings at Durham University Observatory from first recording to 1947*. Durham: University Office.

Anon. 1958. *Temperature, rainfall, sunshine and wind readings at Durham University Observatory from first reading to 1947, supplement 1*. Durham: The Durham Colleges, Science Laboratories.

Baxter, E. F. 1956. Durham University Observatory. *Weather* **11**, 218–22.

Chatfield, C. 1975. *The analysis of time series: theory and practice*. London: Chapman and Hall.

Fox, D. J. and K. E. Guire 1976. *Documentation for MIDAS*. Michigan: Statistical Research Laboratory, University of Michigan.

Harris, R. 1982. Extremes in the Durham University Observatory meteorological record. *Weather* **37**, 287–92.

Lamb, H. H. 1982. *Climate, history and the modern world*. London: Methuen.

Manley, G. 1941. The Durham meteorological record. *Q. J. R. Met. Soc.* **67**, 363–80.

Manley, G. 1974. Central England temperatures: monthly means 1659 to 1973. *Q. J. R. Met. Soc.* **100**, 389–405.

Manley, G. 1980. The northern Pennines revisited: Moor House, 1932–78. *Met. Mag.* **109**, 281–92.

Mason, B. J. 1976. Towards the understanding and prediction of climatic variations. *Q. J. R. Met. Soc.* **102**, 473–98.

Nie, N. H. and C. H. Hull, 1981. *SPSS Update 7–9*. New York: McGraw-Hill.

Plummer, J. J. 1873. On some results of temperature observations at Durham. *Q. J. R. Met. Soc.* **1**, 241–6.

Smith, K. 1970. Climate and weather. In *Durham County and city with Teesside*, J. C. Dewdney (ed.), 58–74. Durham: Local Executive Committee of the British Association.

4

Some aspects of rainfall records with selected computational examples from northern England

ELIZABETH M. SHAW

A review of the compilation and analysis of rainfall records demonstrates how an adequate network of long-term homogeneous records is needed to reflect temporal and a real rainfall variability. Some of the longer series of rainfall data for selected regions of England are described. Attention is focused on the gradual increase in the detail of analysis that may be carried out with the continual advancement of computational facilities, with reference to the rainfall record for England and Wales. Two records for northern England (Manchester and Windermere) have been treated by a selection of standard computer packages in order to demonstrate some of the rainfall analyses that may be applied today to help water engineers in their work.

4.1 Introduction

The birth of orderly scientific enquiry in the years of the Renaissance heralded the growth of reasoned measurement of the variable features of the environment. Air temperature and atmospheric pressure were the noted climatic variables for which precise instruments of measurement were developed at an early stage. However, measurements of rainfall had already been reported, initially from India in the 4th century BC, and then from Palestine in the 1st century AD. Much later, records of rainfall were made in China and Korea, but it was not until the period of rapid scientific advancement in the 17th century that regular measurements of rainfall using carefully designed rain gauges began to be made in Europe. It was only recently (Folland & Wales-Smith 1977) that the possibility of the compilation of a continuous record of rainfall spanning 300 years up to the present time could be contemplated.

Interest in long period instrumental records is being stimulated from two different viewpoints by groups of researchers attempting to examine fluctuations in climatic variables. Firstly, there are numerous purely scientific investigations into changes in the environment; the rapidly expanding knowledge of climatic change providing for greater understanding of the processes involved in the formation of the present physical scene. The interrelationship of instrumental measurements and the observable changes in the present physical environment has enabled scholars to interpret the evidence of past conditions still identifiable on the ground, and thereby to extend backwards in time the results obtained from the instrumental records. Such studies formed a large proportion of Gordon Manley's contribution to environmental science. One example was an interpretation of the glaciological features of the Lake District (Manley 1959), and another was the relationship of snowfall or snow lying to mean air temperature (Manley 1980). The noting of John Dalton's snow drift on Helvellyn (Manley 1952), the analysis of Cruickshank's observations (Manley 1948), and Manley's own annual pilgrimage to the Cairngorms and Ben Nevis each September to examine the extent of the residual snow beds, are examples of the qualitative evidence that must be carefully compiled for comparison with current instrumental observations.

The second viewpoint of long-term records is held by the applied scientists and engineers wishing to appreciate fully the measured vagaries in the weather, and to predict the occurrence of extremes within the proposed timespan of their project design. Additionally, there is often the need to forecast the time of a particular event and, for this, a realisation of the chronological pattern of climatological variables is required within acceptable confidence limits. Gordon Manley also contributed to this aspect of the practical application of studies in climatic change. From his long-term 'Central England' temperature record, and the compiled series of snow-days in the London area, he demonstrated that the tendency for sequences of cold months in the spring should be considered in forecasting fuel requirements for the remainder of the season during which artificial heating was required (Manley 1957).

Long-term precipitation records are essential for efficient water engineering. In the United Kingdom, where the temporal pattern of annual rainfall totals is markedly different over the country (Gregory 1958), a number of long records suitably representative of the distinctive climatic regions is necessary for the nationwide study of water availability. The emphasis in the 1960s on the evaluation of future water resources, for which forecasts of rainfall for many years ahead are a vital requisite, has been replaced by urgent considerations of waste water treatment. The general decline in the rate of increase in water consumption (National Water Council 1982) has encouraged the industry to feel assured of its ability to meet demands from existing supplies and stores. Thus, the designing of further new water resource schemes in the United Kingdom is unlikely in the near future. However, the continued compilation of rainfall measurements is imperative for the successful *management* of existing resources, and, much more seriously, registers of rainfall extremes should be maintained at all existing long-term stations in order to sustain the means for continued improvement in the estimation of the probability of floods and droughts.

The regular measurement of rainfall at one site to ensure the direct assemblage of a continuous homogeneous record over many years is reliably carried out at permanently established meteorological stations. However, with the closure of Kew Observatory, this ideal of permanency is severely challenged. In the early days of weather observations, records were built up by the serious pioneers of the 17th century (Towneley, 30 years; Derham, 19 years), and by many enthusiastic amateurs in Victorian times. Each observer devised his own instrument and method of measurement, and the gauges were invariably set up on high walls or roofs, away from interference, but resulting in the recording of a great range in conditions of exposure. Many of these old records may be consistent within themselves, but cannot be compared one with another without detailed considerations of their composition. Following the famous experiments of Heberden in the 18th century, and the experience of observers in the Lake District in the 19th century, it was gradually deemed essential to establish a uniform procedure for rainfall measurement so that the records could be homogeneous across the country, and meaningful comparisons of rainfall amounts could be made. After many years of experiments and with the co-operation of carefully selected 'volunteer' observers (usually reliable army or church men), Symons compiled the *Rules for rainfall observers* (Symons 1869), and these have provided a 'code of practice' which, with only occasional modification, is still fully operational today. It is to records made under these specifications, with high standards of observational practice, that shorter, irregular series are related and then amended so that they may be incorporated into the basic records to form longer series. An appreciation of the approved procedures for the measurement of precipitation will serve to emphasise the complexities encountered in bringing together several irregular series of measurements to form a satisfactory homogeneous rainfall record covering a long period.

4.2 Precipitation measurement

The nature of the precipitation variable in climatology differs fundamentally from that of the principal variable, temperature, or of the weather indicator, atmospheric pressure. Precipitation is a discontinuous phenomenon, and as such there are periods of time during which no measurements can be made. This is in contrast to the ever-present characteristics of temperature and pressure, values of which can be sampled at will or recorded continuously. In consequence, the compilation of precipitation data is quite different from the processing of temperature and pressure measurements. The continuous variables sampled at specific times, or continuously recorded, are reduced to averages, as, for example, the daily mean temperature may be derived from the averaging of the maximum and minimum values for a particular day. Thence further averaging may be carried out to derive monthly or annual mean temperatures. For the precipitation variable, the occurrence of rain or snow must be awaited, and, when caught, the measured samples give aggregated daily totals, the usual sampling period for basic records. Then daily values, many registering a value of nil rainfall, are summed to provide a monthly total. It is only at this stage that precipitation may begin to resemble a continuous variable and only in areas enjoying rainfall all the year round.

For many applications, knowledge of the incidence of precipitation, the timing and the quantities involved, is vitally important. When using the apparent continuous records of monthly totals, the original composition of those totals must be understood, appreciated, and in some circumstances taken into account.

As with the continuous variables, the measurement of precipitation is a sampling procedure, but more particularly it pertains to the point sampling of the precipitation over an area. Gauge sites, on level ground, must be chosen to be representative of a local neighbourhood, and the instruments need to be exposed so that the catch is not affected by strong winds. A balance must be kept between such over-exposure and undue shelter from adjacent buildings or vegetation. Specific details to be observed in selecting a site are given by the Meteorological Office (1982) in the current version of Symons' rules. Advice on the recognised types of instruments is also provided. The salient features of approved gauges are the turned brass knife-edged rim forming the measuring orifice and the vertical cylinder, about 100 mm in depth above the funnel (the Snowdon funnel), designed to hold snow and hail and prevent splashing rain being blown out by the wind. The main body of the gauge is usually made of copper, but glass-fibre laminate gauges have found approval in recent years. The rain is collected in a glass bottle, with a narrow neck to inhibit evaporation loss, and this is placed in an inner can to collect overflow in heavy rains. In making a daily observation, the water in the glass bottle is poured into a graduated glass measure to obtain the depth of rainfall representative of the volume of precipitation that has fallen over the area of the orifice.

The installation of the rain gauge in the ground has also a bearing on the validity of the measurements. Following the experiments organised by Symons, it was decided to recommend the uniform setting of the gauge rim at 1 ft (30 cm) above the ground surface. To avoid splash into the gauge, the ground surface must be rough; short grass is the usual way of ensuring a non-splash surround, but a covering of gravel or loose chippings would serve the same purpose. With the rim set at a greater height above the ground, there is an increase in the loss of catch due to the wind effects. The setting of the gauge rim at ground level with surrounds of appropriate anti-splash devices has been recommended by hydrologists and shown to give an increase in catch of 6% to 8% for daily rainfall totals, compared with the values at the standard height of 30 cm (Rodda 1967, Green 1969). However, the increase in cost and added difficulties in the management of such an installation for the ordinary observer have precluded its adoption as a recommended standard.

Approximate errors in precipitation measurement have been suggested for different instrumental properties and for exposure (Kurtyka 1953). By far the greatest source of error, with estimated deficiencies as high as 80%, is the exposure of the rain gauge. It is this feature of a rain-gauge station that needs careful consideration when attempting to compare rainfall records. Changes in the exposure of a gauge through site modifications, due to the growth of vegetation, or building demolition, affect the consistency of a record. The record must be checked by correlation with reliable neighbouring records before attempts are made to correct the measurements.

The compilation of most daily rainfall records is made from the simple storage gauge measurements made regularly each day at 09.00h GMT. Measurements of precipitation over shorter durations are made by recording instruments. Many devices have been invented for registering rainfall directly, but in the United Kingdom, two instrumental types are now recognised as reliable and have been approved by the Meteorological Office: the Dines Tilting Syphon Rain Recorder and the Tipping-bucket Rain Gauge. Both gauges are described fully in the latest *Handbook of meteorological instruments* (Meteorological Office 1981). For the student of rainfall records, interest lies in the recording mechanisms of the two instruments. The Dines Tilting Syphon Rain Recorder produces a pen trace on a daily chart (or, more satisfactorily, on a monthly strip chart), and information on the timing of the rainfall and the rainfall depths must be read from the charts. Some observers complete regularly hourly tabulations of depths and durations, and valuable records are available in this form. More usually, only data on high rainfall intensities are abstracted for major storm analyses. At well maintained stations, details of all high intensity falls may be entered on appropriate 'Duration and Rate of Rainfall' forms, which are kept in data archives. The Tipping-bucket Rain Gauge provides the rainfall measurements in a form more readily processed by digital computers. Each tip of the bucket (usually holding the equivalent of 1 mm of rainfall) causes a magnet to complete momentarily an electric circuit; the resultant pulse is recorded on a counter or magnetic tape. With regular time periods entered into the recording process, details of the occurrence of rainfall and intensities can be produced as required. The standard daily totals are also readily evaluated. At all automatically recording rainfall stations, it is essential to have a daily storage gauge to ensure continuity of observations in the event of instrument failure. It is rare to find automatic records maintained with 100% efficiency at the rain-gauge stations established specially for the provision of short duration intensities.

4.3   Compilation of long records

The inspired activities of Symons and his successors in the British Rainfall Organisation, which made the United Kingdom pre-eminent in most aspects of precipitation measurement, were transferred to the Meteorological Office in 1919. The drive and initiative shown in the earlier years were continued in the government department by equally dedicated leaders. The annual volumes of *British Rainfall* appeared regularly in each following year, and included regular reports of further research and of rain-gauge inspections, in addition to the standard presentation of all the station totals with greater details of storm events and periods of drought. Requests for information on precipitation from many parts of the country, most often from water supply authorities and land drainage boards, stimulated the production of county maps of the average annual rainfall for a standard period 1881–1915. When information on the variations of precipitation with time was requested, an assemblage was made of all available data to provide a continuous rainfall record to be representative of conditions over England and Wales. From the year 1727, there have always been at least two rain gauges maintained somewhere in England. The annual totals, expressed as percentages of the standard period average 1881–1915, were published for the years 1727–1927 (Glasspoole 1928). A comparable series for monthly rainfall values, expanded to include Wales, was introduced as a series of 'statistics (which) may be regarded as a history of the rainfall over England and Wales since 1727' (Nicholas & Glasspoole 1932). The serial values of monthly rainfall were calculated by averaging the percentage distributions of rainfall through the months of a year from the records of available stations. Each of these average monthly proportions for the year was then compared with the corresponding 12 standard period (1881–1915) proportions. The resulting 12 percentages were then adjusted so that their mean corresponded to the annual percentage of the 1881–1915 standard for that year. The validity of this series is therefore dependent on the accuracy of the estimated annual values, and in the early years these were based on few gauges, hardly representative of the general rainfall over the country. It was not until 1870 that it was felt that a satisfactory number of records distributed well enough over the area was available to provide nationwide values. However, this 205-year series brought up to date regularly has served well over the years since its compilation.

After a long period of quiescence, disturbed only by occasional individual studies of single station records, a frontal attack has been made on the complexities of assembling historic rainfall records. Recognising the importance of regional variations, Craddock has grouped all the known old records to provide annual rainfall totals representative of 11 regions covering most of England (Craddock 1976). The London area (from 1725) and the East Midlands (from 1726) can provide unbroken records to the present day. The observations of Dr James Jurin (Crane Court) and Thomas Barker (Lyndon, Rutland) are just two examples of renowned chronicles which have contributed to knowledge of the rainfall of those early days. A catalogue of the old records held in the archives of the Meteorological Office (Craddock 1976) now provides a basis for future work on the extension of the regional studies. The discovery of further records in other archives could greatly assist in the delicate task of interrelating data between different rain-gauge sites to fill the present gaps in the regional series. For example, additional records are required to link the 28 years of observations by Dr Huxham at Plymouth (1725–52) with the Exeter record from 1820 to make a satisfactory long series for Devon and Cornwall. This logical framework (set out by Craddock) should give a much needed incentive to other scholars to embark on the detailed scrutiny of the shorter lesser-known records, and also to search for new sources of rainfall data in other parts of the country. For the well documented regions, the next step is the evaluation of monthly values to provide information on the seasonal variations during the centuries.

One of the earlier compilations of a single long-term rainfall record was undertaken in the Meteorological Office (Wales-Smith 1971). The long, continuous record from the Royal Observatory at Kew from 1871 was extended back to 1697, thus making, at the time of its publication, a 274-year sequence of monthly rainfall totals. The basic data were assembled from a wide variety of sources in the London area, and an unfortunate gap in instrumental measurements (1717–24) was filled from a weather diary kept at Richmond. The diary summaries of each month's weather overlapped rain-gauge records both before and after the gap, and thus a scale of estimated monthly rainfalls was established for the diary entries. This was used to fill in the missing eight years. Seven rain gauges, representing an area extending from Upminster to Tonbridge and kept under different conditions of exposure, provided the monthly totals up to 1814. The period 1815–70 was filled directly by the observations at Greenwich Observatory, since for the 10 years of overlap with Kew (1871–80) the minor discrepancies were not worth considering. The methods used in adjusting the contributing short records to the Kew standard are described in detail (Wales-Smith 1971). In this study, the monthly percentages of the annual average for Kew for the later standard period (1916–50) were used in disaggregating annual totals to give monthly values. Overlapping records were tested by regression of the common annual totals; the records for those stations differing from the acceptable values for Lambeth and Soho Square were adjusted by applying appropriate factors.

Meanwhile, following his successful and famous temperature series for 'Central England', Manley was continuing his investigations into old instrumental records with the compilation of a homogeneous long period record of rainfall in the Manchester area (Manley 1972). The problems of comparing the observations at different locations were more acute in Manchester than in the London area, since the rainfall distribution is affected to a greater degree by orographic influences. This is seen in the north–south gradient of the average annual isohyets across the city of about 250 mm (10 ins). The records of a dozen observers were needed to bring the series from 1765 to the period when consistent observations by Manchester Waterworks began at several reservoir sites. A gap of 16 years (1770–85) exists in the rainfall measurements; records of the monthly number of days with rain only were available, and there were no overlapping data to provide a scale of rainfall quantities that was possible for the Kew series. The final series of annual percentages of the 1916–50 average, together with the monthly percentages of each individual annual total, continuous from 1786, has been rendered to be representative of a site in Central Manchester having an annual average of 900 mm (35.4 ins). The record after 1950 may be regularly updated at the Meteorological Office by using the records of six reliable stations in the Manchester area.

A third example of the compilation of a single long-term rainfall record is the homogeneous Pode Hole series (Craddock & Wales-Smith 1977) which may be taken as representative of the East Midlands (Craddock 1976). The amalgamation of the four principal records – Southwick (Oundle), Lyndon (Rutland), South Kyme and Pode Hole – was helped by the small differences across generally flat country demonstrated by a maximum range of 24 mm (0.94 in.) in their average annual rainfalls. The series of monthly and annual rainfall totals runs from 1726 to the present day, and is continued by the observations at Pode Hole, which is centrally placed in relation to the contributing stations for the years before 1829. The factors used to relate observations to the current Pode Hole standard have been modified after further investigation of the rain gauge sites, following the discovery that the records pre-1870 were out of step with other long records (Tabony 1980).

The long record of meteorological observations at Oxford is well known, and although a description of the rainfall is not included here, a detailed appraisal of the Oxford rainfall measurements has been made (Craddock & Smith 1978). The Radcliffe Observatory record from 1815 has been extended back to 1767 by incorporating several shorter series of observations in the Oxford area (Craddock & Craddock 1977).

There are too many long records beginning in the 19th century for them to be cited individually. There is still a lot of work to be done in assessing their quality before analysis can provide reliable information on temporal and areal variations. In order to supplement the regional data series outlined by Craddock (1976), there are many short records that may be combined to enhance the understanding of the causes of varying fluctuations in rainfall quantities. One such study for two regions in the Northern Pennines should be a valuable contribution to the limited knowledge of the precipitation patterns in upland areas (Jones 1981). A wealth of old records exists for parts of Scotland (Manley 1977), but their amalgamation will prove extremely difficult due to the greater distances between rain-gauge sites in the sparsely inhabited regions.

4.4 Analysis of long records

As the years went by, and the number and length of rainfall records over the United Kingdom increased in Symons' archives, a start was made on attempting to answer some of the questions posed by the water engineers responsible for developing potable piped water supplies to the growing towns and industries. Symons himself undertook many of the data analyses, identifying the wettest and driest places in the country, and evaluating the relationships between the wettest and driest years in a record, and the mean annual rainfall (Symons 1884). His findings for the British Isles – that the wettest year has nearly 50% more rainfall than the mean, and that the driest year has one third less than the mean – were supported by the results of a wider study carried out by a consulting engineer (Binnie 1892). Having been concerned with designing waterworks in India, Binnie had collected together records from many different climates, and in his studies of variability he adopted the practice instigated by

Symons of relating the annual rainfall totals to the record mean. He noted that 'the mean does not fall evenly between the extremes, there being a greater divergence of 10% on the side of the wettest years'. This led him to investigate the significance of the mean of a whole record and average values taken over different numbers of years within the record. Using the 97 years of the Padua record (1725–1821) for successive determinations of average values for 5, 10, 15 . . . n years, Binnie found that there was little reduction in the variation from the overall mean after the 35-year average. This result held good for another eight records of 70 years and over from Europe and India. This suitability of a period of 35 years for a standard average was confirmed over 50 years later by a more rigorous statistical analysis (Carruthers 1945).

The concern for water supplies and the provision of adequate storage always leads to the study of sequences of wet or dry years. The average fall of the three driest years was found to be 0.75 of the record mean, taking the values from 153 widely scattered records together (Binnie 1892). Departures from 0.75 for individual stations did not exceed 2% where the mean annual rainfall was over 508 mm (20 in), but below that, much greater departures resulted. The average fall of the three wettest years from the same records was 1.27 of the gross mean, showing that the fluctuations of the three-yearly extremes are virtually equal in both directions. Binnie's identification of similar variations from the annual means occurring in such widely differing climates as those of India and England was the forerunner of the recognition that the frequency pattern of many annual rainfalls resembles that of the normal statistical frequency distribution. The application of the normal distribution to annual rainfalls to determine confidence limits of receiving specified falls was demonstrated by Manning (1956), who also used the famous Padua record (1725–1951). However, while the normal distribution was a satisfactory fit for the Padua record (mean 859 mm, 33.8 in), for stations in low rainfall areas it was less acceptable, and a critical lower limit of about 508 mm (20 in) was implied again. It is noted that while three-yearly rainfall totals are well represented by the normal distribution, many series of annual totals themselves tend to produce a slight positive-skew distribution (Brooks & Carruthers 1953); the goodness-of-fit should always be tested before using related statistics.

In this age of the fast electronic computer which can accomplish a wide range of calculations at the push of a button, it can be rewarding to reflect a while on the achievements of the previous generation of rainfall analysts who were experts at mental arithmetic or at manipulating a hand calculator. One of those prodigious workers was Ashmore, a schoolmaster of Wrexham. In his classical paper to the Royal Meteorological Society, Ashmore (1944) set out a logical analysis of the long period record at Pentrebychan, a reservoir site 2 miles from Wrexham, with a 63-year record (1880–1942). Ashmore was conversant with the analytical techniques adopted by water engineers, such as the derivation of the mass–residue curve and the plotting of ranked annual totals on log–normal probability paper. In analysing the annual rainfall, he also followed the practice of the Meteorological Office scientists, and estimated the driest and wettest periods of up to 10 years. These computations would also have contributed to his production of the five-year running means. Among the statistics of the record calculated by Ashmore, the record mean annual rainfall was 850 mm (33.45 in), his value for skewness was 0.42, and he also discovered that the skewness of the three-yearly totals was zero (indicative of a probable normal distribution). The evaluations of the probability of occurrence of specific annual totals and of the number of minimum and maximum rainfalls to be expected in selected numbers of years were also demonstrated. Similar analyses were undertaken for the monthly rainfall records, such as the derivation of the monthly means over a selection of numbers of years and over the total record, of the wettest and driest months, and of the distribution of the monthly values about their respective means. For manual calculation, the most tedious analysis was surely the identification of the sequences of wettest and driest consecutive months for periods of up to 120 months, but Ashmore made full use of the figures he obtained by relating them to theoretical considerations. After dealing briefly with seasonal rainfall values, Ashmore continued his considerations of the monthly rainfall pattern during the year by fitting Fourier series to the monthly means as percentages of the annual mean (1880–1942). By repeating this procedure for records from other stations in the region, he joined other investigators of the day in attempting to define the rainfall type with the Fourier coefficients, which differed according to station location. Even more tedious data abstraction was involved in the analysis of the daily rainfall. The mean number of rain-days, and the average fall per rain-day in each month and for the year were presented for the full 63 years of the record; however, the detailed study of droughts and rain spells was confined to the period 1926–43. Heavy falls on rain-days (occasions of 25.4 mm (1 in) or more), the wettest n days in each year for various values of n, and the frequency of mean maximum rainfall over numbers of days in wet spells were all statistics contributed with practical applications in mind. It would be superfluous now to quote or comment on Ashmore's results, since the information relates to a limited area and to what is now a relatively short record, but the extent of such a study undertaken by a private individual at that time (1943–4) needs to be recalled and appreciated by present students.

Ashmore's plotting of the five-year running means of the Pentrebychan rainfall record for comparison with the annual number of sunspots is a remainder that climatologists had been anxious over many years to explain observed cyclical variations in natural phenomena by reference to the regular 11-year fluctuations in the number of sunspots. The lack of correlation between rainfall values and the sunspot cycle was demonstrated by Alter (an American from the University of Kansas), who came specially to study the long English records (Alter 1933). He copied the notable data series from the archives in the Meteorological Office, gaining valuable advice on their validity from Brooks and Glasspoole. From correlation periodograms calculated from quarterly values of 15 records spanning the period 1727–1929, Alter summarised the following major results.

'(a) Long cycles do not exist in these data.
(b) Strictly periodic terms do not exist.
(c) At least one and probably more cycles do exist which vary their phase in step with the sunspot numbers.
(d) Nothing has been developed to give long-range prediction of any commercial value.'

These early, largely negative, findings will be recalled when more sophisticated recent time series analyses are described.

The lean years for rainfall analysis following Ashmore's exhaustive example were finally terminated by a sharp stimulus from the long-term weather forecasters. In the late 1950s, the development in digital computers had made short-term weather forecasting by numerical modelling of the atmosphere a routine operational procedure. There was then considerable pressure on the research teams to extend the forecasting timescale to months, and even to seasons ahead. This entailed a climatological approach to average weather sequences, rather than the dynamic atmospheric modelling on a day-to-day basis. Historic records of monthly and seasonal temperatures and rainfalls were fed into the electronic computer for analysis. By the mid-1960s, there was also further encouragement from the Water Resources Board which was anxious to assess available water resources to the year 2001. Results of some of the rainfall studies were published by Stephenson (1967) and Murray (1967).

Stephenson used the England and Wales record from 1727. He compared sequential six-monthly totals, January to June with July to December, then February to July with August to January, and three-monthly totals with the following three-monthly totals, and then in successive three-month intervals up to a year ahead. The various data series so compiled were classified into terciles, dry, average and wet, and analysed in six different periods defined by the two predominant atmospheric circulation types, westerly regimes or blocking regimes, identified by the synoptic climatologists. An example of the many contingency tables produced is given in Table 4.1 for the summer to winter relationships for the whole of the record, 237 years. When these absolute values

Table 4.1 Contingency tables of six-monthly rainfall terciles for England and Wales, 1727–1963 (from Stephenson 1967).

April–September related to following October–March				October–March related to following April–September			
		October–March				April–September	
		Dry	Average	Wet			Dry	Average	Wet
April	dry	28	28	24	October	dry	27	27	29
	average	28	30	21		average	24	29	24
September	wet	27	19	32	March	wet	30	22	25

are converted into probabilities of occurrence, transition probability matrices are obtained (Table 4.2). To evaluate the chances of dry summers being followed the next year by dry, average or wet summers, the two matrices may be multiplied together to give a single transition probability matrix for the sequential summer seasons (April to September) shown in Table 4.3. A similar matrix could be derived for the winter sequences. From the even distribution of all these probabilities, it is evident that there is an equal chance of a rainfall total in one of the three terciles. Similarly, from all the other contingency tables, there was no strong evidence of correlation between compared seasonal rainfall totals, and the general conclusion was that this approach to seasonal forecasting of rainfall was not likely to prove fruitful.

Murray also used the England and Wales record, but confined his analysis to the last 100 years (1866–1965) when overall national values were obtained from a good network of gauges. Tercile values were again used, but in this study, sequences of individual months classified as dry, average or wet were abstracted from the data series. An example of the number of sequences of dry (and wet) months up to seven months in length is shown in Table 4.4. It will be noted that the longest sequence was of seven wet months, beginning in a July, and that there were two sequences of six dry months, beginning in an April and a September. The average length of the dry spells was 1.54 months, and of the wet spells 1.55 months. Sequences of dry or average and average or wet months were also abstracted. However, the coarse tercile classification of the average monthly rainfalls over the whole country cannot be related closely enough to the variable rainfall-producing weather patterns to make these results useful for forecasting purposes.

The divergence of some annual and many monthly rainfall totals at stations in the United Kingdom from complete randomness, signified by the non-fitting of

Table 4.2 Transition probability matrices for summer/winter rainfalls for England and Wales.

		October to March					April to September		
		Dry	Average	Wet			Dry	Average	Wet
April	dry	0.35	0.35	0.30	October	dry	0.33	0.33	0.34
	average	0.35	0.38	0.27		average	0.31	0.38	0.31
September	wet	0.35	0.24	0.41	March	wet	0.39	0.29	0.32

Table 4.3 Transition probability matrix for the summer season.

		Summer II		
		Dry	Average	Wet
Summer I	dry	0.34	0.33	0.33
	average	0.34	0.33	0.33
	wet	0.35	0.32	0.33

Table 4.4   Frequencies of runs of 1,2,3 etc. dry (wet) months (from Murray 1967).

Starting month	Runs greater than or equal to number of months						
	1	2	3	4	5	6	7
January	20(19)	6(5)	2(2)	1(2)	0(1)	0(0)	0(0)
February	19(26)	6(10)	3(3)	1(1)	0(0)	0(0)	0(0)
March	25(23)	8(10)	1(3)	1(0)	0(0)	0(0)	0(0)
April	23(21)	9(8)	3(4)	2(2)	1(2)	1(1)	0(0)
May	23(16)	9(5)	3(1)	0(0)	0(0)	0(0)	0(0)
June	22(24)	6(9)	5(2)	1(1)	0(0)	0(0)	0(0)
July	22(24)	8(8)	3(6)	1(2)	1(1)	0(1)	0(1)
August	19(21)	7(4)	3(0)	0(0)	0(0)	0(0)	0(0)
September	22(19)	7(9)	2(4)	1(1)	1(1)	1(0)	0(0)
October	21(22)	6(9)	2(3)	2(2)	1(0)	0(0)	0(0)
November	23(20)	10(5)	2(4)	0(1)	0(0)	0(0)	0(0)
December	19(23)	8(6)	4(1)	0(0)	0(0)	0(0)	0(0)

the normal probability distribution, calls for specifically designed analysis of rainfall quantities to provide the required information. Wales-Smith (1973) presented some results of computations carried out in the Meteorological Office to help solve water engineering problems. Using his long Kew record (1697–1970), he produced frequency diagrams of seasonal and annual rainfall totals and, from the six wettest and six driest months in the 274 years for each calendar month, extreme value analysis resulted in plots of monthly rainfalls likely to be equalled or exceeded and those not to be exceeded for return periods up to 200 years. The extreme value computations were also carried out for selected monthly sequences, and for the defined seasons. With judicious extra-polation of the return period graphs, rough estimates of the 500- and 1000-year events were possible. An appraisal of general trends in the Kew rainfall series was provided by plotting the maximum and minimum values in each decade, the 'second extremes', and the decadal means for each of the seasons. For the annual totals, decadal averages were low at the beginning of the 18th century, but rose to a peak in the early 19th century, since when there have been several long period fluctuations with the two annual extremes occurring in the present century (wettest 1903, driest 1921).

A more extensive study of long period records was made by Tabony (1977), who used the reliable monthly rainfall totals from 1911 for stations covering England and Wales, as well as the long England and Wales record from 1727 and the Kew and Manchester series. He worked primarily with the coefficients of variation and skewness (C_v and C_s respectively) and examined their different values for each month over 60-year blocks of the records. Overall, C_v tends to a maximum in the spring and a minimum in autumn, to be expected from generally dry springs and wet autumns. However, the marked variability in the seasonal patterns of the C_v between the different blocks of years is indicative of large random components in the rainfall amounts. The C_s values show no regular seasonal pattern. The coefficients were also determined for sequences of months with varying starting months; an example of the results averaged over all the starting months is given in Table 4.5.

Table 4.5 Coefficients of variation and skewness of monthly rainfalls over a range of periods for the years 1911–70 (from Tabony 1977).

		Periods (months)					
		1	3	6	12	24	36
Kew	C_v	0.553	0.327	0.239	0.176	0.128	0.106
	C_s	0.670	0.300	0.010	−0.180	−0.110	−0.030
Manchester	C_v	0.485	0.274	0.198	0.145	0.100	0.078
	C_s	0.600	0.240	0.170	−0.080	0.050	0.070

The values show a general decrease with increase in the number of months in the sequences. This study of long-term monthly rainfall then considered the interrelationships of C_v and C_s with period length in months, mean rainfall, and area. These resulted in a method for calculating the return periods of the rainfall in sequences of wet or dry months for a range of months starting in any month of the year for anywhere in Great Britain. The comprehensive model developed used the three parameter log-normal frequency distribution for calculating the rainfall–return period relationships, and Tabony's detailed monograph provides the means for determining the parameters necessary for evaluating the values of C_v and C_s over the required period of months for any location in Great Britain.

A later study (Tabony 1979) investigated in great detail the incidence of periodicities by the latest techniques of spectral analysis. To the three long records used in the earlier work, was added the monthly series for Pode Hole from 1726. Several analytical methods were compared, and for the annual rainfalls at Kew, all identified significant peaks at the periods of 2.1, 2.4, 3.9 and 12.0 years. The starting month for an annual total can affect the incidence of short periodicities and, in comparing the spectra obtained from the 12-months total starting in January with those from the 12-months total starting in July, the 2.1-year periodicity becomes insignificant in the second data series. Analyses of the four long records were made for 12-monthly totals starting at 3-monthly intervals, for the winter and summer halves of the year, and for the four seasons. A selection of the results is given in Table 4.6. In addition to the differences seen between the stations, great variations in the spectral patterns were found between three separate blocks of 80 years for the winter half-year rainfall over England and Wales. The 5-year periodicity was scarcely identifiable in the earlier years, but completely dominated the 1896–1975 period. While periodicities of 2.1, 2.4, 3.9, 5 and 6 years were all identifiable with at least 5% significance over the complete length of the four records, they only account for very small proportions of the total variance of the data. The incidence of the several series of rainfall totals examined was therefore largely random, which is again seen to make forecasting of monthly or seasonal rainfalls very difficult in the United Kingdom.

Table 4.6 Significant spectral peaks in the long rainfall records (from Tabony 1979).

	J	F	M	A	M	J	J	A	S	O	N	D
12 months starting in January												
England & Wales	51.2	.			3.90				2.39		2.11	2.00
Kew		12.6		6.11	3.90				2.40		2.11	
Pode Hole			9.61				2.93				2.11	
Manchester	41.6				3.85		2.90	2.56	2.41			2.00
Summer half-year												
England & Wales			10.3	6.05	3.78				2.40		2.13	
Kew		48.7		6.09					2.42		2.12	
Pode Hole			7.96	4.64	3.74						2.12	
Manchester	39.2			6.18				2.55	2.41			2.00
Winter half-year												
England & Wales					4.97		2.75					2.00
Kew				8.84	4.92					2.19		
Pode Hole	133.2	21.5			4.95	3.93	3.57	3.35				
Manchester				8.36	4.97		2.93					2.00

The need for forecasting seasonal and annual rainfalls is much more critical in South Africa, where the incidence of rainfall is variable and there are considerable seasonal differences over the country. Seven major rainfall regions have been defined (Dyer 1977), and for two of these regions, Dyer has demonstrated the application of stochastic models to the rainfall data series from 1910–72. He showed that although the regional rainfalls could be related to the sunspot cycle by a regression model, the sunspot numbers could not be used satisfactorily in forecasting, since the model only accounted for a small proportion of the rainfall variance. A trigonometric model of the rainfall series gave a satisfactory fit to the observed rainfall values, and was considered worthy of further investigation.

In the United Kingdom, the identification of rainfall regions in a statistical sense by using principal component analysis is a prime requisite, to aid the compilation of the regional long-term series (Craddock 1976). This alternative approach would seek to complement Tabony's comprehensive model derived to assess the variability of long-term rainfall over Great Britain.

4.5 Two northern England rainfall records

In reviewing the compilation and analysis of rainfall records, without necessarily endorsing every contribution emphasis has been placed on the advisability of having an adequate network of long-term homogeneous records to reflect the temporal and areal rainfall variability across the country. Attention has also been drawn to the general characteristics of rainfall occurrence, and to the gradual increase in the detail of analysis that may be carried out with the continual advancement of computational facilities. Many of the findings of the early rainfall analysts remain unchallenged today; indeed the more sophisticated techniques have verified original results. However, there is a contemporary requirement to improve upon precision in measurement and in the estimate of rainfall statistics.

To demonstrate some of the rainfall analyses that may be applied on a routine basis to fulfil requirements of water engineers today, two different records in northern England have been treated by a selection of standard computer packages (Table 4.7). The first record has already been introduced: the Manchester monthly rainfall record 1786–1976 compiled by Gordon Manley to be representative of a site in Central Manchester (2292 data values). The second is the daily record for Windermere (Holehird) from 1911 to 1973 (23 011 data values). These two records illustrate the contrast in rainfall data series described earlier; the monthly totals for Manchester constitute a continuous series of discrete data values while the daily observations at Windermere form a discontinuous series of discrete rainfall measurements. The basic data sets are treated quite differently, but clearly the continuous monthly totals for Windermere may be analysed similarly to those for Manchester.

Table 4.7 Particulars of the northern England records.

	Manchester	Windermere (Holehird)
location (National grid reference)	(SJ 845 990)	NY 411 007
height above mean sea level (m)	(45)	149
period	1786–1976	1911–73
number of years	191	63
record average annual rainfall (mm)	905	1807
average annual rainfall (mm) (1916–50)	900	1769

The most commonly required statistics of the individual monthly and annual totals are given in Table 4.8. The average values for the two stations cannot be compared strictly since they have been derived over different periods. However, for each of the stations, the maximum annual total is 150% of the annual mean, in agreement with Symons' findings. The driest years, representing 61% and 64% of the annual means of Manchester and Windermere respectively, are more than the third less than the means noted by Symons. The relatively small standard deviations produce low values for the coefficients of variation. The positive coefficient of skewness indicates a tendency to have more annual values below than above the mean. Nevertheless, the annual data provide a reasonably good fit to the normal distribution, as seen in Figure 4.1. The annual totals are plotted as proportions of the mean against probability of

Table 4.8 Rainfall statistics Manchester (191 years) and Windermere (63 years).

	J	F	M	A	M	J	J	A	S	O	N	D	Year
Manchester													
mean (mm)	71	59	57	53	63	71	90	94	83	94	85	87	905
std dev.	34	32	30	26	33	37	43	41	44	39	37	44	136
C_v	0.48	0.55	0.52	0.49	0.53	0.52	0.48	0.44	0.53	0.42	0.44	0.50	0.15
C_s	0.76	0.51	0.77	0.52	1.02	0.70	0.86	0.34	0.58	0.44	0.47	0.50	0.40
max. (mm)	200	163	146	129	204	184	288	240	232	230	183	231	1359
year	1948	1848	1827	1843	1792	1882	1828	1799	1918	1870	1825	1792	1792
min. (mm)	8	7	8	8	6	0	8	7	7	9	14	7	549
year	1861	1921	1808	1851	1826	1925	1868	1959	1959	1922	1805	1844	1887
Windermere													
mean (mm)	197	140	117	101	101	103	127	153	159	183	189	201	1769
std dev.	79	85	60	44	45	50	53	67	86	97	98	92	288
C_v	0.40	0.61	0.51	0.44	0.45	0.49	0.42	0.44	0.54	0.53	0.52	0.46	0.16
C_s	0.50	0.09	0.72	0.52	0.38	0.28	0.22	0.33	0.60	0.69	0.34	0.26	0.44
max. (mm)	476	326	256	207	209	220	269	302	383	468	399	387	2652
year	1928	1915	1912	1913	1920	1938	1938	1917	1918	1967	1929	1951	1928
min. (mm)	16	4	20	22	17	7	27	5	23	24	12	44	1134
year	1963	1932	1944	1938	1935	1911	1911	1947	1972	1946	1945	1933	1973

exceedence (P) where $P=(m - 0.375)/(n + 0.25)$, with m the descending ranked value and n the number of years record. The probability of exceedence relates to the return period. The straight line of the distribution fitted by the method of moments passes through the mean value at the central frequency point on both graphs. From such plots, estimates of the standard deviations and hence coefficients of variation can be made. Further calculated statistics of probability of annual totals are shown in Table 4.9.

Table 4.9 Probability estimates of annual rainfall.

Exceedence probability	Return period	Normal variate	Manchester (mm)	Windermere (mm)
0.990	1.01	−2.33	588.6	1099.0
0.980	1.02	−2.05	625.7	1177.6
0.950	1.05	−1.64	681.4	1295.4
0.900	1.11	−1.28	730.8	1400.0
0.800	1.25	−0.84	790.7	1526.8
0.500	2.00	0.00	905.3	1769.3
0.200	5.00	0.84	1019.8	2011.8
0.100	10.00	1.28	1079.7	2138.6
0.050	20.00	1.64	1129.2	2243.3
0.020	50.00	2.05	1184.8	2361.1
0.010	100.00	2.33	1221.9	2439.6
0.005	200.00	2.58	1255.9	2511.5
0.001	1000.00	3.09	1325.9	2659.7

The monthly data for Manchester and Windermere show the expected seasonal variations, with a tendency for maximum values in the winter and minima in the spring and early summer. The range of variation is much greater than for annual values, shown by the higher coefficients of variation, and the wide range in the coefficients of skewness is indicative of very variable frequency distributions for the monthly values. The maximum monthly totals may be as high as 2.5 times the monthly mean; at Windermere the October maximum in 1967 is 256% of the mean, but at Manchester, an exceptionally wet July in 1828 is 320% of the mean. The minimum totals reach nil in only one month, June 1925 at Manchester, but 4 mm in February 1932 and 5 mm in August 1947 both represent a mere 3% of the respective monthly means at Windermere.

In considering the sequential time series of the two records, the annual totals as percentages of the respective record means are plotted in Figure 4.2. Superimposed on the bar graphs are broken curves joining the five-year running means plotted at each central year. The running means smooth out irregularities in the annual totals and, for the Manchester record, they emphasise distinctive groups of wet and dry years in the first 40 years of the record and in the period 1885–1930. During the other years, the incidence of the annual values is highly irregular. There is no apparent regular fluctuation in the annual rainfall at Manchester. The comparable plot for Windermere over the much shorter

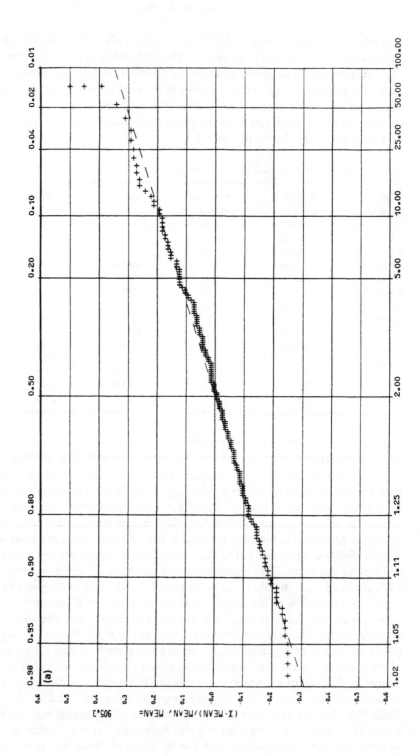

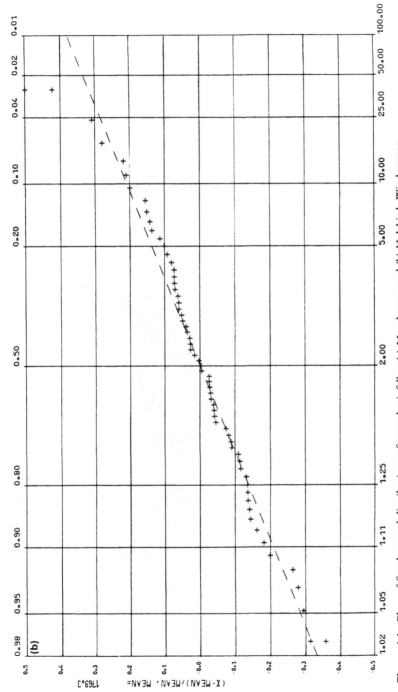

Figure 4.1 Plots of fitted normal distributions of annual rainfall at (a) Manchester, and (b) Holehird, Windermere.

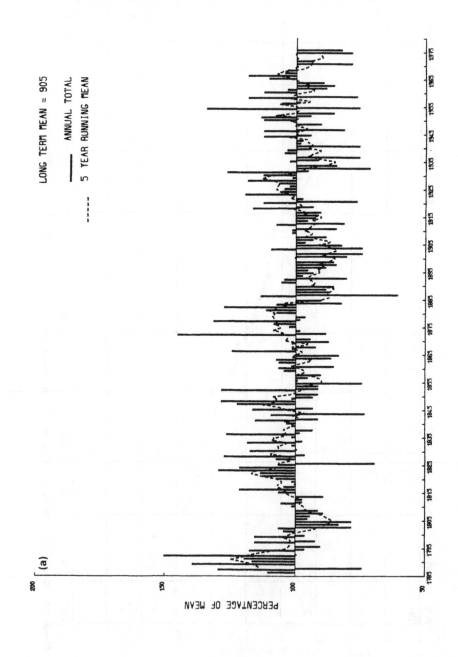

(a)

LONG TERM MEAN = 905

—— ANNUAL TOTAL

----- 5 YEAR RUNNING MEAN

PERCENTAGE OF MEAN

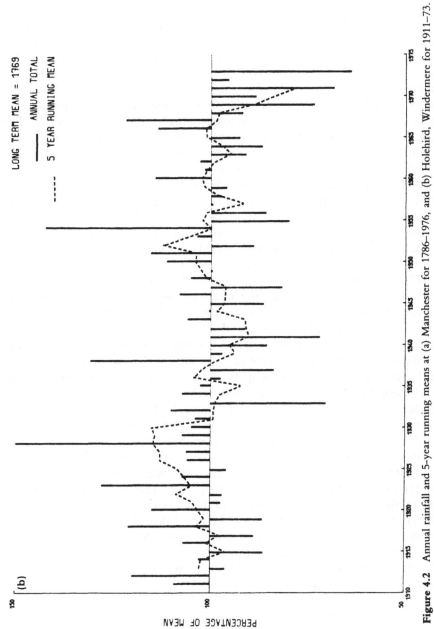

Figure 4.2 Annual rainfall and 5-year running means at (a) Manchester for 1786–1976, and (b) Holehird, Windermere for 1911–73.

period demonstrates a similar irregular pattern, but the fluctuations of the five-year running means compare favourably with those of Manchester over the same years – 1911–73. The serial correlation coefficients up to lag 50 form the correlograms of the annual rainfalls for Manchester and Windermere in Figure 4.3. There is a complete lack of any regular cyclical variation in the correlograms, and few of the individual coefficients show any distinctive exceedence of the 95% confidence limits. These negative findings are supported by the lack of any outstanding spectra shown in the results of a smoothed spectrum analysis (Fig. 4.4). There is only the slight indication of a weak 2.8-year cycle. The comparable analysis for Windermere shows a weak 3.8-year cycle. (These bear comparison with Tabony's more exhaustive analysis.)

When the time series of the monthly totals are analysed, a very different pattern of the serial correlation coefficients appears in the correlograms (Fig. 4.5). Waves with a consistent length of 12 months are seen for both records, and there are regular occurrences of significant serial correlation coefficients. It will be noted that the highest correlation coefficients are found at the wetter station, Windermere. The dominance of the annual cycle of monthly rainfalls is shown in the plots from the spectrum analysis (Fig. 4.6).

The 23 011 pieces of data comprising the Windermere daily record over 63 years have been subjected to a much more complex analysis to demonstrate the powerful potential of the digital computer. The days have been classified into seven groups bounded by the rainfalls as follows: 0 and ≤10 mm, >10 and ≤20 mm, >20 and ≤30 mm, >30 and ≤40 mm, >40 and ≤50 mm, >50 and ≤60 mm, and >60 mm. By inspecting the sequential occurrence of rainfalls in different groups, the relationship between one day and the next has been evaluated, and is given in the contingency table (Table 4.10). Here, the numbers of days in different rainfall groups for a single day having the following day in one of the seven rainfall groups are presented. For example, there are 1675 days with >10 and ≤20 mm followed by a day with 0 to ≤10 mm, and 12 days with >30 and ≤40 mm followed by a day with >30 and ≤40 mm. The contingency table is then converted to a state transition matrix which shows the probabilities of occurrence of the various rainfall group days (Table 4.11). Thus there is a probability of 0.704 of the occurrence of the first example, and a probability of 0.038 of the second example. It will be noted that after the first group, the probability values do not vary very much. This is due to the inclusion of dry days in the first group and the tendency to have a dry day followed by another dry day. The probability of any day being in one of the groups is given by the probabilities of occurrence derived from the sum of the rows in the contingency table. This is called the steady state probability, and is given by the values in Table 4.12. To determine the probability of daily falls two or more days ahead, given the rainfall group on the first day, a two- or more step ahead transition matrix can be evaluated by multiplying the one-step transition matrix by itself two or more times. The two-step ahead transition matrix is shown in Table 4.13, and it will be seen that the probabilities beginning with a day in any of the rainfall groups is tending towards the steady state. From the comparison of the

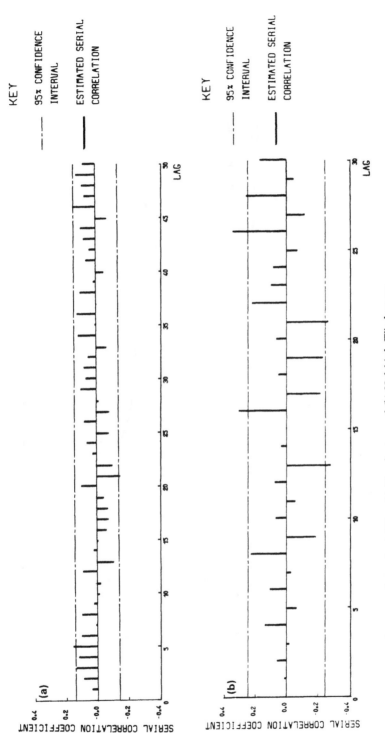

Figure 4.3 Correlograms of annual rainfall at (a) Manchester, and (b) Holehird, Windermere.

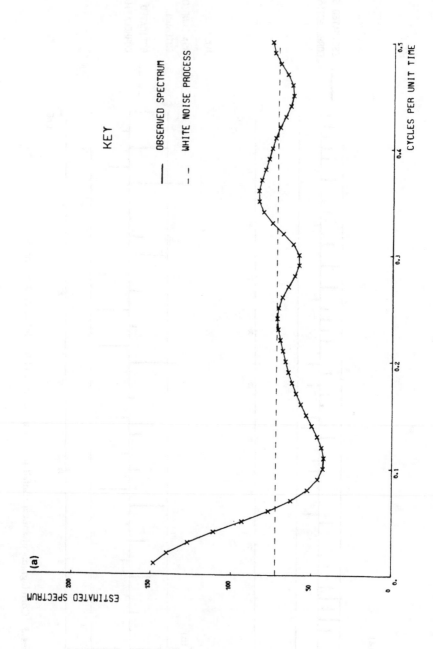

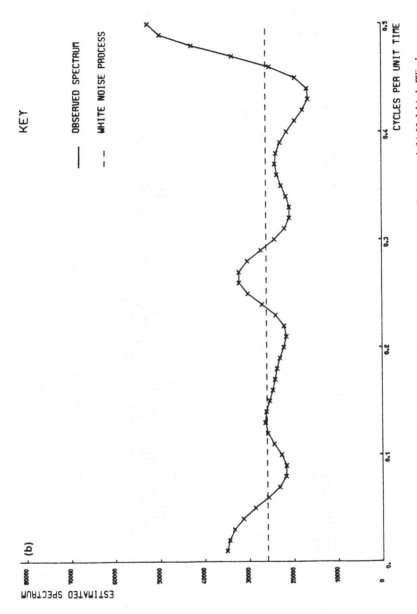

Figure 4.4 Smoothed spectra of annual rainfall using parzen window at (a) Manchester, and (b) Holehird, Windermere.

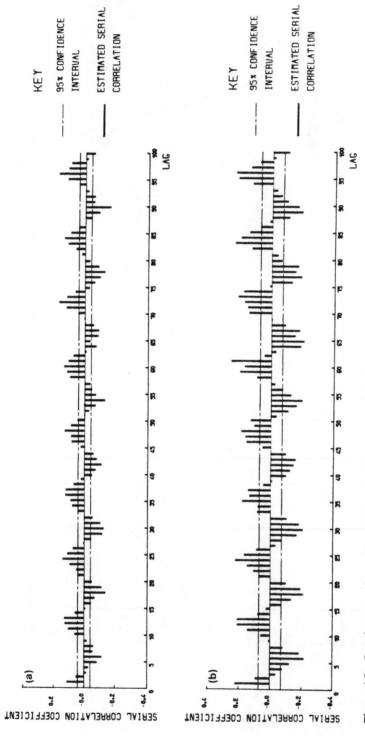

Figure 4.5 Correlograms of monthly rainfall at (a) Manchester, and (b) Holehird, Windermere.

Table 4.10 Daily rainfall contingency table.

Group	0–10	10–20	20–30	30–40	40–50	50–60	>60
0–10	16616	1643	608	217	89	41	24
10–20	1675	454	139	67	27	10	8
20–30	590	173	51	20	8	2	2
30–40	202	71	27	12	4	0	4
40–50	92	24	11	3	1	0	2
50–60	37	8	3	1	3	0	1
>60	26	7	7	0	1	0	0

Table 4.11 State transition matrix.

Group	0–10	10–20	20–30	30–40	40–50	50–60	>60
0–10	0.864	0.085	0.032	0.011	0.005	0.002	0.001
10–20	0.704	0.191	0.058	0.028	0.011	0.004	0.003
20–30	0.697	0.205	0.060	0.024	0.010	0.002	0.002
30–40	0.631	0.222	0.084	0.038	0.013	0.000	0.013
40–50	0.692	0.181	0.083	0.023	0.008	0.000	0.015
50–60	0.698	0.151	0.057	0.019	0.057	0.000	0.019
>60	0.634	0.171	0.171	0.000	0.024	0.000	0.000

Table 4.12 Steady state probabilities of daily rainfall at Windermere.

Group	Probability
0 to ≤ 10	0.836
> 10 to ≤ 20	0.103
> 20 to ≤ 30	0.037
> 30 to ≤ 40	0.014
> 40 to ≤ 50	0.006
> 50 to ≤ 60	0.002
> 60	0.002

Table 4.13 Two-step ahead transition matrix.

Group	0–10	10–20	20–30	30–40	40–50	50–60	> 60
0–10	0.841	0.100	0.035	0.013	0.006	0.002	0.002
10–20	0.814	0.118	0.041	0.016	0.007	0.002	0.002
20–30	0.813	0.119	0.041	0.016	0.007	0.003	0.002
30–40	0.800	0.126	0.044	0.017	0.007	0.003	0.002
40–50	0.811	0.119	0.043	0.016	0.007	0.002	0.002
50–60	0.812	0.118	0.044	0.016	0.007	0.002	0.003
>60	0.804	0.126	0.031	0.017	0.007	0.003	0.002

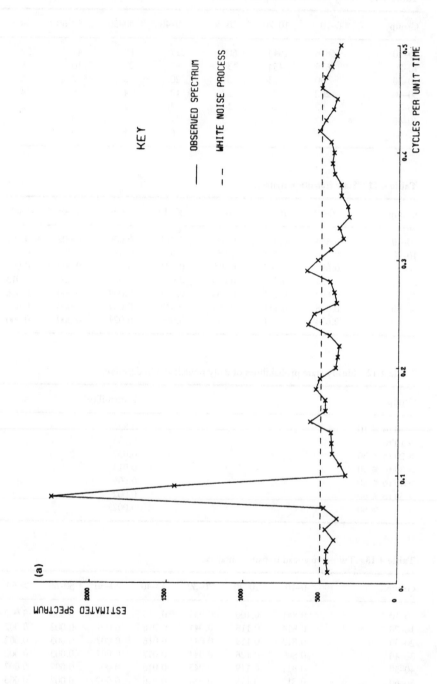

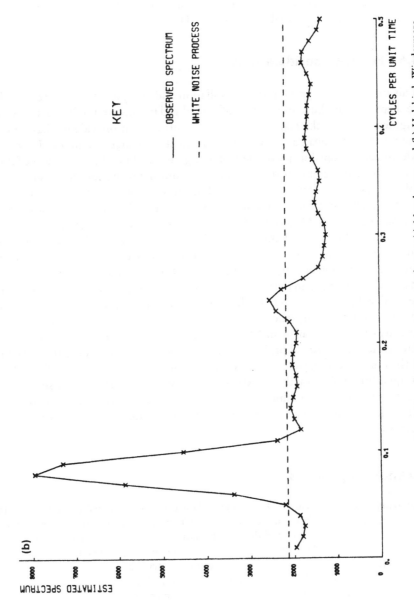

Figure 4.6 Smooth spectra of monthly rainfall using parzen window at (a) Manchester, and (b) Holehird, Windermere.

one-step and two-step ahead transition matrices, it is evident that there is a rapid convergence towards the steady state with increasing steps. This is due to a very weak correlation in rainfall from day to day.

4.6 Further considerations

This outline of progress in the compilation and analysis of rainfall records has concentrated on the manipulation of annual and monthly rainfall totals. In water engineering, it is the statistics of these quantities that provide the basic information for the evaluation and management of water resources. However, additional responsibilities of the water engineer include the prediction of flood levels, the forecasting of their occurrence, and the provision of design criteria for storm-water drainage systems. For these applications, information on rainfall quantities over shorter durations is required. The standard measurement period of 24 hours, 09.00 h to 09.00 h GMT, at the majority of rainfall stations throughout the United Kingdom produces an extensive archive of daily data. For large drainage areas, daily rainfall values may suffice for calculating particulars of flood events, and many relevant studies of daily records have been made. Only a single analysis of the Windermere daily values has been included here. The digital computer is able to process a large number of data items, and thus much more comprehensive analyses than formerly can now be produced. The discontinuous daily rainfall series can be represented by a combination of three different statistical probability distributions: one describing the incidence of the rain (wet or dry days); a second, the duration of wet spells; and the third, the quantities falling on the wet days. Similarly, and in particular for small urban areas, the modelling of rainfall events over shorter durations in terms of hours or minutes can help to provide the essential information for engineering design purposes.

Acknowledgements

The daily rainfall record for Windermere (Holehird) was originally supplied by the Meteorological Office in computer-compatible form, on punched cards; the record was recently completed by the addition of the last five years. The Manchester record, supplied by Professor Manley, had been transferred to punched cards. I am most grateful to Mr K. Guganesharajah for organising the analysis and running the computer packages, and to Sir M. MacDonald and Partners, consulting engineers, for time on their computer. Miss Anna Hikel very kindly typed the manuscript.

References

Alter, D. 1933. Correlation periodogram investigation of English rainfall. *Monthly Weather Rev.* **61**, 345–50.
Ashmore, S. E. 1944. The rainfall of the Wrexham District. *Q. J. R. Met. Soc.* **70**, 241–69.

Binnie, A. R. 1892. On mean or average annual rainfall and the fluctuations to which it is subject. *Proc. Inst. Civil Engrs* **109**, 89–133.

Brooks, C. E. P. and N. Carruthers 1953. *Handbook of statistical methods in meteorology.* London: Meteorological Office, MO 538, HMSO.

Carruthers, N. 1945. The optimum period for a British rainfall normal. *Q. J. R. Met. Soc.* **71**, 144–50.

Craddock, J. M. 1976. Annual rainfall in England since 1725. *Q. J. R. Met. Soc.* **102**, 823–40.

Craddock, J. M. and Eveline Craddock 1977. Rainfall at Oxford from 1767 to 1814 estimated from the records of Dr. Thomas Hornsby and others. *Met. Mag.* **106**, 361–72.

Craddock, J. M. and C. G. Smith 1978. An investigation into rainfall recording at Oxford. *Met. Mag.* **107**, 257–71.

Craddock J. M. and B. G. Wales-Smith 1977. Monthly rainfall totals representing the East Midlands for the years 1726 to 1975. *Met. Mag.* **106**, 97–111.

Dyer, T. G. J. 1977. On the application of some stochastic models to precipitation forecasting. *Q. J. R. Met. Soc.* **103**, 177–89.

Folland, C. K. and B. G. Wales-Smith 1977. Richard Towneley and 300 years of regular rainfall measurement. *Weather* **32**, 438–45.

Glasspoole, J. 1928. Two centuries of rain. *Met. Mag.* **63**, 1–6.

Green, M. J. 1969. *Effects of exposure on the catch of rain gauges.* Reading: Water Research Association TP.67.

Gregory, S. 1958. Some aspects of the variability of annual rainfall over the British Isles for the standard period 1901–30. *Q. J. R. Met. Soc.* **81**, 257–62.

Jones, P. D. 1981. A survey of rainfall recording in two regions of the Northern Pennines. *Met. Mag.* **110**, 239–52.

Kurtyka, J. C. 1953. *Precipitation measurement study.* Report No. 20. Urbana, Illinois: State Water Survey Division.

Manley, G. 1948. Records of snow cover on Scottish Mountains. *Met. Mag.* **77**, 270–72.

Manley, G. 1952. John Dalton's snowdrift. *Weather* **7**, 210–12.

Manley, G. 1957. Climatic fluctuations and fuel requirements. *Scott. Geogr. Mag.* **73**, 19–28.

Manley, G. 1959. The late-glacial climate of north-west England. *Liverpool & Manchester Geol. J.* **3**, 188–215.

Manley, G. 1972. Manchester rainfall since 1765. *Mem. & Proc. Manchester Lit. Phil. Soc.* **114**, 1–20.

Manley, G. 1977. Annual rainfall in England since 1725: some comments. *Q. J. R. Met. Soc.* **103**, 820–22.

Manley, G. 1980. The northern Pennines revisited: Moor House, 1932–78. *Met. Mag.* **109**, 281–92.

Manning, H. L. 1956. The statistical assessment of rainfall probability and its application in Uganda agriculture. *Proc. R. Soc. Lond.* B. **144**, 460–80.

Murray, R. 1967. Sequences in monthly rainfall over England and Wales. *Met. Mag.* **96**, 129–35.

Meteorological Office 1982. *Rules for rainfall observers.* Meteorological Office Leaflet No. 6. Bracknell: Meteorological Office.

Meteorological Office 1981. *Handbook of meteorological instruments* Vol. 5, 2nd edn. London: HMSO.

National Water Council 1982. *Annual report and accounts 1981/82.* London: National Water Council.

Nicholas, F. J. and J. Glasspoole 1932. General monthly rainfall over England and Wales, 1717–1931. *Brit. Rainfall 1931*, 299–306.

Rodda, J. C. 1967. The systematic error in rainfall measurement. *J. Inst. Water Engrs* **21**, 173–7.

Stephenson, P. M. 1967. Seasonal rainfall sequences over England and Wales. *Met. Mag.* **96**, 335–42.

Symons, G. J. 1869. Rules for rainfall observers. *Brit. Rainfall 1868*, 102–4.

Symons, G. J. 1884. On the limits of fluctuation of total rainfall. *Brit. Rainfall 1883*, 29–32.

Tabony, R. C. 1977. *The variability of long-duration rainfall over Great Britain.* Meteorological Office Scientific Paper No. 37. London: HMSO.

Tabony, R. C. 1979. A spectral and filter analysis of long-period rainfall records in England and Wales. *Met. Mag.* **108**, 97–118.

Tabony, R. C. 1980. A revised rainfall series for Spalding, Lincolnshire. *Met. Mag.* **109**, 152–7.

Wales-Smith, B. G. 1971. Monthly and annual totals of rainfall representative of Kew, Surrey from 1697 to 1970. *Met Mag.* **100**, 345–62.

Wales-Smith, B. G. 1973. An analysis of monthly rainfall totals representative of Kew, Surrey from 1697 to 1970. *Met Mag.* **102**, 157–71.

5

A critical assessment of proxy data for climatic reconstruction

HERMANN FLOHN

Long proxy data series from Switzerland, southern Germany, and France are carefully compared and cross-correlated. Calibration correlations with long temperature series are investigated; a reliable reconstruction of individual summer temperatures extending back some 500–600 years is possible, as recently carried out by Pfister. The high spatial correlation of many proxy data series hints at their potential capability for climatic reconstruction, if their response to climatic factors can be ascertained.

5.1 Introduction

Gordon Manley devoted most of his time and his indefatigable zeal to reconstructing climatic records extending as far back as possible; his classic article of 1974 is now one of the most quoted papers of the past decade. The unselfish devotion of this great scientist inspires an investigation of the calibration of proxy data for climatic reconstruction. Already, Chinese colleagues are promising an even longer unbroken series of rainfall estimates from the Far East (Wang & Zhao 1981), and Swiss records going back to 1525 will be published soon (Pfister 1984). The time is ripe for a review of the value of proxy data.

The number of reliable observation series for climatic reconstruction before about 1850 is limited, and in many areas of the globe there are no series for this period. Only a few temperature records are available for the 200 years before that date, and far fewer precipitation records. In the Netherlands and in Italy, investigations are in progress to evaluate further long series, some dating back to the construction of meteorological instruments around 1650. Before that date, proxy data from weather diaries (Flohn 1949, Lenke 1960, Pfister 1977, Brumme 1981), tree rings (Fritts 1976, Frenzel 1977, Schweingruber et al. 1978 & 1979, Fritts et al. 1979, Hughes et al. 1982), wine harvest (Le Roy Ladurie & Baulant 1980), and yield data and many other quantifiable sources may be used to reconstruct climate for individual seasons or months. An example has been given by Pfister (1984) who, in his thesis at the University of Bern, employed phenological data available in the unusually complete Swiss archives, and used empirical correlations with climatic records in order to derive reliable regression equations.

During the past seven years, test correlations have been obtained between proxy data and temperature records, and between different kinds of proxy data to test their reliability, representativity and stability using records available from southern Germany, east-central France, Switzerland and Austria. The results of this research are only preliminary: the calibration of proxy data from the past can certainly be improved, and high cross-correlations between different proxy data series will allow an improvement in the accuracy of reconstruction. Seasonally varying cross-correlations between climatological parameters (temperature, precipitation, sunshine duration) are required to derive reconstructions of such important parameters as soil moisture (Wigley & Atkinson 1977, Brumme 1981) and global radiation (Wacker 1981) for the instrumental period.

5.2   Sources

The following series of proxy data were used.

WD = Grape harvest data from northeastern France, western Switzerland and southwestern Germany, area-averaged corrected for Dijon for the period 1484–1879 (Le Roy Ladurie & Baulant 1980).

SL =   Annual maximum density of dead wood from Lauenen, northern Swiss Pre-alps for the period 1269–1977 (Schweingruber et al. 1978 & 1979).

SO =   Annual maximum density of late wood, raw data from Ötztal (Tirol) for the period 1370–1970 (Lambrecht 1978).

YR and QR =   Yield and quality of wine 1719–1950 from Schloss Johannisberg, Rheingau (Weger 1952).

YS =   Yield of grape harvest (1540–1825), central Switzerland (Pfister 1981).

CS =   Cereal harvest data (tithe auctions), central Switzerland (Pfister 1979).

The Oberrhein temperatures of Basel, Strasbourg and Karlsruhe (von Rudloff 1955–6) were first used in the experiment, but when a lack of homogeneity was detected in this record (see Appendix A), individual long records were used (München, Hohenpeieszettenberg, Basel and Geneva, mostly available since 1780).

Earlier investigations on spatial correlations between distant stations have often been hampered by the lack of stability of correlations. In this investigation, long series have been split into smaller sections, with a length of around 50 years, to test the stability. A good example of such a reconstruction has been given in a dendroclimatic reconstruction of annual precipitation amounts in Iowa (Duvick & Blasing 1981). Rainfall statistics showed weak positive correlations with ring width for each of the 12 months preceding August (the end of the ring-forming season): the sum yielded, for a 60-year calibration period, a correlation coefficient (ccf) $r = 0.74$, with very weak autocorrelation in both series. A preceding 46-year verification period yielded $r = 0.72$ for the same

rainfall period August-July. These results indicate stability and a high contribution of the tree-ring signal to the total variance of rainfall ($r^2 > 0.5$).

Cross-correlation and other statistical parameters can be used to test the
available proxy data; high correlations are found for the period 1781–1825, with
its extreme temperature deviations. However, a critical reconstruction of
climatic series has been postponed until Pfister's data are fully available for
comparison (Pfister 1984).

5.3 On the representativity of proxy data

Investigations on climatic variability are only meaningful if they are based on
data series which are homogeneous and representative. Many series lack these
qualities; great sources of error are the growth of cities and the transfer of
stations within cities, for example from an urban garden area to an airport.
Proxy data may also show an urban effect: this is almost certainly the case for
data from a series of horse-chestnut trees in Geneva (situated along the walls of
the old fortress), which have shown a marked warming trend during the last
decades (Lauscher & Lauscher 1981).

Homogeneity and representativity of proxy data series can be tested by
comparisons with a uniform climate. Spatial correlations between different
series, split into shorter periods to test the stability, allow an estimate of the
spatial (and time) coherency of these series. Tree-ring data have been frequently
proposed as usable proxy series for climatic reconstruction; many examples
from the western USA with its strongly seasonal climate (with a high degree of
variability), have been more or less successfully investigated (Fritts 1976, Fritts
et al. 1979, Hughes et al. 1982). In this case an immediate reconstruction of
pressure anomaly patterns from tree-ring series, omitting intermediate steps
(precipitation, seasonal temperatures), may be too optimistic because of the risk
of a multiplication of stochastic errors. Tree-ring data from humid Western and
Central Europe do not allow very convincing climatic reconstructions, mainly
because of their long climatic memory (e.g. oaks), which may respond to the
12–24 months' impact of both temperature and precipitation (Hughes et al.
1982). Representative results can be expected from trees at the alpine or polar
tree limit (Schweingruber et al. 1978 & 1979), where the temperature of the short
vegetative period mainly controls the growth rate. Significant progress has been
made through the Roentgen density measurements of wood (Schweingruber et
al. 1978), which allow the evaluation of a quantitative parameter such as the
density of late wood, which is produced mainly during late summer.

Phenological data (Schnelle 1981) have also been used for climatic reconstruction (Pfister 1979 & 1984). One of the most representative series has gradually
been built up by Le Roy Ladurie and Baulant (1980); their area-averaged wine
harvest data from eastern/central France, western Switzerland, and a few
villages from southwestern Germany seem to be very valuable. Older wine
records available in Germany, at the northern limit of vineyards, do not always

distinguish between quantity of yield and quality of wine. While the latter, measured as density of the grape juice, depended strongly on global radiation (and indirectly on temperature) of the summer (Table 5.1e), the correlation between yield and quality varied with time, harvest date and types of grapes (Trenkle 1970) (Table 5.1b). Eighteenth-century series of yield are available from south-west Germany (Weger 1952, Tisowsky 1957), and Switzerland (Pfister 1981). Their significant correlation (Table 5.1a) seems to indicate that they can be used cautiously as climatic indicators, in spite of some negative results (Table 5.1d). Only data relating to the period before the epidemic of *Peronospora* in the 1880s are meaningful. The high correlation between wine quality and harvest data (Table 5.1b) in spite of the distance, is quite remarkable, because the harvest dates before 1879 were not postponed to improve the quality, but were fixed by local authorities immediately after maturity of the grapes.

One of the surprising results of these experiments was the high correlation (Table 5.1c) between cereal harvest date (in July) and grape harvest date (in early October). This is due to the high correlation of the latter with the temperatures of late spring and early summer (Table 5.2), as indicated by Pfister

Table 5.1   Correlation between proxy data series (unit 0.01).

(a)   Yield of grape harvest: central Switzerland (YS), Johannisberg–Rheingau (YR) and Iphofen–Franken (YF)

   YS versus YF: 1670–1729: +46, 1730–1794: +52
   YS versus YR: 1719–1780: +53, 1781–1825: +62
   YR versus YF: 1719–1756: +45, 1757–1794: +54

(b)   Quality of wine harvest, Johannisberg–Rheingau, estimated* (QR)

	1719–80	1781–1825	1826–75
QR versus YR	−38	−49	−40
QR versus WD	+63	+72	+71

   * Estimates are given from 1 (best) to 9 (worst); high quality is thus correlated with increasing yield and early harvest (maturity). 1787 and 1788 omitted, probably not representative.

(c)   Date of cereal harvest (tithe auctions, CS) in central Switzerland

	1611–60	1611–1710	1711–60
CS versus WD	+76	+71	+77
CS versus SL	−46	−56	−52

(d)   Correlations of YS versus WD and SL are either variable or statistically insignificant.

(e)   Representative data from an experimental vine-culture near Freiburg im Breisgau allow a check of the present-day relation between the sugar content of grape juice (Oe), the degree-days above 5°C (DD), and sunshine duration (SS), all during the growing season (Trenkle 1970; additional data by courtesy of Dr H. Trenkle). For the period 1954–76 the ccf Oe–SS amounts to 0.90, Oe–DD to 0.78 and SS–DD only 0.70; all ccfs are significant above the 0.1% level. For red grapes (Late Burgundy), increasing Oe is correlated with decreasing yield. Gordon Manley himself would probably also have enjoyed this approach to historical climatology.

Table 5.2 Calibration correlations for proxy records, unit 0.01 (I = 1782–1831, II = 1829–78, III = 1879–1928).

	SL (Lauenen) –Temp. Basel			SO (Otztal) –Temp. HP.			WD Dijon –Temp. Basel		–Temp. Geneva	
	I	II	III	I	II	III	I	II	I	II
April	25	46	14	14	35	06	−21	−33	−34	−37
May	43	23	05	46	41	−09	−52	−51	−45	−53
June	36	02	−05	28	00	−02	−58	−42	−50	−48
July	37	33	17	28	26	20	−63	−48	−64	−43
August	55	23	55	60	52	71	−36	−14	−35	−13
September	41	65	65	31	45	62	−25	−50	−26	−48
April–June	56	45	10	48	51	−04	−69	−77	−70	−77
May–July	55	35	11	48	43	05	−83	−83	−81	−78
June–August	59	31	35	54	42	46	−73	−55	−70	−54
July–September	68	61	62	59	65	71	−65	−56	−64	−51
August–September	68	60	71	61	73	78	−43	−43	−41	−43
April–September	76	66	59	68	75	56	−83	−83	−82	−80
July–September	49	59	64	with temperature						
April–September	59	60	63	Hohenpeieszettenberg (HP)						

(1984). This coincidence is a convincing test for the usability of proxy records for climatic reconstruction.

The interannual correlations between the two proxy series WD and SL (Table 5.3a) are (with minor exceptions in the mid-18th century) fairly stable and statistically significant. In both series, autocorrelations with lag 1 are statistically insignificant; the significance limits of a population of 50 pairs are 0.28 (5% level), 0.36 (1% level) and 0.45 (0.1% level). The series SL (corrected) and SO (raw, unadjusted data) are also well correlated, while SO and WD are less well correlated, partly due to the increasing distance. It has been shown (Flohn 1981) that the wine harvest data from Geneva and Vienna are still significantly correlated (two consecutive subperiods between 1560 and 1779, with 85 and 87 pairs of data, $r = 0.51$ and 0.54 respectively), in a distance of 800 km.

A further test has been made to check the long-term homogeneity of these records. Non-overlapping 11-year averages have been correlated (Table 5.2b). The high correlation between SL and the raw data of SO indicates only a minor rôle of adjustment; the correlation between SL and WD is also highly significant. However, with $r < 0.7$, the variance of one series is represented by less than 50% in the variance of the other series. This limits the use of even good proxy series for climatic reconstruction.

Some results from European tree-ring series are given in Appendix B. Here, the high autocorrelations (in oak trees) substantially limit their use for a reconstruction of interannual climatic variability. The correlation betwen non-overlapping 50-year series from Lamb (1977) is remarkable.

Table 5.3 Interannual and interdecadal correlations between proxy series (unit 0.01).

(a) Interannual correlations (50-year periods)

	SL–SO	SL–WD	SO–WD
1500–49	80	−61	−48
1550–99	71	−60	−56
1600–49	51	−60	−46
1650–99	67	−62	−46
1700–49	74	−56	−52
1750–99	38	−42	−27
1800–49	79	−66	−50
1730–79	50	−37	−13
1780–1829	81	−73	−54
1830–79	66	−52	−32

(b) Correlation between non-overlapping 11-year averages (n = number of pairs)

SL versus SO: +73   (1424–1907, n = 44)
SL Versus WD: −62 ⎫
SO versus WD: −44 ⎭ (1490–1863, n = 34)

5.4   Calibration experiments with proxy series

The use of proxy data series for climatic reconstruction depends first on a careful comparison with simultaneous observations from a truly representative climatic station. An area-averaged series such as WD can be compared with several long climatic series; Table 5.2 gives two examples with the temperature of Basel and Geneva. The greatest ccfs (correlation coefficients) are found with the months of May, June and July; late summer months contribute only little, but increase the total ccf for the summer half-year April–September to quite promising values of between 0.80 and 0.83.

Results with the SL series are not as good; during June, the ccfs drop in two subperiods to zero, but satisfactory results are obtained for late summer. Basel (Bider *et al.* 1959) was used for a comparison since no mountain station was available (see, however, Schweingruber *et al.* 1978). Table 5.4 indicates that during the summer half-year there is little difference between stations in the lowland north of the Alps and the two mountain stations Hohenpeieszettenberg (996 m) and Zugspitze (2968 m) (Paesler 1970, Attmannspacher 1981); all data from which can be used as a first approximation for the Lauenen record (at about 1900 m). The rôle of precipitation should be rather small, because of the negative ccf between summer temperature and precipitation. For further studies, methods of multivariate statistics should be employed, or several proxy data series should be used to reconstruct representative series of summer temperatures and precipitations.

If proxy data are sufficiently homogeneous, they can even be used for testing the homogeneity of old climatic records. Appendix A gives an example from a

Table 5.4 Temperature correlations for summer months, 1900–49, (unit 0.01). In several 50- or 100-year periods between 1781 and 1957, ccfs are essentially the same.

	April	May	June	July	August	September
Hohenpeieszettenberg–München	99	99	99	98	97	98
Hohenpeieszettenberg–Basel	94	95	93	98	95	98
Hohenpeieszettenberg–Zugspitze	95	92	96	93	93	95
München–Zugspitze	94	93	94	95	90	94
Basel–München	93	93	93	96	92	96
Basel–Genf	96	97	94	97	99	99
München–Genf	88	89	85	93	97	97

combined record: this result should be used to indicate the great sensitivity of climatic records.

During careful examination of all historical series – as seen from the viewpoint of 'classical' climatology – the following checks should be made:

(a) Check, if possible, the homogeneity of all records by comparison with adjacent records; check also the stability of cross-correlations.
(b) Check the spatial representativity of all kinds of climatic records: the famous 'central England' record or the Greenland ice-core records are unrepresentative for far distant areas. There are many cases with negative spatial correlations, such as between Greenland and north-west Europe, or between Western and Eastern Europe.
(c) Use of phenological records necessitates a careful examination of the time and type of response to climate parameters which they may represent.
(d) Area-averaged series (Pfister 1979 & 1981, Le Roy Ladurie & Baulant 1980) are preferable if their coherency can be warranted.

5.5 Conclusion

The increasing interest in historical climatic records has been largely due to the life-long efforts of such pioneers as G. Manley, C. E. P. Brooks and H. Arakawa, who have demonstrated the possibilities of climatic reconstruction. They have indicated the occurrence of rare extreme anomalies in individual years or (even more interesting) of climatic patterns significantly deviating from the present. The coincidence of climatic anomalies in individual years with historical events has also given rise to increasing interest (Wigley *et al.* 1981).

Recent controversy has focused on the cause of climatic change with respect to the 'Little Ice Age'. Was this a consequence of increased volcanic activity, or of the Maunder minimum of solar activity, which has not been repeated since 1715? It is obvious that a reliable, well-calibrated reconstruction of climate is needed far back into the pre-instrumental period (Groveman & Landsberg

1979). Some of the approaches of 'classical' climatology are still of the highest value, in addition to the up-to-date statistical techniques now used for dendro-climatic reconstruction (Fritts 1976, Hughes *et al.* 1982). A careful comparison of quite different proxy series may lead to greater confidence. Furthermore, a transition of individual months to seasons (Le Roy Ladurie & Baulant 1980, Hughes *et al.* 1982), in a biological–phenological sense, greatly improves the climatic response function.

Appendix A   Proxy data used to check the homogeneity of temperature records

If proxy data are sufficiently homogeneous, they can even serve to test the homogeneity of early temperature records, at least before the establishment of governmentally supervised climatic networks. An example of this can be given, using a combined temperature record (*Oberrhein*) published by H. von Rudloff (1955/6). This record consists of three independent series: Basel (1755–1954), Strasbourg (since 1806), and Karlsruhe (since 1834), and therefore gives quite representative data after 1834. The use of this combined record indicated some doubts about the homogeneity of the *Oberrhein* record. This indicated that another experiment should be done separately for the periods 1755–1805 (Basel), 1806–34 (Basel and Strasbourg), 1834–79 and 1880–1929 (three stations). Table 5.5 contains the ccfs between TO (Oberrhein, April–September) and the proxy records, together with the relevant averages (indicated with a bar).

Table 5.5 Test correlations between TO (deviation from a given average, °C), WD and SL for four different periods.

	TO/WD	TO/SL	$\overline{\text{TO}}$	$\overline{\text{WD}}$	$\overline{\text{SL}}$
1755–1805	−0.64	0.49	+0.47	29.87	85.9
1806–1833	−0.71	0.70	+0.12	32.58	81.2
1834–1879	−0.87	0.62	−0.19	29.19	83.4
1880–1929	—	0.50	+0.16	—	84.6

While the ccfs indicate no serious inhomogeneity, a careful look at the averages indicates that the changes between the two first periods are internally consistent, as are those between the last two periods. However, the inclusion of Karlsruhe since 1834 is accompanied by an inconsistency: while $\overline{\text{TO}}$ becomes apparently colder, the wine harvest ($\overline{\text{WD}}$) is more than 3 days earlier, and $\overline{\text{SL}}$ increases more than two units (0.01 g cm^{-3}). Using the two linear regression equations for WD/Basel and SL/Hohenpeieszettenberg, $\overline{\text{TO}}$ should increase by 0.23°C (0.21°C) instead of decrease; the inhomogeneity should thus be in the order of 0.5°C.

Manley was infinitely careful to homogenise the early temperature records, and to calibrate the lost early thermometers. Brumme (1981) has used two statistical relationships between non-instrumental data (frequency ratio between snowfall/total precipitation versus temperature, frequency of rainy days during summer versus temperature) for this purpose. At Hohenpeieszett-enberg (1879–1978), the latter relation gives ccfs between − 0.56 and − 0.69 for each month between May and September (Table 5.6). Unfortunately, the Hohenpeieszettenberg temperature data up until 1900, as given by Attmann-spacher (1981), lack homogeneity (Keil 1983).

Table 5.6 Correlation between temperature and precipitation (unit 0.01).

	J	F	M	A	M	J	J	A	S	O	N	D
Hohenpeieszettenberg 1879–1978												
temp.–precipit. amount	−04	02	−27	−44	−35	−40	−48	−35	−44	−40	−08	01
temp.–precipit. days	−23	−07	−46	−54	−59	−58	−69	−61	−56	−52	−08	−01
precipit.: amount–days	78	86	81	75	68	69	62	66	71	86	78	78

Appendix B Two oak tree-ring series from western Germany

In his well known treatise on climatic change, Lamb (1977) reproduced two very long tree-ring chronologies, from western Germany, evaluated by Holl-stein (Rheinland) and by Giertz-Siebenlist (Spessart). Both have been combined from many overlapping trees – however, without some of the corrections used by Fritts (1976) and Fritts et al. (1979) – covering more than 1100 years at a distance of not more than about 150 km, in quite similar climates. Unfortunately, the climatic interpretation of oak tree rings under the climatic conditions of Central Europe has not, until now, been sufficiently successful for a climate reconstruction.

Some experiments to test their usability for this purpose have been made. First decadal averages from both series have been compared: the ccfs between the two series, for the periods 840–1199, 1200–1569, and 1570–1949, are 0.64, 0.28 and 0.64; an intermediate period (1400–1799) also gave 0.58. Thus during the first and the last third of the record (i.e. for the medieval warm period and for the Little Ice Age and recent periods) the interdecadal fluctuations coincide quite well, with r signficant for the 0.1% level. During the last medieval period, however, the decadal fluctuations are incoherent. Between twenty-two 50-year averages of the total 1100 years, r = 0.58, significant to the 1% level.

Within each of these 50-year periods, the interannual fluctuations are mostly coherent: in 18 of the 22 half-centuries, r varies between 0.56 and 0.85. However, such ccfs are not necessarily significant, because of the well-known high persistency of ring-width indices (Fritts 1976). As an example, the auto-ccfs for lags 1–3 in the Rheinland series of 1650–99 are, respectively, 0.55, 0.33

and 0.17, thus indicating Markovian behaviour. From the 4 half-centuries with $r<0.5$, two (1350–99 and 1550–99) are hampered by obvious inhomogeneities, while during the 19th century the Spessart series shows a distinct downward trend not observed in the Rheinland series.

Acknowledgements

The data in Tables 5.2–5.5 have been selected from a greater number of correlations provided by Dr R. Kreuels; the manuscript has been carefully prepared by Mrs I. Lockwood. Support has been provided by the *Rheinisch-Westfälische Akademie der Wissenschaften*, Düsseldorf, and in earlier years by the Rockefeller Foundation, New York.

References

Attmannspacher, W. (ed.) 1981. 200 Jahre meteorologische Beobachtungen auf dem Hohenpeineszettenberg. *Ber. Dt. Wetterdienst* **155**, 1–84.

Bider, M., M. Schüepp and H. von Rudloff 1959. Die Reduktion der 200-jahrigen Basler Temperaturreihe. *Arch. Met. Geophys. Bioklim.* **B 9**, 360–412.

Brumme, B. 1981. Methoden zur Beobachtung historischer Meeszett- und Beobachtungsdaten (Berlin und Mitteldeutschland 1683–1770). *Arch. Met. Geophys. Bioklim.* **B 29**, 191–210.

Duvick, D. N. and T. J. Blasing 1981. A dendroclimatic reconstruction of annual precipitation amounts in Iowa since 1680. *Water Resources Res.* **17**, 1183–9.

Flohn, H. 1949. Klima und Witterungsablauf in Zürich im 16. Jahrhundert. *Viertelj. Schrift. Naturf. Zürich* **94**, 28–41.

Flohn, H. 1981. Short-term climatic fluctuations and their economic role. In *Climate and history*, T. M. L. Wigley, M. J. Ingram and G. Farmer, (eds.), 310–18. Cambridge: Cambridge University Press.

Frenzel, B. (ed.) 1977. *Dendrochronologie und postglaziale Klimaschwankungen in Europa.* Wiesbaden: F. Steiner.

Fritts, H. C. 1976. *Tree rings and climate.* London: Academic Press.

Fritts, H. C., G. R. Lofgren and G. A. Gordon 1979. Variation in climate since 1602 as reconstructed from tree rings. *Quatern. Res.* **12**, 18–46.

Groveman, B. S. and H. E. Landsberg 1979. Reconstruction of Northern Hemisphere temperature 1579–1880. *Univ. Maryland Met. Prog., Coll. Park., Md. Publ.* **79**, 181.

Hughes, M. K., P. W. Kelly, J. R. Pilcher and V. C. La Marche Jr (eds) 1982. *Climate from tree rings.* Cambridge: Cambridge University Press.

Lamb, H. H. 1977. *Climate: present, past and future,* vol. 2. London: Methuen.

Lambrecht, A. 1978. *Die Beziehungen zwischen Holzdichtewerten von Fichten aus Subalpinen Lagen des Tirols und Witterungsdaten aus Chroniken in Zeitraum 1370–1800 AD.* Dissertation. Zürich: University of Zürich.

Lauscher, A. and F. Lauscher 1981. Vom Einflueszett der Temperatur auf die Belaubung der Roeszettkastanie nach den Beobachtungen in Genf seit 1808. *Wetter und Leben* **33**, 103–12.

Lenke, W. 1960. Klimadaten von 1621–1650 (Landgraf Hermann von Hessen). *Ber. Dt. Wetterdienst* **63**, 1–51.

Le Roy Ladurie, E. and M. Baulant 1980. Grape harvests from the fifteenth through the nineteenth centuries. *J. Interdisc. Hist.* **10**, 839–49.

Paesler, W. 1970. Die Temperaturmessungen in München 1781–1968. *Wiss. Mitt. Met. Inst. München* **19**, 1–92.

Pfister, C. 1977. Zum Klima des Raumes Zürich im spaten 17. und frühen 18. Jahrhundert. *Viertelj. Schrift. Naturf. Ges. Zürich* **122**, 447–71.

Pfister, C. 1979. Getreide-Erntebeginn und Frühsommertemperaturen im schweizerischen Mittell and seit dem frühen 17. Jahrhundert. *Geog. Helvet.* **34**, 23–35.

Pfister, C. 1981. Die Fluktuationen der Weinmosterträge im schweizerischen Weinland vom 16. bis ins frühe 19. Jahrhundert: Klimatische Ursachen und sozioökonomische Bedeutung. *Schweiz. Zeitschr. f. Geschichte* **31**, 445–91.

Pfister, C. 1984. *Das Klima der Schweiz von 1525–1860 und seine Bedeutung in der Geschichte von Bevölkerung und Landwirtschaft.* Habil. Schrift. University of Bern.

von Rudloff, H. 1955/6. Die Klimapendelungen der letzten 120 bis 200 Jahre im südlichen Oberrheingebiet. *Ann. Met.* **7**, 12–34.

Schnelle, F. 1981. Beiträge zur Phänologie Europas IV: Lange phänologische Beobachtungsreihen in West-, Mittel- und Ost-Europa. *Ber. Dt. Wetterdienst* **158**, 1–35.

Schweingruber, F. H., H. C. Fritts, O. U. Bräker, L. G. Drew and E. Schär 1978. The X-ray technique as applied to dendroclimatology. *Tree-Ring Bull.* **38**, 61–91.

Schweingruber, F. H., O. U. Bräker and E. Schär 1979. Dendroclimatic studies on conifers from Central Europe and Great Britain. *Boreas* **8**, 427–52.

Tisowsky, K. 1957. Häcker und Bauern in den Weinbaugemeinden am Schwanberg. *Frankfurter Geogr. Hefte* **31**, 1–94.

Trenkle, H. 1970. Die Verwendung phänologisch-klimatologischer Beobachtungen bei der Gütebewertung von Weinberglagen. *Die Wein-Wissenschaft* 327–38.

Wacker, U. 1981. Untersuchungen langfristiger Schwankungen der Sonnenscheindauer. *Arch. Met. Geophys. Biokl.* **B 29**, 269–81.

Wang, S. W. and Z. C. Zhao 1981. Droughts and floods in China, 1470–1979. In *Climate and history*, T. M. L. Wigley, M. J. Ingram and G. Farmer (eds.), 271–88. Cambridge: Cambridge University Press.

Weger, N. 1952. Weinernten und Sonnenflecken. *Ber. Dt. Wetterdienst US-Zone* **38**, 229–37.

Wigley, T. M. L. and T. C. Atkinson 1977. Dry years in south-east England since 1698. *Nature (Lond.)* **265**, 431–4.

Wigley, T. M. L., M. J. Ingram and G. Farmer (eds) 1981. *Climate and history.* Cambridge: Cambridge University Press.

6

The Little Ice Age period and the great storms within it

HUBERT H. LAMB

The introductory section of this chapter gives a brief description of Gordon Manley's derivation of the monthly mean temperatures in central England, and their reliability, back into the climax period of the Little Ice Age in the 17th century and presents it in an updated diagram. The later sections describe various other probes with which the writer has been concerned, and which have to do with the causation, or mechanism, of the cold climate period of recent centuries: changes in the sea temperature and ocean current situations, the mean atmospheric circulation patterns prevailing, and the individual daily synoptic pressure and wind maps during several great storms in the period between the late 16th and 18th centuries are diagnosed, and the analyses tested, by techniques described in the chapter.

6.1 Introduction: air temperatures

Gordon Manley's work (carried out over 30 years) on the production of his now famous series of monthly mean temperatures in central England from 1659 to 1973 has provided a landmark in the development of our knowledge of the past record of the climate. No doubt, he was inspired by the earlier efforts of Labrijn (1945), in producing a 200-year series of temperatures and rainfall in the Netherlands, and of Birkeland (e.g. 1925 and 1949), who published long series, in some cases extending back over 150 to nearly 200 years, of values of the atmospheric pressure and temperature at Bergen, Oslo, Trondheim and Vardø in Norway. It is said that Birkeland had a team of assistants who filled a large room and were not allowed to speak or to disturb the work, even during a lunch break by rustling their sandwich papers. But Manley did the work himself, without the benefit of modern computing aids and virtually without assistants. Among the most valuable parts of his effort are his meticulous accounts in a series of articles (Manley 1946, 1953, 1959, 1974) of the sources he used and of the methods by which the overlapping records from different places were scrutinised and applied to the construction of a single series of monthly values for representative lowland sites in central England. The resulting historical temperature series, the longest based directly on thermometer readings for anywhere in the world, is widely quoted in the literature and treated as the firmest piece of our knowledge of climatic events since the middle of the 17th century.

The occurrence of low thermometer readings in winter could be checked against reports of frost phenomena, and so indications were obtained of the magnitude of the effects of instrumental errors, and of the odd positions in which the thermometers were exposed. It was harder to get reliable checks of the highest temperatures measured by the early thermometers in summer, and Manley was still hopeful that some way – some chemical method, he was inclined to think – of improving the summer values from the earliest years may yet be found. Perhaps this will come from oxygen or carbon isotope measurements from tree rings, but these have their own error margins and difficulties of interpretation. Estimates of the error margins of the monthly values can be broadly gauged by comparisons between the series and the most carefully worked out temperature series for other places in England and in neighbouring countries, back to some time in the first half of the 18th century. Pending such investigation, the impressions gained from use of the central England series, together with consideration of Manley's accounts of the thermometers and their observers, and his own comparisons among the different original series available, suggest that the monthly mean values are probably within 0.1°, or at most 0.2°C, of the true value back to about 1730. Around the 1720s, the standard error of the individual monthly values may be of about this magnitude, and back as far as 1680 the standard error is probably rather less than ± 0.5°C.

Indications of the reliability of Manley's published values (Manley 1974) in the earliest decades, when comparisons with other series are not available, can be gleaned from the monthly summary maps for the Januarys and Julys, showing reports of winds and weather, that have been compiled. These support Manley's decision that it was worth suggesting monthly values to the nearest 1°C back to 1659, although occasional greater errors than this may be present before about 1680 at any time of the year. The standard error of the individual monthly values given by Manley in those first decades seems, when judged from this test, to be around ± 0.6°C or very little more.

Weather mapping – even the mapping of such fragmentary collections of reports as can be gathered in the mid-17th century and earlier – has a positive value, not only in providing a more complete picture of the situation than can be obtained in any other way, but also as a consistency check on the reliability of individual reports, and an outline check on values estimated from the readings of the earliest instruments of different kinds.

The estimates of standard error suggested above are based on comparison of the 42 earliest monthly mean temperature figures given by Manley for the Januarys and Julys before 1680, with the summary maps referred to for those months. These maps display descriptive reports for 5 to 15 places in Europe, including points in the British Isles. From examination of this sample, with a view to estimating the greatest errors possibly present in Manley's figures, only one month (January 1678) looked doubtful: its temperature figure may need correcting (upwards) by up to 3°C. (The published monthly mean temperature figure of + 1.5°C is too low for a month in which a ship in port on the south coast experienced 61% westerly surface winds, even though the implied gradient wind would be from about WNW during a central part of the Little Ice Age period.) In only four other months did it seem possible that an error as great as 1°C (and in none of these cases significantly over that) could be present, whereas 69% of the monthly figures published seemed probably within ± 0.5°C of the true value. With these estimates of the greatest possible error margins applying to the figures in the earliest 22 years of Manley's series – where the original data presented most difficulties – it stands as a remarkably reliable probe into the middle of the period of the so-called Little Ice Age climate.

Manley's most famous diagram summarising the temperatures prevailing at a typical site in central England in each season of the year by 10-year running means, here continued up to 1982, is presented in Figure 6.1. Since the standard errors of the 10-year means can be estimated as no more than ± 0.2°C in the 1660s and 1670s, and little over ± 0.1°C in the 1690s (which appears as the coldest decade), this diagram gives us a fully acceptable measure of the depression of the prevailing temperatures in this part of the world in the climate of that time.

Moreover, the deduced course of the prevailing winter temperatures in central England is now seen to be closely paralleled by that derived for the Netherlands by comparing the records of freezing of the Dutch canals back to 1634 with their thermometer records which go back as far as 1735 (van den Dool et al. 1978).

Manley himself became interested in the problem of the causation of temperature variations and, in particular, what modifications of the atmospheric and ocean-current circulation characterised the conditions then prevailing.

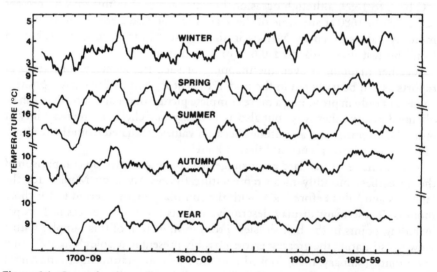

Figure 6.1 Seasonal and overall average temperatures in Central England from 1659 to 1982: 10-year running means.

6.2 Ocean surface temperatures and their effects on the air temperature pattern

The longest series of sea surface temperature measurements in the world at something approximating to a fixed point comes from the Faeroe Islands and goes back to 1867. With the aid of proxy data from the records of the fisheries and vagaries of the Arctic sea ice, this can be extended in outline back to the period 1675–1704, to compare with the air temperatures in central England. Several general observations showed in advance that this comparison was likely to be interesting.

The Faeroe Islands lie in the midst of the broad channel between Europe and Iceland through which the main arm of the North Atlantic Drift (with its warm, saline water of Gulf Stream origin) normally passes in the present epoch on its way to the Arctic. Yet it is well known that a branch of the polar water current from east Greenland sometimes rounds the east coast of Iceland, and there are suggestions apparent in old atlases that this was commonly present in the last century. Moreover, an aircraft survey with modern remote sensing equipment took place on 9 April 1968, when a tongue of the polar ice itself had advanced south down the east coast of Iceland, and discovered another tongue of the polar water reaching to within about 100 km of the Faeroe Islands (Fig. 6.2). By pursuing the matter by means of local inquiries in the islands, it was learnt that in 1888 the polar ice itself had extended south on both sides of the Faeroe Islands, and the fishing fleet had had to turn back to port. Over the 115-year long record

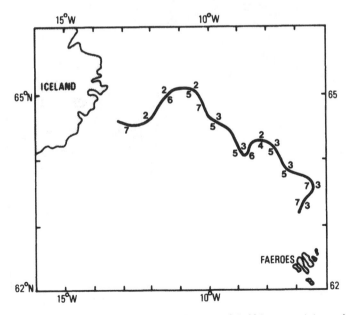

Figure 6.2 Boundary between polar water and water of Gulf Stream origin, and sea-surface temperatures (°C) observed by an airborne sensor on 9 April 1968.

of sea surface temperature measurements at the Faeroes, the overall average sea surface temperature has been 7.7°C, but the coldest years (1867–9 and 1965–9) produced mean temperatures of 6.5°C to 6.9°C, and the warmest years (1894 and 1951) produced means of 8.5°C and 8.9°C respectively. The warmest and coldest 5-year means at the Faeroe Islands since 1867 have differed by 1.0°C, or twice the range of the 5-year means of the air temperatures in central England. Thus, the ocean surface temperatures in this region cannot be seen as a stabilising influence on the climate, but are subject to great variations as the ocean currents vary.

An expedition (West 1970) visited the Faeroes in the summer of 1789, which was a slightly warmer than normal summer for those times, and found sea temperatures there about 1.3°C below the modern average. A map (Lamb 1979) of the sea temperatures measured by the same expedition over a 5-month period in 1789 in northern waters (Fig. 6.3) indicates a more meridional (north–south)

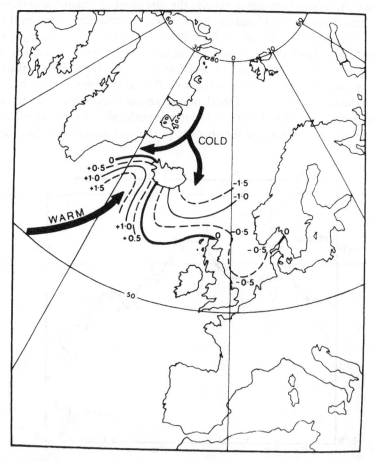

Figure 6.3 Sea-surface temperatures observed in the summer of 1789, and the implied ocean current anomalies.

orientation of the ocean surface currents at that time, and an active east Iceland branch of the polar current supplying cold water as far south as the North Sea.

Because the range of the cod seems to be effectively limited by the 2°C isotherm, the records of the Faeroe Islands cod fishery, which go back to the beginning of the 17th century, can give some indications of the intrusions of the polar water back to that time. The cod are finely tuned to water temperatures. They are at their best in waters between 4° and 7°C, and cannot survive for long in water temperatures below 2°C, which results in kidney failure. It is recorded by Svabo in his report on the Faeroe Islands (1782) that from verbal interrogations he learnt that the cod fishery began to fail in the early 17th century, and that in 1625 and 1629 there were no fish – something that had never been known to happen before. After some better years, the cod disappeared altogether from Faeroese waters for 30 years from 1675 to 1704. The worst year (1695) saw the cod also disappear from the whole coast of Norway, and become scarce at Shetland, while the Arctic sea ice itself surrounded Iceland. There is little doubt therefore that the disappearance of the cod marked an enormous advance of the polar water south over the ocean surface and, hence, that between 1675 and 1704 this water mass dominated the ocean as far south-east from Iceland as to about 61–62°N 0°E. This means that the ocean surface in that region in that period must on average have been 4–5°C colder than the 100-year mean of the period 1867 to around 1970–80.

An anomaly of this magnitude in the ocean surface in that region means that the air temperatures in the northern half of Scotland and in south Norway must have been lower when the winds blew from that part of the ocean than they are now. The region of the North Atlantic Drift water farther south – as far north as Biscay, the coasts of Ireland and south-west England and the English Channel – probably experienced water temperatures no more below modern levels than the air temperatures in central England. So a greater variability of air temperatures than now would be expected in all parts of the British Isles as the winds veered between SW and NW or N. The greatest variability would be in the northern half of Scotland, especially on the eastern side remote from the influence of the warm-water areas of the Atlantic. As a result, the overall mean air temperatures in the last quarter of the 17th century in the northeastern half of Scotland should have been lowered much more below 20th century values than the temperatures in England as shown by Manley's series (1979).

Verification of this in broad terms seems to be forthcoming from the following.

(a) English travellers who described visits to eastern Scotland between the years 1600 and 1770 or after, mostly referred to permanent snow on the tops of the Cairngorms (1200–1300 m above sea level), whereas Manley (1949) calculated the mid-20th century theoretical snowline for the (western) Highlands (Ben Nevis) as about 1600 m. The difference of 300–400 m implies an average temperature level in the 17th and 18th centuries, 2–2.5°C below the mid-20th century level.

(b) There are a couple of reports known of small lakes (lochans or tarns) in the eastern and northern Highlands at heights greater than about 750 m above sea level which bore ice even in the summer time. One report, by Mackenzy (1675), is of a 'little lake in Straglash (Strathglass) at Glencannich on land belonging to one Chisholm . . . the lake is in a bottom between the tops of a very high hill so that the bottom itself is very high. This lake never wants ice on it in the middle, even in the hottest summer, tho' it thaws near the edges.' (Identification is uncertain: the description would fit either Loch Tuill Bhearnach at 57° 21.6′N 5° 2.5′W 750 m, or Loch a' Choire Dhom-hain at 57° 18.5′N 5° 5.5′W 925 m above sea level.) The other report is by Thomas Pennant, in his account of his first visit to Scotland in 1769, that under Ben-y-Bourd (Aberdeenshire) is 'a small loch, which I was told, had ice the latter end of July'. (This can be clearly identified as Dubh Lochan, 57° 4.5′N 3° 29.5′W 942 m above sea level, though the report may refer just to that particular year of 1769.) As the present annual mean temperature at these positions is from 2°C to 3°C, these reports also suggest a temperature lowering at least in the mid- to late 17th century of about 2.5°C in the area.

(c) Reconstruction by Matthews (1977) of the temperatures in central southern Norway from tree-ring studies pointed to temperatures prevailing there between 1700 and 1725 about 1.6–1.7°C below recent norms, and to even lower values in the preceding quarter century (1675–1700). A similar temperature lowering of about 2°C seems to be implied by the formation of small new glaciers on the open plateau of Hardangervidda at heights between 1460 m and 1570 m above sea level (e.g. Omnsbreen 60° 39′N 7° 30′E) reported by Liestøl (personal communication 1977).

Thus, there seems to be evidence of an enhanced thermal gradient around 1700 on the one hand between central England and the Atlantic south and west of the British Isles, and on the other hand, a region from Iceland extending south-east to the northern North Sea and the lands closest to it on either side.

6.3 Atmospheric pressures and the general wind circulation pattern

Use has been made of the available barometric pressure measurements from past centuries to elucidate the characteristics of the wind circulation. The barometer (invented by Torricelli in 1643) was developed at an early stage into a reliable instrument. By 1648, Blaise Pascal was already experimenting with its use to discover the differences between the weight, or pressure, of the atmosphere on mountain tops and on the plain below, nearer sea level. Only a decade or so after that, Robert Boyle was using the instrument for his experiments. Exposing the barometer so as to obtain representative measurements of the pressure of the atmosphere involves fewer problems than does the use of the thermometer or rain gauge. The desirability of recording the temperature of the instrument, or of the room that it was in, was soon appreciated. So some good barometric pressure measurement starting from dates between 1700 and 1750, and even a few that begin earlier are available. Corrections to sea level and to standard gravity (i.e. gravity as at latitude 45°) are, of course, needed.

It was found to be possible to derive maps of monthly mean atmospheric pressure at sea level covering most of Europe and the fringe of the Atlantic Ocean for every year back to 1750 (Lamb & Johnson 1959, 1961, 1966). Series of January and July maps going back to that date were completed and published after a testing procedure had shown the extent of the region for which reliable isobars with predetermined limits of error could be drawn.

Because even these map series did not go back far enough to reveal the situation in the main period of the Little Ice Age in this part of the world, its sharpest phase having occurred in and around the 1690s, compilation of maps for the Januarys and Julys back to 1680 was attempted. This meant exploring the situation with maps which showed barometric pressure values, some of which were evidently rather rough, for six places in Europe back to 1745, and just two to four places back to 1724. (The network of point values that can be assembled changes from year to year, owing to some short runs of observations, but, for example, in 1730 it included London, Berlin and three places distributed through the Baltic.) Values from barometric measurements in or near London, and occasionally just one or two other places, were available for some of the earlier years. Clearly, the information on the maps had to be eked out with every available kind of supplementary information. Wind roses for each particular January and July were plotted wherever such data could be obtained, including British naval ships in ports as far away as Lisbon and Leghorn (Livorno, Italy). (Thus, the January map for 1694 shows wind roses for London, Plymouth, Lisbon, Ulm, and Livorno. It was usually possible to include somewhere in the western or central Mediterranean.) Entries were also made describing the weather of the month, and included particular attention to summaries giving the number of days with rain, snow, frost, fog or fine weather etc. wherever these details were known. It is quite certain that much more information of these kinds could be found if time and funds permit.

The maps were analysed when information for about 10 to 20 points in Europe (the number actually ranged from 7 to commonly 15 to 20 places) had been compiled. Reports seem to be most abundant when the weather was most dramatic: within this period, most detail is known regarding the winters, and specifically the Januarys, of 1684 and 1709, with reports in 1709 of conditions in nearly 40 places in Europe. Some reports from Iceland were commonly available and reports from one or two points in North America.

The analysis procedure used was first to draw 'free-hand' the distribution of high and low pressure, by means of hypothetical or 'putative' isobars, controlled when possible by an actual pressure reading, and so as to fit the known winds and weather best. Next, a similar meteorological analysis of the Januarys and Julys of the 1880s (actually 1879–88) – a decade just 200 years later, known to have had a rather cold climate in Europe and not familiar to the analyst – was drawn by the same analyst, using a restricted compilation of information as nearly as possible equivalent to that used for the 1680s and following decades on to the 1720s. Later comparison with an isobar analysis of the Januarys and Julys of 1879–88 based on full information from Europe and the nearer Arctic revealed the distribution of errors in the analysis of only primitive data (Fig. 6.4). This was basically the same test technique as had been used earlier for the published map series from 1750 to modern times. However, in the case of the maps for the Januarys and Julys from 1680 to 1749, a further stage seemed appropriate, namely the development of a second approximation analysis. In other words, the findings of the error survey were used to 'correct' the first analysis, considering the pattern of errors in the 10-year mean as a measure of the analyst's unconscious bias in analysing situations of the type occurring in the 1880s, and in the partly similar situations occurring between 1680 and 1749. Areas of the map where the analyst's error in the 10-year average from the first analysis exceeded 5 millibars were discarded.

The error distributions shown by the pressure values, wind directions as represented by the isobar directions in the main airstreams, and latitude and longitude errors of the main features of the pressure pattern, on the individual maps of the 10-year series of January and July maps of the test series 1879 to 1888 were also studied. The results are summarised in Table 6.1. The tests indicated

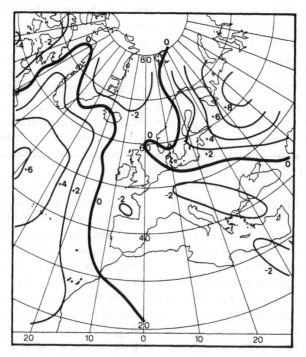

Figure 6.4   Test charts for the Januarys 1879 to 1888 analysed from observational data as in 1680 to 1720: mean error in millibars.

Table 6.1 Errors on the test series of mean pressure maps for the Januarys 1879 to 1888 analysed on the basis of primitive data equivalent to those available for the 1680s. The values are the number of cases in each error class.

| Error class | Isobar direction errors main airstreams | | | | | Longitude errors at 45°N | | Latitude errors eastern Atlantic |
| | North Atlantic Westerlies | | European winds | | Polar Easterlies | West Atlantic trough axis | Atlantic high axis | 'Iceland' low-pressure axis |
	West of 30°W	30–0°W	0–10°E	c. 30°E	East of 30°W			
little error	2	7	4	4	4	1	1	
1–15°	5	3	6	2	0	5	5	
16–30°	3			0	4	2	1	
31–45°				2	0			
over 45°				2	2			
ill-defined cases						2	3	
average error	−9°	+2.5°	−1.5°	−16.5°	0°	−11°	−5°	+2.0°

(as had been expected) that the area which could be analysed with some approach to reliability on the maps for the individual years was quite narrowly restricted to the region for which the most specific reports were available, whereas (owing to the evening out of random errors) the 10-year means gave a reasonable approximation to the true distribution over a considerably wider area.

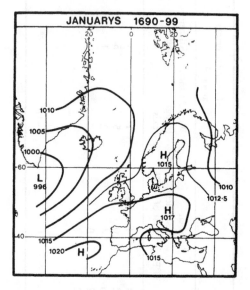

Figure 6.5   Derived average MSL pressure map for the Januarys 1690–99 (with largely putative pressure values).

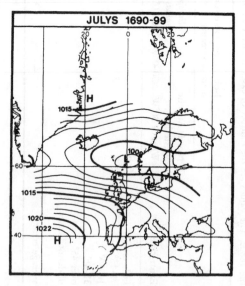

Figure 6.6   Derived average MSL pressure map for the Julys 1690–99.

Therefore, the last stage of the analysis was to derive the 10-year mean maps shown in Figures 6.5, 6.6, 6.7 and 6.8 for the coldest and warmest decades (the 1690s and 1730s respectively) in the period between 1680 and 1749. These maps were arrived at by averaging the putative pressure values indicated on the January (and similarly on the July) maps for the individual years of the decades named. The pressure values attached by the analyst to the isobars on these maps

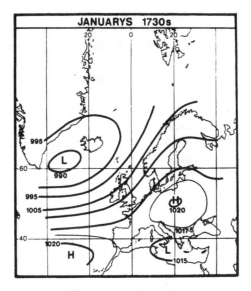

Figure 6.7   Derived average MSL pressure map for the Januarys 1730–39.

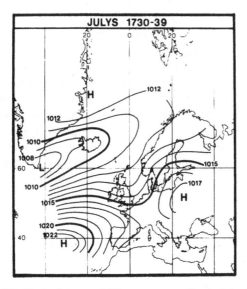

Figure 6.8   Derived average MSL pressure map for the Julys 1730–39.

should, of course, be disregarded. But the test procedures suggest that the patterns over the region shown on these maps are likely to be sound.

The main features revealed by the maps are:

(a) a displacement south from 20th-century positions of both the low and high pressure maxima over the Atlantic in both decades;

(b) the displacement mentioned seems to be associated with higher pressure over the Arctic (or a greater southward thrust of high pressure from the Arctic) particularly in the Greenland sector;

(c) between the 1690s and the 1730s, the low pressure zone seems to have deepened and the pressure gradients south of it to have intensified; at the same time, the subtropical high pressure system seems to have become orientated so as to spread over central Europe in both winter and summer, while the low pressure development turned more to the north-east, tending to penetrate the Arctic towards the Barents Sea;

(d) in the winters of the 1690s, as represented by the Januarys of that decade, there seems to have been much development of 'blocking anticyclones' over Europe with the 10-year mean indicating a separate centre of high pressure over Scandinavia. This is the only decade in which this feature seems to have been dominant enough to appear on the 10-year average.

6.4 Studies of great storms in the Little Ice Age period

Indications of the occurrence of occasional storms of outstanding severity during the colder climate regime in recent centuries and about the time of its first symptoms in the 13th and early 14th centuries, were perhaps first noticed by C. E. P. Brooks (1949, 1st edn. 1926). He pointed out that there seemed to have been two periods, the first broadly around 500–200 BC and the second between about AD 1300 and some time between 1550 and 1800, when the coasts of Western and Northern Europe had been altered at a number of low-lying points by extraordinary movements of blown sand. He also brought to notice the high frequency of great storms affecting Europe in the 13th century (Fig. 6.9), particularly between about AD 1210 and 1250 (Lamb 1977). Many of these were accompanied by coastal flooding and some more permanent inroads of the sea, including the formation of the Jadebusen basin in north-west Germany and the beginnings of the Zuyder Zee in the Nertherlands. He drew appropriate attention to the storm of exceptional violence, described by Defoe (1704), which crossed Southern England in early December (Gregorian or New Style calendar) 1703. In fact, these events may be rough indicators of the beginning and time of climax respectively of a long-term surge of storm development.

It seems likely that there was an increase of storm activity in these latitudes over the North Atlantic and Europe soon after AD 1200, when the Arctic sea ice reappeared in increasing amounts off the coast of Greenland as far south as latitude 60°N (Lamb 1982). There is some evidence that this occurred at a time

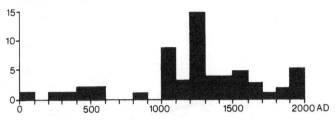

Number of reported SEVERE SEA FLOODS per century

20th century total estimated as ⁴⁄₃ (1900-75) total

Figure 6.9 Numbers of severe sea floods causing much loss of life or land in the North Sea and Channel coasts per century.

when sea-level had been somewhat raised by several centuries of warmer climate in the latitudes coinciding with the main Northern Hemisphere glaciers and ice sheets on land. It is known that there had been glacier recession in Greenland, as well as in the Alps, and it is probable that there had been recession also in Scandinavia, Baffin Island, and Alaska (Grove 1979). Hints of a slightly raised sea-level – a reasonable calculation suggests (Lamb 1980) that world sea-level in the 13th and 14th centuries may have been around 50 cm higher than it was around AD 700, and somewhat less than that above today's level (Tooley 1978, Coles & Funnell 1981) – may be seen in the wet state of the Fenland in eastern England south of the Wash, and the existence of a channel between Thanet and Kent. This diagnosis, if right, indicates that the low-lying coasts of north-west Europe, and particularly the islands of low relief in the North Sea, should have been particularly vulnerable to storms and storm erosion in the 13th and 14th centuries.

Because of the hints, also implied, of long-term variations in the amount, and perhaps also the intensity, of storm activity over north-west Europe and the North Sea, it seems highly desirable to attempt a synoptic meteorological analysis of some of the great storms reported in the Little Ice Age period.

Some outstanding storms can be identified by the nature and extent of the reported damage to buildings and to shipping, and from the reported erosion, or alternatively the alteration, of the coast by massive transport of sand. The last-named cases may sometimes involve exceptionally low tides (affected by the direction of storm wind as well as by astronomical variations) and the exposure of unusually great expanses of estuarine sand or of sand that had been previously moved by storm waves. Some outstanding storms may also be identified from the records of exceptional sea floods, produced by storm surges leading to abnormally high tides, on low-lying coasts. None of these methods can be expected to give a complete list of the severest class of storms on the North Sea and other exposed regions of Europe and the European seas. These lines of discovery can yield a useful sample selection of the most outstanding storms for study.

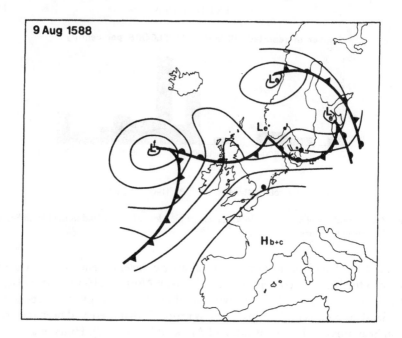

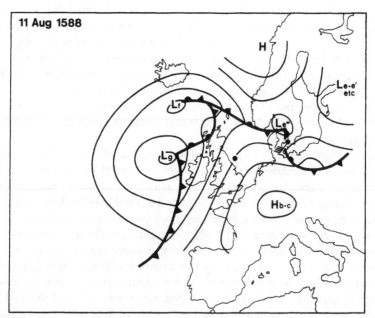

Figure 6.10 Synoptic weather maps for four dates in August 1588 derived from weather observations made on the ships of the Spanish Armada and in Denmark. Positions of main observation data are shown.

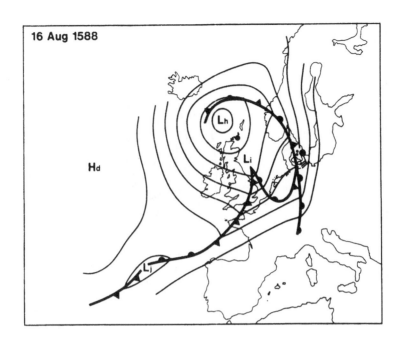

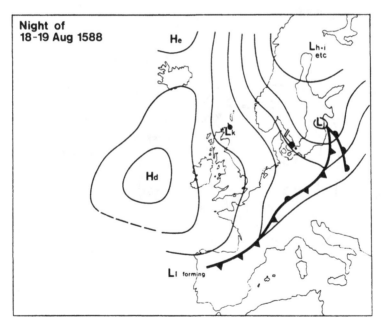

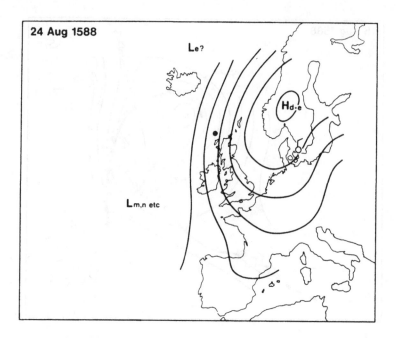

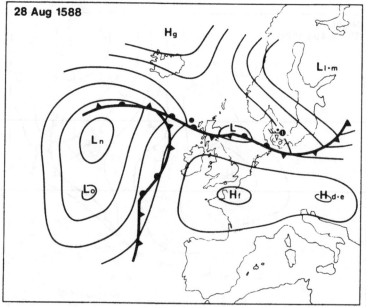

Figure 6.11 Synoptic weather maps for dates in late August and September 1588, derived as for Fig. 6.10.

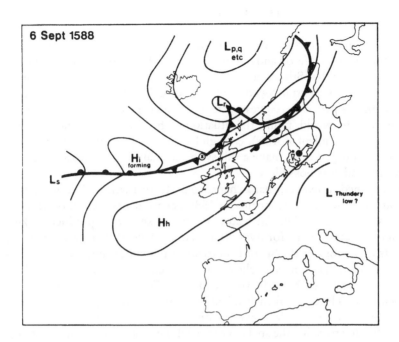

6 Sept 1588

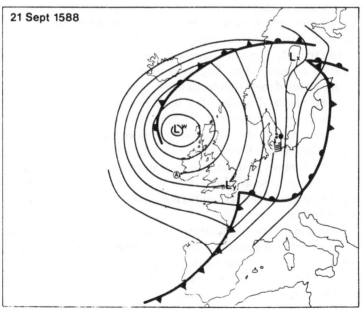

21 Sept 1588

The first main effort was to concentrate on a run of severe storms in the 1790s, including the worst flood and highest tide level of the River Elbe near Hamburg that occurred in the 18th century, and a storm in 1795 which felled so many trees in the forests of central Sweden that it took 25 years to clear the damage. Although this period is 100 years after the coldest phase of the Little Ice Age in this part of the world, unlike the situation in the 1690s (when there were some possibly more violent storms), enough barometric pressure measurements can be gathered together to make a full synoptic analysis and measurement of the pressure gradients possible.

It happened, however, that a chance to perform a worthwhile analysis of the famous Spanish Armada storms in 1588 occurred first (Douglas *et al.* 1978, Douglas & Lamb 1979). An historian's own exercise in producing sketched synoptic weather maps for over 30 individual days of that summer, from the reports in the logs of the Spanish ships and a few other sources, was presented for comment. The analysis shown was tested by its ability to explain the weather observed each day by the astronomer Tycho Brahe on the island Hven in Denmark. The original analyst had been unaware of this Danish observation series. Since the test result showed a satisfactory agreement on 72% of the days, it was decided to refine the analysis of the whole series of daily maps, after entry of the observations made in Denmark and with every care for logical continuity of the development and movement of the wind circulation systems, from one day to the next. Two sample sequences from the resulting maps are shown in Figures 6.10 and 6.11. The dates have been corrected to the modern (Gregorian or New Style) calendar. The August sequence shows the deduced weather situations that enabled the Armada to escape northwards through the North Sea to the Orkney–Shetland region after its harassment by fire in the battle of Gravelines on the 7th of that month. A great storm approached on the 15th, and from the 16th to the 18th the Spanish ships reported continual squalls, rain and heavy seas in the region of the Northern Isles, as well as some fogs. The later maps show the Armada held in northern waters by almost continual southerly winds, until, on the 21 September, the severest storm of that season beat many of the ships to pieces on the rocks and in the harbours of the west coast of Ireland and in the Hebrides.

Although this map series pre-dates the invention of the barometer, so that no pressure levels can be suggested, the data are sufficient to indicate the position of many of the individual cyclone centres on the daily maps. Hence, the 24-hour movements of these systems can be measured. And, because there are statistical associations between the distances travelled by the systems in the young, open frontal wave and occluded states respectively, and the strengths of the gradient wind (Palmén 1928) and of the jet stream (Douglas *et al.* 1978, using Chromow 1942), deductions can be made about the probable wind speeds. Study of the movements of the depression centres between 22 July and 1 October 1588 indicates that on six occasions in that one late summer–early autumn there were jet streams probably reaching or exceeding 200 km/h or 108 knots. The strongest was probably about 155 knots on 8–9 August 1588, and in mid-

September 130 knots seems to have been again reached. This means that on six occasions towards the end of the summer season, the jet stream reached a strength at or beyond the limit of what might be expected at that time of year from the statistics of 20th-century observations.

The methods used for analysing the situations occurring in the 1790s can be very close to those applicable to modern maps. Observation data were available from 50 places in Europe, ranging from northern Iceland to Rome, and from Barcelona to St Petersburg and northern Sweden and Finland. Additionally, occasional reports from ships at sea, or in port, at useful positions, though without instruments, strengthen the analysis. Through information about the positions and practices of the observing stations and their instruments supplied with much of the archival data, and from the experience gained by Mr J. A. Kington of the Climatic Research Unit in the University of East Anglia in analysing the daily weather situations over the six years from 1781 to 1786, a gazetteer of the stations and their specific details has been built up. From this, the barometric pressures reported at a network of key stations, for which one can apparently be confident in the details known, can be corrected to standard temperature and gravity and to sea-level. On the basis of this network, a monthly mean pressure map can be drawn covering Northern and Central Europe as far south as northern Italy and as far west as the British Isles. It is then possible to establish the average total correction required by the reported barometric pressures at other stations, for which detailed specification of the station position is lacking, simply by the adjustment needed to make their mean pressure for the month fit the map.

This technique is bound to lead to some rough fits of the barometric pressures reported at some of the relatively higher level stations, and at places in valleys and on plains when very low surface temperatures are observed beneath an inversion layer, but it has the virtue of making use of many stations where careful observations were made, which would otherwise have to be discarded, and which can therefore contribute to a fuller analysis on most days. Finally, for every station from which barometric pressure readings were used, the reliability of the values could be explored by calculating the standard error of the pressure values (corrected and reduced to sea-level) revealed by comparisons with the values indicated for the place by the final and most careful analysis of the daily synoptic pressure maps. These standard error, or reliability, measurements are a valuable addition to knowledge of the old observing stations compiled in the station gazetteer. (The results may be somewhat flattering to stations at the edge of the mappable area, which tended to be relied on because of their position, but this can hardly be so where the network of observations was dense so near the limit of the area covered. From this it was apparent that the Swedish stations in the 1790s and Gordon Castle in northern Scotland, as well as the French stations of that date, were among the most reliable, with standard errors of about one millibar or less. It seems, from inspection of the data, that the more isolated station at Trondheim, Norway, was also among the best.)

A separate test of the reliability of the synoptic maps with regard to pressure gradients and the storm winds over the North Sea and surrounding regions was performed by first analysing the maps for March 1791 with only four stations (Trondheim, Stockholm, St Petersburg and Copenhagen) in Northern Europe north of 55°N apart from the British Isles stations. These places were as far apart as the places on opposite sides of the North Sea:

e.g.	Stockholm – Trondheim	600 km
	Copenhagen – Trondheim	830 km
	Stockholm – St Petersburg	720 km

compared with

Edinburgh – Hamburg	600 km
Norfolk – Hamburg	600 km
North–east Scotland – Trondheim	1000 km

Subsequently, the observations made at six more places in Sweden and Finland, in the gaps in the network used for the first analysis, were entered. Unfortunately, the barometer used at Gothenburg was not one of the best, its record being marred by failure to respond fully to the quickest changes of pressure. Eliminating the minority of days, 8 out of 31 in that stormy month, when the Gothenburg barometer readings were obviously badly wrong because of its slow response, the standard error of the pressures indicated by the isobars of the first analysis at that place was 1.24 mb, part of which may still be due to relatively poor behaviour of the Gothenburg barometer. Overall, the test suggested standard errors of 1–1.5 mb at places in the midst of the biggest gaps in the network used for the test analysis, but bigger errors at no great distance beyond the limits of that network (e.g. standard error of the isobars 3.7 mb at Härnösand, 250 km beyond the northern limit of the network). The corresponding errors of the pressure gradients over the North Sea would be about 1 mb/200 km, or barely 10 knots in the gradient winds indicated.

To illustrate the results obtained from this analysis the synoptic maps for 14.00 h on 20, 21 and 22 March 1791 are reproduced in Figures 6.12, 6.13 and 6.14.

The first warning of disturbance to the anticyclonic situation with light winds which had been dominating Northern Europe since 16–17 March, including the British Isles and with light southerly winds as far north as Iceland, came with the passage of a cold front southwards over Iceland on the night of 18–19 March followed by three days of hard northerly winds with bitter frost and blowing snow there. A strong northerly (or in the south, northwesterly) gale reached the British Isles on the 20th, after falls of pressure of the order of 20–30 mb in 24 hours, followed by equally rapid rises. Temperatures fell low for the season in the strong winds, with snow and hail showers as far south as Hampshire. And at Hamburg, where the pressure had fallen 37 mb in 24 hours between the afternoons of the 20th and 21st, at 0.5.00 h on the 22nd, after a night of storm,

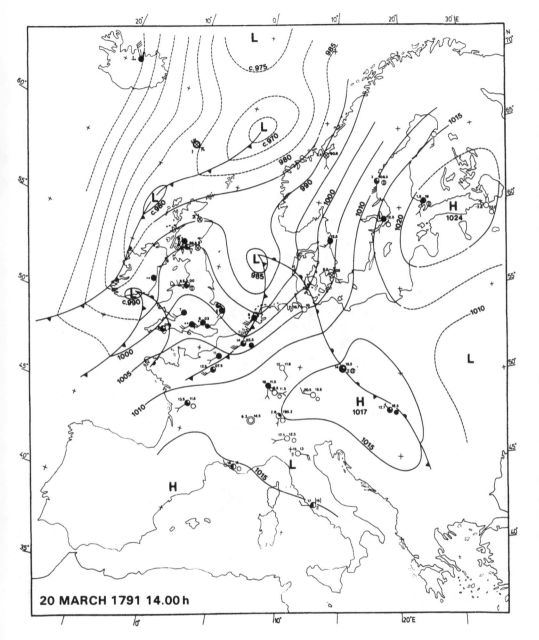

Figure 6.12 Synoptic weather map for 14.00 h on 20 March 1791.

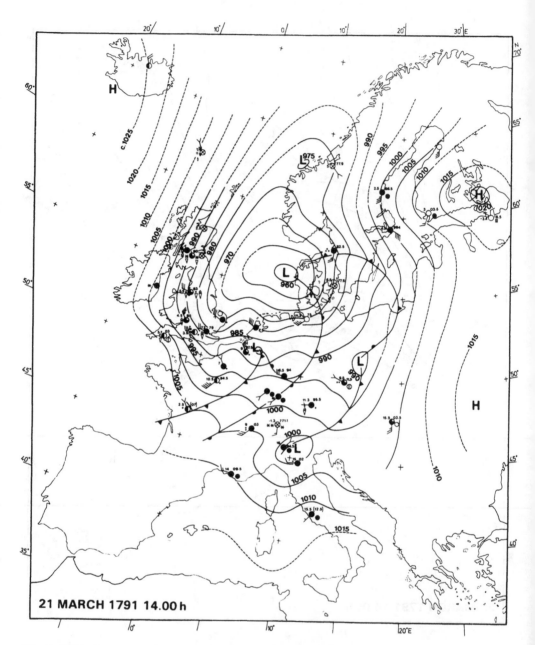

Figure 6.13a Synoptic weather map for 14.00 h on 21 March 1791, showing a severe North Sea storm developing.

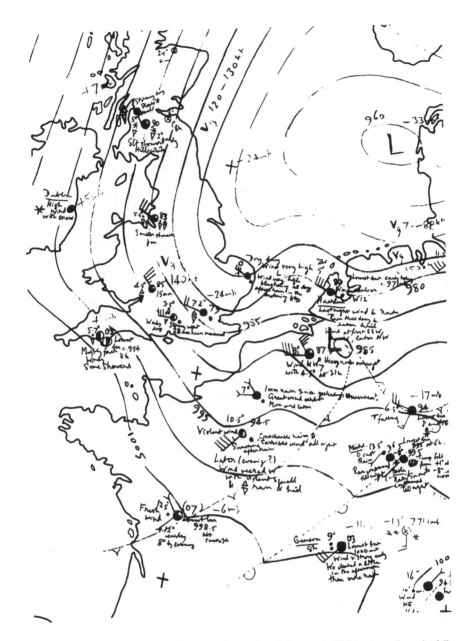

Figure 6.13b Excerpt of the original manuscript version of the map for 21 March, to show the full amount of data, much of it relating to additional hours of the day.

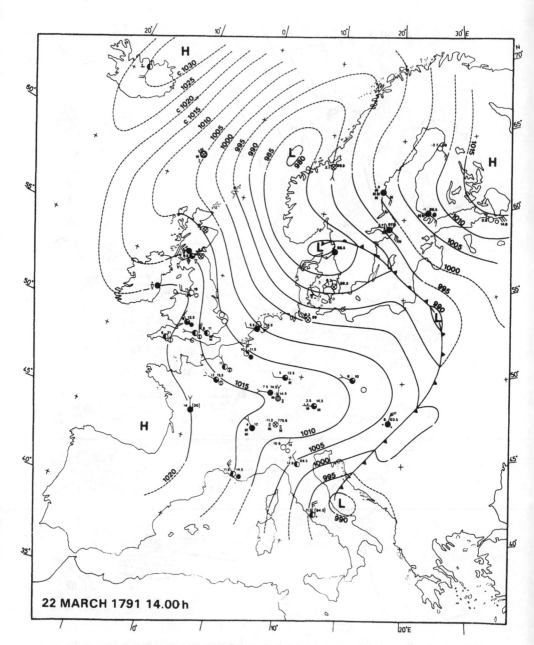

Figure 6.14 Synoptic weather map for 14.00 h on 22 March 1791.

the tide in the River Elbe reached the level of two previous high floods in 1751 and 1756, a level exceeded significantly only once in the 18th century, on 11 December 1792. This storm in March 1791, which maintained gale strength for three days over some parts of the North Sea, was probably the longest in duration of any of the severe storms of that century, owing to the rather slow movement of the cyclonic system from the 20th onwards in the increasingly meridional situation.

On this occasion, as also during three separate storms in the month of December 1792, geostrophic winds reaching 130–150 knots at some points over the North Sea and neighbouring coasts were indicated by the surface pressure map. Since in all these cases 'straight' or great circle windstreams occurred in the regions where the pressure gradients were strongest, the gradient winds were presumably also of about this strength. Strongest gusts at the surface over the sea and exposed areas near to the coasts may also have been over 100 knots, perhaps by a wide margin. The gradient winds measured in these storms, and in one other case in February 1825 which produced the severest sea flood of the 19th century on the coasts from Holland to Denmark, exceed the strongest gradient winds measured in any of the 15 greatest storms of the 20th century analysed over the same region, though a gust of the surface wind amounting to 154 knots was measured in Shetland during the storm of 16–17 February 1962, when the gradient wind was estimated at only 80–90 knots. (An even stronger gust of 163 knots was measured by a standard anemometer on the mountainous island of Jan Mayen (71°N) in the Norwegian Sea on 9 April 1933.)

So far, the investigation which has compiled analyses of 29 of the severest storms known to have occurred in the region from 1588 to 1981 supports the impression that several of the severest storms in the Little Ice Age period between the mid-16th and early 19th centuries exceeded the violence of any that have occurred since. There are indications that the severest cases occurred during periods when, with the Arctic sea ice in advanced positions near Iceland and the polar water extending farther south, the thermal gradient between latitudes 50°N and 60–65°N in the eastern Atlantic sector was increased compared with most later situations.

Acknowledgements

The author expresses his gratitude to Shell Exploration Limited for funding the research on the severest North Sea storms of the past reported on in this chapter.

His thanks are also due for gifts of weather observation data to the late Professor Gordon Manley for obtaining the records for Cambuslang near Glasgow, for Liverpool, for Lyndon, Rutland, for Stroud near Gloucester, for London, and for Modbury near Plymouth; to Sjöfn Kristjánsdottir for the record from Eyafjörður in Northern Iceland, to Dr Harold Nissen of the University Library, Trondheim, for the Trondheim record; to Dr K. Frydendahl and the Danish Meteorological Institute for the observations from Copenhagen and from Danish ships at sea; to Dr E. Hovmøller, formerly of the Swedish

Meteorological and Hydrographic Institute, for observations from ten places in Sweden and Finland; to Dr P. K. Rohan, formerly Director of the Irish Meteorological Service, for the observations made in Dublin; to Mr J. A. Kington of the Climatic Research Unit for the observations which he obtained from Paris, Poitiers and Montdidier in northern France and from Milan; to the Director of the *Schweizerische Meteorologische Anstalt*, Zürich, for the observations from four places in Switzerland and for Mühlhausen in Alsace; to Professor Puigcerver for the observations in Barcelona; and to Dr G. Zanella of the University of Parma for the observations made at Parma in northern Italy.

Apart from these places, most of the observations mapped came from the collection in the yearbook published by the then existing *Societas Meteorologica Palatina*, Mannheim, for 1791. A valuable record for Gordon Castle in North-east Scotland came from the UK Meteorological Office, and Gilbert White's record at Selborne, Hants, was made available for copying by the Trustees at Selborne. The writer's thanks go to the relatives and Trustees of persons who kept weather diaries in the 1790s at Kemnay in Aberdeenshire, and at three places in Norfolk (Weston Longville, Letheringsett, and Hoveton). The agreement between the weather reported in these three diaries meticulously kept at places 20–30 km apart is remarkable and lends confidence to the use of such records.

References

Birkeland, B. J. 1925. Ältere meteorologische Beobachtungen in Oslo. Luftdruck und Temperatur in 100 Jahren. *Geofysiske Publikasjoner* 3.

Birkeland, B. J. 1949. Old meteorological observations in Trondheim. *Geofysiske Publikasjoner* 15.

Brooks, C. E. P. 1949. *Climate through the ages*, 2nd edn. London: Ernest Benn.

Chromow, S. P. 1942. *Einführung in die synoptische Wetteranalyse*, 2nd edn. Vienna: Springer.

Coles, B. P. L. and B. M. Funnell 1981. Holocene palaeoenvironments of Broadland, England. *Special Publs. Int. Ass. Sediment* **5**, 123–31.

Defoe, D. 1704. *The Storm: or a collection of the most remarkable casualties and disasters which happen'd in the late dreadful tempest both by sea and land.* London: G. Sawbridge.

van den Dool, H. M., H. J. Krijhen and C. J. E. Schunrmans 1978. Average winter temperatures at De Bilt (the Netherlands) 1634–1977. *Climatic change* **1**, 319–30.

Douglas, K. S. and H. H. Lamb 1979. Weather observations and a tentative meteorological analysis of the period May to July 1588. *Climatic Research Unit Research Publication CRU RP 6a.* Norwich: University of East Anglia.

Douglas, K. S., H. H. Lamb and C. Loader 1978. A meteorological study of July to October 1588: the Spanish Armada storms. *Climatic Research Unit Research Publication CRU RP 6.* Norwich: University of East Anglia.

Grove, J. M. 1979. The glacial history of the Holocene. *Prog. Phys. Geog.* **3**, 1–54.

Labrijn, A. 1945. Het klimaat van Nederland gedurende de laatste twee en een halve eeuw. *Mededelingen en verhandelingen* **49**, 1–102.

Lamb, H. H. 1977. *Climate: present, past and future*, vol. 2. London: Methuen.

Lamb, H. H. 1979. Climatic variation and changes in the wind and ocean circulation: the Little Ice Age in the northeast Atlantic. *Quatern. Res.* **11**, 1–20.

Lamb, H. H. 1980. Climatic fluctuations in historical times and their connexion with transgressions of the sea, storm floods and other coastal changes. In *Transgressies en Occupatiegeschiedenis in de Kustgebieden van Nederland en België*, A. Verhulst and M. K. E. Gottschalk (eds.), 251–84. Ghent: Rijksuniversiteit.

Lamb, H. H. 1982. *Climate, history and the modern world*. London: Methuen.

Lamb, H. H. and A. I. Johnson 1959. Climatic variation and observed changes in the general circulation (Parts I and II). *Geog. Ann.* **41**, 94–134.

Lamb, H. H. and A. I. Johnson 1961. *Ibid*. (Part III). *Geog. Ann.* **43**, 363–400.

Lamb, H. H. and A. I. Johnson 1966. Secular variations of the atmospheric circulation since 1750. *Geophysical Memoirs* **14** (110), 1–125. London: HMSO.

Mackenzy, Sir G. 1675. Report and notes. *Phil. Trans. R. Soc. Lond.* **10**, 307; also in *Hutton's abridgment of the philosophical transactions of the Royal Society* **2**, 1672–85.

Manley, G. 1946. Temperature trend in Lancashire, 1753–1945. *Q. J. R. Met. Soc.* **72**, 1–31.

Manley, G. 1949. The snowline in Britain. *Geog. Ann.* **36**, 179–93.

Manley, G. 1953. Mean temperature of central England. *Q. J. R. Met. Soc.* **79**, 242–61.

Manley, G. 1959. Temperature trends in England, 1698–1957. *Arch. Met. Geophys. Bioklimat.* **B9**, 413–33.

Manley, G. 1974. Central England temperatures: monthly means 1659 to 1973. *Q. J. R. Met. Soc.* **100**, 389–405.

Manley, G. 1979. The climatic environment of the Outer Hebrides. *Proc. R. Soc. Edin.* **77B**, 47–59.

Matthews, J. A. 1977. Glacier and climatic fluctuations inferred from tree-growth variations over the last 250 years, central southern Norway. *Boreas* **6**, 13–24.

Palmén, E. 1928. Zur Frage der Fortpflanzungsgeschwindigkeit der Zyklomen. *Met. Zeitsch.* **45**, 96–9.

Pennant, T. 1771. *A tour of Scotland 1769*. Chester.

Svabo, J. C. 1782. *Indberetninger fra en Reise i Faerøe 1781 og 1782*. Copenhagen (republished 1959 by Selskabet til Udgivelse af Faerøiske Kildeskrifter og Studier).

Tooley, M. J. 1978. Interpretation of Holocene sea-level changes. *Geol. För. Stockholms Förh.* **100**, 203–12.

West, J. F. 1970. *The journals of the Stanley expedition to the Faroe Islands and Iceland in 1789.* Vol. I: *Introduction and diary of James Wright*. Torshavn: Førøya Froðskaparfélag.

7

The timing of the Little Ice Age in Scandinavia

JEAN M. GROVE

Field studies and documentary evidence of glacier advances in southern Norway during the Little Ice Age are examined. In northern Scandinavia field evidence for glacial advances is scanty, but lichenometric dating provides some useful information. It seems that glacial expansions in northern Scandinavia were not always contemporaneous with those further south, and possible meteorological explanations are suggested.

7.1 Introduction

It is generally accepted that glaciers reached their Little Ice Age maxima during the 17th, 18th and 19th centuries, and that most were in an enlarged state at least from the final decades of the 16th century until sometime in the 19th century (e.g. Porter 1981a). Manley (1966) pointed to the association of minor advances within this period (Fig. 7.1), with decades suitable for glacier growth indicated by his Central England temperature series, which he took to be of general European significance: 'quite fairly representative of the course of events between Basle and Bergen'. There is no doubt that the glaciers of the European Alps had already expanded by the late 16th century. The main advances of many Swiss glaciers occurred between the mid-1580s and the turn of the century (Messerli *et al.* 1978) while glaciers of the French *versant* of Mont Blanc, in a more responsive maritime position, were probably affected a decade or so earlier, and phenological historical studies in Switzerland have made possible unusually detailed reconstruction and understanding of the meteorological circumstances of the 16th and 17th centuries which caused these responses (Pfister 1980, 1981). The climatic history of the 13th and 14th centuries is known in less detail, but it is now clear that the deterioration following the medieval warm period was such as to invoke marked extension of Alpine glaciers, including the Grosser Aletsch and the two Grindlewald tongues (Delibrias *et al.* 1975, Messerli *et al.* 1978).

The traditional view of Norwegian glaciologists, recorded by Hoel and Werenskiold (1962) and based on historical evidence, has been that the glaciers in Norway were very small in the 16th century and enlargement began sometime in the latter part of the 17th century, reaching a climax in the mid-18th century. No suggestion of 13th-century advances is to be found in the historical

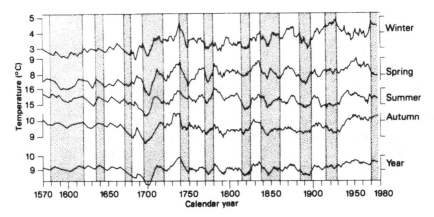

Figure 7.1 The relationship between periods of marked advance of European glaciers (shaded) and decadel running averages of seasonal and annual mean temperatures from Central England 1559–1979. The periods of glacial advance – based on Le Roy Ladurie (1971) and Bray (1982) – are from alpine sources (after Manley (1966), and further data from the chart kept in his study up till the time of his death).

literature (e.g. Eide 1955). In contrast to this, several recent papers reporting the results of field investigations have suggested that glaciers in both Norway and Sweden had, like those in the Alps, reached forward positions by 1590 (e.g. Denton & Karlén 1973, Karlén & Denton 1976, Karlén 1979, 1981). Karlén (1979) concluded that in the Svartisen area of northern Norway, the ice margins were well advanced in the early 14th century and, after a period of retreat, expanded again towards the end of the 16th century and in the mid- as well as the late 17th century. He also cited field evidence from southern Norway suggesting enlargement in the 14th century as well as in the last half of the 15th or first half of the 16th century.

Good documentary data can provide the most exact information about the timing of glacial fluctuations during the Little Ice Age period (Porter 1981a), but by no means all documentary data are reliable (Bell & Ogilvie 1978, Ingram *et al.* 1981). In the absence of satisfactory archival material, glacial fluctuations can be dated approximately if the ages of the associated moraines can be determined. The methods available include determination of the radiocarbon age of organic material underlying, overlying or included within moraines, and lichenometric dating of moraine surfaces, which has been particularly useful in Scandinavia. Each of these methods has its difficulties and limitations. In addition to possible contamination of samples, laboratory error and the problems of soil dating (e.g. Karlén 1979, Matthews 1980, 1981, 1982), radiocarbon dating is ambiguous for the period between about AD 1500 and 1950, as any one value for a sample may indicate a number of possible calendar dates (Stuiver 1978). Therefore, radiocarbon dates cannot be used for Little Ice Age inter-regional comparisons. Under ideal circumstances, lichenometry provides a much more precise tool for this period (Porter 1981a,b), but only if the method is used with proper precaution, and knowledge of the dates of exposure of a substantial number of lichen-bearing surfaces allows an adequate growth curve to be constructed (Benedict 1967, Webber & Andrews 1973, Matthews 1974, 1977, Porter 1981b).

If the meteorological characteristics of the Little Ice Age period are to be properly understood, the existence or otherwise of widespread 13th-century glacial advances, or of a lag of a century or more between the onset of glacial expansion in the Alps and in Scandinavia, is of substantial interest. Resolution of the present conflicts of view must begin with a critical review of the evidence, bearing in mind that glacial expansions in northern Scandinavia cannot be assumed to have been contemporaneous with those further south, any more than synchronicity in the Alps and Scandinavian mountains can necessarily be expected.

7.2 Southern Norway

In southern Norway, farming communities have lived close to glaciers for hundreds of years, but no unambiguous documentary evidence about the extent of the ice in the 13th and 14th centuries has been found. However, there is reliable evidence that farms in Olden and Loen parishes immediately north of Jostedalsbreen (the largest ice cap in continental Europe, Fig. 7.2) suffered some major disaster in 1339. A special court held in January 1340 registered many farms as deserted, and reduced the taxes on the others substantially. Unfortunately, the records (*Diplomatarium Norvegicum* 1847) do not specify the cause of the devastation in 1339, but tax reductions of this sort on individual farms were made only in cases of major physical damage. It is noteworthy that these farms in Olden and Loen were amongst those which were to suffer severely from glacial advances and associated avalanches, rockfalls and landslides in the late 17th and 18th centuries (Grove 1972). It has been suggested (Professor A. Holmsen, personal communication 1970) that the earlier damage might also have been connected with an episode of glacial expansion. On the other hand, the area could have been shaken by an earthquake; Pálsson (1795) mentioned that many communities in Norway had been destroyed by an earthquake in Iceland in 1339, though no reliable confirmation has yet been found in the literature. (Western Norway is certainly not immune to earthquakes. Erik Iversen Nordal, parish priest at Leikanger, Sogn, recorded that there was an earthquake at midnight on Christmas Eve 1660: 'in Sogn which was felt as far away as Sunnfjord' – Finne-Grønn 1945.) An earthquake triggering a series of mass movements may be a more likely explanation than glacier enlargement in the early 14th century. Gordon Manley (personal communication, undated) found 'it difficult to run across any hint of climatic support for a glacial advance in Norway about 1340', but considered that if, as seemed to him quite possible, there was a minor advance in Scandinavia between 1250 and 1300 and 'after the (very wet) early 14th century the glaciers were still relatively 'forward', a series of mild but rainy seasons could produce serious summer flooding'.

There is field evidence to indicate rapid climatic deterioration in the late

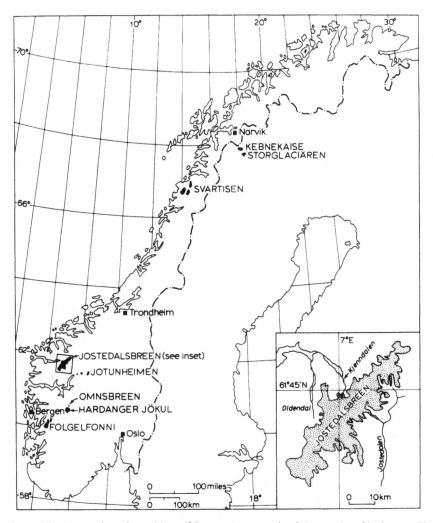

Figure 7.2   Map to show the positions of the most important localities mentioned in the text. The dot shading on the inset shows the extent of the ice cap.

Middle Ages in southern Norway. Andersen and Sollid (1971) examined the moraines of Tverrbreen (now known as Tuftebreen), one of the eastern tongues of Jostedalsbreen, and described the protrusion of an older moraine from beneath the main outer ridge generally accepted as representing the maximum Little Ice Age extension of the glacier during the 18th century (but see Matthews 1982). A date of 410 ± 60 BP (T-799) on a layer of peat partly distal to the outer moraine has been taken as indicating an age of *c*. AD 1540 (calibration according to Damon *et al.* 1974) and as possible confirmation of a legend that Nigardsbreen, a neighbouring outlet of Jostedalsbreen, had advanced some time earlier than the late 17th-century expansions which culminated in 1748 (Karlén 1981). Anderson and Sollid had, however, noted that a 16th-century advance of Jostedals-breen 'is not immediately compatible with the historical records'. (Matthews (1982) suggested that both ridges may have been formed close to AD 1750, explaining the well vegetated nature of the outer ridge in terms of its high peat and soil content.)

Melting away of the central part of Omnesbreen, a small glacier lying north of Hardangervidda (Fig. 7.2) in the 1970s, revealed an extensive area of fresh tills with humus and plant remains, which have been dated to 550 ± 110 (T-1485), 440 ± 120 (T-1486a), 430 ± 100 (T-1578) and 430 ± 100 (T-1479) (Elven 1978). The undisturbed nature of the plant remains appears to indicate rapid climatic deterioration causing the plants to be embedded in increasingly permanent snowbeds. It is particularly interesting that the species involved grow at somewhat lower altitudes today. Elven (1978) very reasonably concluded that the change in climate was too swift for vegetation to respond before being overwhelmed. He placed the deterioration in the 14th century, having calibrated the ^{14}C ages of the plant fragments according to Ralph et al. (1973).

Evidence from Omnesbreen, and dates from moss beneath a moraine fronting Storbreen in Jotunheimen (Fig. 7.2) of 664 ± 45 (SRR. 1083) and 532 ± 40 (SRR. 1084) (Griffey & Matthews 1978), were considered by Karlén (1979). He calibrated these radiocarbon dates according to Damon et al. 1974, and suggested 'a glacial expansion in southern Norway at 500–400 ^{14}C years BP, i.e. in the last half of the 15th century or the first half of the 16th century'. The initiation of Omnesbreen must be presumed to have occurred at the end of a period somewhat warmer or with less accumulation of snow than nowadays, while the moraine fronting Storbreen was deposited when the glacier was substantially larger than it is now. It does not seem probable, therefore, that the two events were contemporaneous.

Recent collaboration aimed at rationalisation and experimental quantification of the uncertainties involved in ^{14}C dating and estimation of the extent of inter-laboratory variability has led to the conclusion that quoted laboratory errors should be multiplied by a factor of between two and three (International Study Group 1982). The nature of the uncertainty involved in radiocarbon dating of samples from the last few centuries can be deduced, given the combination of Stuiver's (1978) work on the relationship between radiocarbon ages and calendrical ages for the period 1500–1950, the International Study Group's cautious approach to the extent of laboratory error and Klein et al.'s (1982) main calibration tables. Taking the most recent of the Omnesbreen dates of 430 ± 100, the extreme date limits are between AD 1025 and AD 1950 for $\sigma = 30$. However, all the dates from Omnesbreen and Storbreen are dates from material that grew before the climatic deterioration and glacial expansion set in, and the most recent are not necessarily the most accurate. It must be concluded that the radiocarbon evidence is inadequate to determine conclusively when the initiation of Omnesbreen took place, or when the smaller outer moraine of Tverrbreen was formed. Of the two dates from Storbreen falling outside the period covered by Stuiver's work, the date from the uppermost of the two moss layers sampled is the older, and Griffey and Matthews (1978) state that 'no importance can be attached to the difference between the two dates'. Therefore, although these ^{14}C dates are 'believed to antidate closely the glacier maximum', they give us no more than an approximate maximum age for the moraines, and certainly do not reveal in which century the advance began. The field evidence being inconclusive, the extent of the reliable documentary evidence bearing upon the time of initiation of the Little Ice Age in southern Norway must now be investigated.

The earliest definite evidence of direct damage to farmland by advancing ice comes from Jostedal (a valley on the eastern side of Jostedalsbreen) in a brief account, dated 1684, of the arraignment before the local court (or *ting*) of two farmers, Knut Grov and [illegible] Berset for non-payment of land rent by the proprietor of their farms, a widow called Brigette Munthe (*Statsarkivet Bergen MSS a*). They pleaded that they could not pay because the huts (or *støler*) on their high pastures had been covered by advancing ice. The exact position of these huts and the *saeter* grasslands upon which they stood are unknown, but Grov and Berset (or Bergset) farms were, as they still are today, in Krundalen, a right hand tributary of Jostedalen. As it happens, it was the outer moraines of Tverrbreen, the glacier which terminates above Berset farm, which were investigated by Andersen and Sollid. There is not a great deal of space in upper Krundalen, and the implication is that if the land here was suitable for use as pasture, it must have been ice free for a considerable time. It is significant that Brigitte Munthe submitted that the two farmers could use pastures at Kriken and Espe, also her properties, as there was sufficient grass there. Kriken and Espe lie on the eastern side of Jostedalen, that is on the side of the valley away from Jostedalsbreen (Fig. 7.3).

The advance of the ice after 1684 was swift. According to the records of an enquiry held in August 1742, Knut Grov's son Ole, born about 1678, and a neighbour, Ole Bierck, born about 1672: 'men of good repute, explained that they remembered that in their youth, the said glacier [Tverrbreen] had been only high up on the mountain in the narrowest neck of Tufteskar, but it had forced its way through the gap, and down on to the flat fields towards the river, and, according to Rasmus Cronen's explanation it had advanced 100 fathoms [*c.* 200 m] in only 10 years.' They made no mention of temporary retreats or halts between 1684 and 1742, or of earlier advances of the glacier. This is hardly surprising in view of the history of settlement in the valley.

In Norway, the effects of the Black Death were severe (Hovstad 1971), but its incidence was uneven. Many upland areas were left with little or no population for prolonged periods. After the plague year of 1349–50, Jostedalen was deserted. There was still one farm occupied in Krundalen in 1374, but eventually this too lay empty. Survivors commonly migrated to richer lowland areas or to coastal situations. It was not until the population had risen substantially in the 16th century that remote valleys like Jostedalen were taken in once more.

The resettlement of Jostedal can be traced by examination of the tax rolls (Laberg 1944). When a farm was reoccupied or taken in from the wild, a few

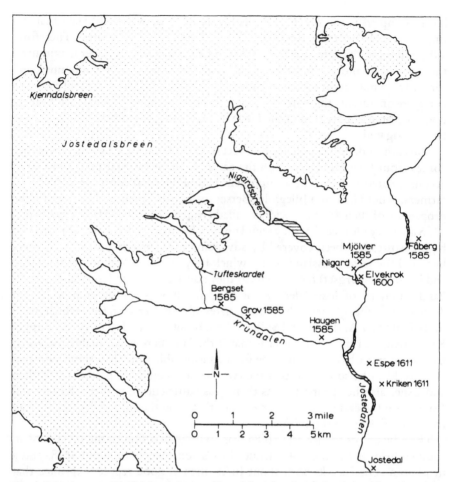

Figure 7.3 A map of Krundalen and part of Jostedalen, showing their relationship to Jostedals-breen. The date at which each farm first appeared in the tax rolls is marked against it. The dot shading shows the extent of the ice cap.

years of grace were allowed before it was assessed for land rent, taxes or tithes. It may be assumed that farms were reoccupied during the decades preceding the dates at which they first appear in the tax rolls. The results of Laberg's investigation are mapped on Figure 7.3. Jostedal was not farmed again after its desertion until the years immediately preceding 1585, when a wave of immi-grants came in and settled down. The Krundalen farms, Grov and Berset, were reoccupied just at the time when the glaciers around Mont Blanc were expanding to cover or damage farms and high villages in the Arve valley (Grove 1966, Le Roy Ladurie 1971). The grasslands at the head of Krundalen must not only have been clear of ice in the late 16th century, but surely for a long period before that, as they were suitable for use as summer grazing. Conditions in Jostedal were sufficiently satisfactory for a further wave of incomers to arrive in the early part of the 17th century. Several farms, including Kriken and Espe, appeared in the tax rolls in 1611.

When Jostedalsbreen eventually expanded in the late 17th century, the damage caused to the farmlands led to on-the-spot investigations (*avtaksforret-ninger*) and courts of enquiry, and hence to graphic contemporary records and descriptions (see *MSS* references). From these it is clear that the advance of Tverrbreen was by no means exceptional in its extent or timing. Foss, vicar of Jostedal, gave an account of his parish in 1744 in which he explained how the ice had pushed forward 'both in Krundalen and in Milverdalen, where in its advance it has widened more and more as the valleys open out' (Foss, in Laberg 1948). Nigardsbreen took longer to reach the grasslands in Milverdalen, but the front moved some 2800 m forward between 1710 and 1735, by which time the ice was within a stone's throw of the farmhouse occupied by Guttorm Johanssen, had carried away the best part of his meadowland, and threatened further damage to what remained (Eide 1955). Witnesses of the glacier's behaviour were not lacking between 1585 and 1684. They may well have noticed minor fluctuations during that period. Other natural events which occurred during the 16th century are also known about. For instance, Kruna, or Kronen, one of the largest farms in Krundalen, passed from one Sjur to his son Klaus in 1553 after the son's farm had been destroyed by flooding (Laberg 1944), but there is no hint of glacial damage to the farmlands to be found in the documents before the late 17th century. It seems clear that the grasslands occupied by Tverrbreen and Nigardsbreen in the late 17th and early 18th centuries had been free of ice since the resettlement of Jostedal. It could be argued that as there were no witnesses available in the 15th and 16th centuries, a substantial advance could have occurred sometime then, although the use of upper Krundalen by 1585 militates against such an assumption.

The likelihood of 16th- and even 15th-century advances diminishes further in the light of historical evidence from Oldendal on the northern edge of Jostedalsbreen (Fig. 7.4). This valley was not completely deserted after the Black Death, a number of farms remaining occupied without interruption. Re-immigration began earlier than in Jostedal, the main resettlement taking place between 1530 and 1620 (Aaland 1973). By 1563 there were 23 farms in Olden parish and a further 16 had been added by 1667. Not only the old deserted farms but also new farms had been taken in, and the valley was fully occupied.

The most significant histories from the point of view of the present enquiry are those of two farms, Aabrekke and Tungøen, which both held grazing land up in Brendalen, a hanging valley tributary to Oldendal. These farms, established before 1563, huddled beneath the rock step which terminates Brendalen at its western end. Aabrekkebreen, an outlet of Jostedalsbreen, closed the eastern end of Brendalen, but according to tradition (Rekstad 1901), after the settlement of the farms the ice was just visible on the skyline and did not reach the floor of Brendalen. Both Aabrekke and Tungøen certainly used Brendalen for pasturing their cattle. Documents from the period 1563 to 1670 do not reflect any anxiety about the future of the farms, and no damage to their lands is recorded. The tax

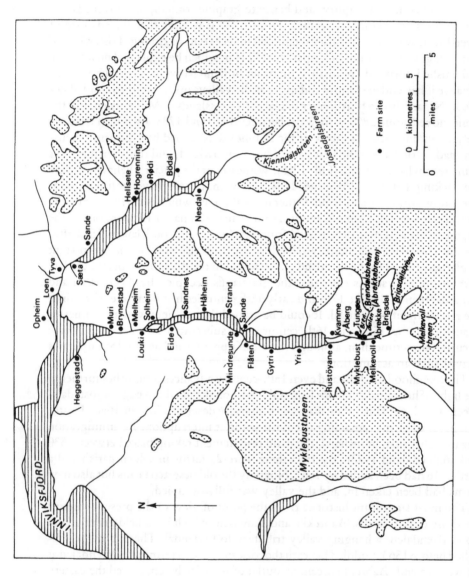

Figure 7.4 Map of valleys at the northern margin of Jostedalsbreen showing farms which were severely damaged by floods or mass movements once or more at dates between 1667 and the late 18th century. The dot shading shows the extent of the ice cap and the horizontal line shading the extent of water. Oldendal extends from Brigsdal to Muri (see inset, Fig 7.2, p. 135).

assessment on Tungøen in 1667 was 2 *laup* 1 *pund* of butter; it maintained 38 head of cattle and 3 horses; and yielded 29 *tønner* of corn and 3 *tønner* of seed corn. (A *laup* was approximately 18 kg. Taxation at this date was in terms of kind, and in the Oldendal area assessments were commonly in butter, 24 *merker* being equal to 1 *pund* and 3 *pund* to 1 *laup*. A *tønner*, was a corn measure, fixed in 1684 as equivalent to 139.4 litres or about 4 bushels. Before 1684, the value varied from district to district.) But by 1685 the ice was spilling over into Brendalen and accumulating below the rock step at its head. A court sitting at Tungøen in 1685 noted that the farm had suffered great damage from the flooding which took place as the enlarged glacier penetrated upper Brendalen. The course of events as it enlarged further is well documented (Rekstad 1901, Aaland 1932, Eide 1955) because of the many courts of enquiry held in connection with the occupants' application for tax relief in 1685, 1695, 1702, 1723, 1728, 1734 and 1744. After 1700, the ice front moved rapidly down valley towards the farms, covering their pastures. In 1728, Tungøen was described as lying 'under a very high and dangerous mountain where a terrible glacier now lies in a large valley where they grazed their cattle in summer in former times'. The report goes on to explain that the 'glacier discharges water, rocks and gravel each year and with such force and power that it has now broken up and undermined the soil many hundreds of *alen* deep and all has burst out and run down into the other large and cruel river which flows down the whole valley [that is Oldendal] from two other great glaciers, and below the farm; and this great river is and has been blocked, so that swollen, it has again burst its banks, flooded and completely carried away all their best and finest meadows'. The ice was now so close that the farm buildings were moved to a site which seemed safer. By 1734 the ice 'has so far advanced that it is to be feared that it will come right down to the main stream in a few years'. It did not in fact reach the main river draining Oldendal, but in December 1743 it carried away the new houses of Tungøen completely, killing all but two of the occupants (Eide 1955). By 1735, Tungøen tax assessment had been reduced to a mere 12 *merker*. After the disaster of 1743, it was removed from the tax roll, and the small remnants that remained transferred to the neighbouring Aaberg. Aabreeke remained, but its inhabitants had been reduced from meagre but reasonable self-sufficiency to penury and begging. The documents leave no doubt that Brendalen was for the most part, if not completely, clear of ice in the second half of the 16th and most of the 17th century. The main advance took place between 1685, when Tungøen's summer pastures had already been damaged, and 1725, when Aabrekkebreen loomed over Aabrekke 'directly above the place where their poor small dwellings formerly stood' (*Statsarkivet Bergen MSS* c), that is when the front had traversed the length of Brendalen and reached the rock step at the Oldendal end so that 'a mass of lumps of ice mixed with gravel and rough stones has often fallen'.

While Tungøen and Aabrekke were not reoccupied until sometime between 1520 and 1563, there were some farmers in Oldendal throughout the period between the Black Death and the main resettlement of the valley. When

Aabrekkebreen reached the entrance of Brendalen and toppled over into Olden-dal, it would have been visible from other farms in upper Oldendal and from the pathways to them. It is noteworthy that this late 17th century advance was accompanied by very serious flooding of both the stream issuing from it and of the main stream in Oldendal, taking the water from other glaciers such as Brigsdalsbreen which were enlarged at the same time. Flooding of this sort which was typical of the enlarged glaciers of the Little Ice Age (Grove 1966) caused a great deal of serious damage to farms and farmland, as did mass movements such as avalanches, rockfalls and landslides which occurred at the same time (Grove 1972, Grove & Battagel 1983). Reading through the reports of the many *avtakforretninger* (special courts of enquiry following applications for tax relief in respect of serious physical damage to farmland), it becomes clear that while the farms in Oldendal had always been subject to occasional avalan-ches or rockfalls, the intensity and scale of these had changed dramatically by the 1690's (*Riksarkivet Oslo MSS*).

Myklebust, facing Tungøen and Aabrekke across Oldendal and reoccupied between 1520 and 1563, was assessed at 4 *laup*, 2¾ *pund* of butter in 1626, and described as rather a good farm. The number of subdivisions of Myklebust, each with its own farmer, increased from 4 in 1556 to 7 in 1602, 8 in 1612 and 9 in the 1640s, but in 1685 the river from Brigsdalsbreen had already caused serious damage. By the 1690s, a good deal of Myklebust land was derelict. It is recorded in an *avtak* report from 1702 that by 1693 'this farm has suffered very great damage from the terrifying river which comes from the glacier [that is, Brigsdalsbreen] which has washed away most of their best arable and pasture, leaving nothing but roots, scree and gravel . . . Further many avalanches fall across the farm in winter and spring . . . Because of the above mentioned circumstances nearly the whole of the said farm Myklebust has lain waste for some years.' The court of 1702 removed 2 *laup* from the farm's assessment. Had the glaciers advanced in upper Oldendal during the 16th century, Mykelbust could hardly have enlarged and prospered as it is known to have done.

The lower part of Oldendal was never entirely deserted, so in this valley, unlike Jostedalen, some witnesses were available even during the 15th century. Documents remain from this period (Aaland 1932), but no trace of events such as those which accompanied the expansion of the glaciers in the late 17th and 18th centuries is to be found in them.

The evidence from Kjenndalsbreen, to the east of Oldendal, is even more striking. The advance of Kjenndalsbreen (Fig. 7.5) was contemporaneous with that of Tverrbreen and Aabrekkebreen. The occupants of Bødal farm in Loen parish suffered accordingly. An enquiry in 1693 found that 'a valley called Tiørdahl, where they formerly owned . . . [illegible] with the Indre Naesdal farmers, has now been wholly lost and completely ruined by the destructive glacier and by the great river running from it, so that where they formerly had five enclosures and six barns, they and their contents have been completely carried away.' As both Bødal and Nesdal farms owned property in 'Tiørdahl' (or 'Tiøndahl') the glacier which swept away the barns in its advance can be identified as Kjenndalsbreen. The barns, enclosures and the farmlands now lost would have been taken into account at the time of the general assessment of 1667. The reduction of the 1667 assessment recommended by the 1693 assessors indicates that Kjenndalsbreen too must have advanced rapidly between 1667 and 1692/3. Bødal and Indre and Ytre Nesdal were the farms in the area most vulnerable to damage associated with glacial enlargement. They were occupied continuously, while others such as Sande and Breng were deserted after the Black Death, and here too no evidence of glacial extensions in the 15th and 16th centuries has been found in the documentary sources.

As the various tongues of Jostedalsbreen have oscillated in harmony with each other since measurement of frontal position began in the late 19th century (e.g. Faegri 1948), the evidence of four of the major outlets having affected the surrounding valleys towards the end of the 17th century may be taken as symptomatic of the whole ice cap. Measurements of the 20th century have demonstrated a general coherence in the behaviour of the glaciers in southern Norway (Kasser 1967, 1973, Müller 1977), although there have naturally been differences in the response times of larger and smaller glaciers, and the role of precipitation has been more decisive in the maritime west than in the more continental east, so that glaciers near the coast have expanded in recent decades, whilst those further inland have continued to retreat (Liestøl 1973).

The only other localities from which there might be contemporary reports of early Little Ice Age advances are the parishes around Folgefonni. Øyen (1899) collected together references to this ice cap, but there are no accounts of marginal fluctuations earlier than the 18th century. A search of *avtak* reports has failed to disclose any mention of damage directly due to glaciers, although the pattern of damage caused by flooding and mass movements is similar to that for parishes near Jostedalsbreen or between Jostedalsbreen and the sea (Grove 1972, Grove & Battagel 1983), except that trouble here began earlier. For example, Broe and Skousell in Strandebarm Skipreide pleaded for land rent reductions on account of 'great and irreparable damage' to pasture and hayfields by flooding and river bursts in 1661 (*Riksarkivet Oslo MSS*). Hoel and Werenskiold (1962), in the acknowledgement section of their book, referred to a 1677 document dealing with an advance of Buarbreen 'in the period immediately preceding that year', but it has not proved possible to locate such a manuscript in the *Statsarkivet* in Bergen. A court of enquiry held in July 1667 found serious damage to Buar farms; 'a cruel mountain had fallen on the pasture . . . and the best of it taken away, besides which the river had broken out of its course and carried away all the arable.' This report (*Statsarkivet Bergen MSS* b) is fragmentary and reconstituted, but the missing parts are small and there is neither reference to Buarbreen nor any other reference to Buar in 1677 to be found in the *Tingbok*. The failure of Buarbreen to reach the farmhouse in the 18th century, despite its proximity, suggests that the late 17th and 18th century advances of Folgefonni may have been more muted than those of Jostedals-breen. Buarbreen and Bondhusbreen, the best-known ice tongues, reached their historical maxima in the late 19th century, as did some of the Icelandic glaciers

(Thorarinsson 1943) and Blomsterskardbreen around 1940 (Tvede & Liestøl 1977). Although the fluctuations of Folgefonni are recognised to have deviated from those of other glaciers in Norway and Sweden in the 20th century, there is no evidence at all that its tongues advanced earlier than those of Jostedalsbreen.

7.3   Northern Scandinavia

Glaciers and ice caps in northern Norway and Sweden are generally remote and far from farming settlements, and consequently very few records are to be found relating to their positions at times earlier than the late 19th century. Little Ice Age history must be pieced together almost entirely from field evidence of one sort or another.

The somewhat scanty evidence from northern Scandinavia during the period of observation was surveyed by Karlén (1979). Mikkajekna was practically stationary between 1896 and 1900, advanced in fluctuations between 1901 and 1916, and then receded. It appears that this behaviour is typical of those glaciers in both the Kebnekaise and Sarek mountain areas (see Fig. 7.2) in which much Swedish work has been concentrated. Glaciers in the Svartisen and Okstinden areas of Norway expanded in the early 1900s, reaching advanced positions between 1910 and 1920. A subsequent slow retreat was broken in some places by advances around 1928–32, and was followed by a more precipitate retreat. Engabreen, an outlet of Svartisen, began to readvance in 1965. This enlargement has persisted in most years since then. Precipitation declines from the more oceanic western mountains eastwards into Sweden. Glaciers such as Engabreen therefore respond more sharply and sooner to climatic oscillations than do those further east. The dividing line between glaciers increasing in mass in the west and those declining in mass in the east during the last decade has been nearer the coast in the north than in the south of Norway (Wold 1982). Nonetheless, an underlying coherence of behaviour is discernible from the observational data available; such a coherence may be assumed to have characterised events earlier in the Little Ice Age.

Evidence of glacial events in the Middle Ages so far available depends upon ^{14}C and lichenometric dating. Karlén (1979) extracted a sample consisting of 70% wood fragments from the top 2 cm of a peat deposit underlying the outermost moraine of Fingerbreen, an eastern tongue of Svartisen. This gave a date of 695 ± 75 (I-10364) as a maximum for the deposition of the moraine. Another date of 600 ± 100 (St-6757) was obtained from the upper 2 cm of a peat layer under the forest beds of a delta formed in a lake dammed by the expansion of one of the tongues of the western section of Svartisen into Glomdalen (Svartisen consists of two ice caps separated by Glomdalen which runs north–south). This sample was considered to pre-date the glacial expansion closely. However, Karlén (1979), having pointed out that peats may be deposited over a long period, illustrated the considerable disparities which may occur between dates obtained within the top 4 cm of a peat layer, and noted that the uppermost sections of a peat layer could be missing. In view of these points, maximum dates obtained from peat layers cannot be expected to be very accurate. The two ^{14}C dates from Svartisen could be compatible with 13th-century advances, but do not prove their existence decisively. Unfortunately, there is no documentary evidence available from northern Scandinavia about the state of the glaciers at that time.

Karlén (1973, 1979, Karlén & Denton 1976) mapped and dated numerous sets of moraines in the Kebnekaise and Sarek mountains of Sweden between 67° and 68°N, as well as in the Svartisen area of Norway. The lichenometric techniques employed were described in detail (Karlén 1975). The measurements made on both *Rhizocarpon geographicum* and *Rhizocarpon alpicola* thalli were consistent, drift units known to be older on a basis of geomorphological mapping bearing larger maximum diameter thalli, and maximum diameters on different subsections of the same drift units almost always giving similar results. Maximum lichen diameters on control surfaces of known age in the Sarek area agreed closely with those of similar age in the Kebnekaise mountains. Thus, in both regions, surfaces dating from 1900 to 1916 bore lichens 21–27 mm in diameter, while surfaces from the 17th century gave maximum values of 66–85 mm. Only a few surfaces of known age were available in the Sarek Park, and so, in view of the close agreement found in data from the two study areas, a lichen growth curve was constructed using 21 basic control points drawn from both the Kebnekaise and Sarek mountains and adjoining areas. Twelve of these were 20th century, mostly drift surfaces exposed at times known from historical records and photographs. The older control points were provided by copper and silver mine tips formed in the 17th and 18th centuries. Most of these were dated to a certain span of years, and two of them to alternative times. Thus one of the copper-mine tips at Sjangeli was formed between 1894 and 1902, and another either between 1745 and 1751 or between 1699 and 1702.

Similar studies made in the Svartisen region revealed that here too there have been a number of periods of glacial growth or stationary periods causing moraine formation during the last few centuries. The growth curve used to date the moraines of the Svartisen tongues and the little glaciers of the Okstinden and Saltfjell areas was based on the ages of 14 surfaces of known age, the great majority of them being late 19th-century or 20th-century mine tips. Only one pre-19th-century surface could be found, a tip at Nasa silver mine which was in operation between 1635 and 1659. (Karlén noted that lichen growth here may be slightly slower than it is around Svartisen because of the more continental location.) Karlén (1979) gathered lichenometric data from about 125 moraines, and concluded that the glaciers in northern Norway had reached advanced positions in the early 1300s and subsequently retreated. Further periods of glacial expansion were identified towards the end of the 1500s, and in the mid- and late 1600s: 'The results also permit relatively good dating of several of the youngest general periods of glacial maxima. These occurred at AD 1780, *c*. 1800, 1810–1820, 1860, *c*. 1880, 1900–1910 and *c*. 1930.' This summary of results compares closely with that from the Kebnekaise–Sarek region (Karlén &

Denton 1976) where 'individual Little Ice Age advances culminated about 1916–20, 1880–1890, 1850–1860, 1800–1810, 1780, 1700–1720, 1680, 1650 and 1590–1620.'

Karlén has unquestionably demonstrated the complex nature of the Little Ice Age in northern Scandinavia in the course of his impressive series of field investigations. Many of the periods of glacial expansion identified approximate with those identified in the Alps (e.g. Le Roy Ladurie 1971, Bray 1982), such as 1570–80, 1596–1616, 1676–80 and 1694–1724. Unfortunately, such coincidences of timing may only be apparent.

Lichen cannot start to grow on a moraine until recession of the ice has begun and the deposit is sufficiently stabilised. A lichenometric age therefore refers to the end of a period of expansion or a stationary period. Lichenometric dating is insufficiently precise to differentiate between the time of recession and the previous period of glacial expansion, except over periods covered by lichen growth curves so well based that narrow error limits have been obtained. The Swedish curve may be considered to meet this criterion as far as the last century or two are concerned, though no error limits were stated, but it cannot be expected that earlier Little Ice Age episodes can be dated even to the nearest decade or two in view of the lack of precision of the older control surfaces. Porter (1981b) based his growth curve for Mount Rainier on 19 control points, of which the most loosely dated was ± 5 years and, assuming measurement accuracy of ± 1 mm, estimated the accuracy of his dating at between ± 3 and ± 5 years for surfaces between 120 and 200 years old, rising to ± 10 to ± 20 years for surfaces between 200 and 400 years old. Construction of growth curves for northern Scandinavia was more difficult, involving dating from photographic evidence and the use of mining tips. It is questionable whether the Scandinavian curves are as accurate as the Mount Rainier curve, certainly for anything earlier than the 19th century. The record from Engabreen of an ice advance that carried away the farm Storstenørn, according to the general assessment made in 1723, and is also said to have 'come right down to the water's edge' 50 years earlier (Liestøl 1979), is scarcely sufficient to confirm the accuracy of the lichenometric data.

It is necessary to bear in mind that moraines may be formed not only when glaciers advance, but also when there is a halt during retreat. Some moraines are destroyed by later advances, so that the morphological record is incomplete and, as Rekstad (1901) noticed, differing local conditions may cause discrepancies in the number of moraines on the forefields of neighbouring glaciers. In order to obtain a well-based chronology, it is essential not only to obtain data from a very large number of moraine sets, as Karlén has so energetically succeeded in doing, but also to confirm the results using other forms of evidence.

The data presently available for northern Scandinavia suggest the possibility of glacial advances in the 13th, 16th, 17th, 18th and 19th centuries, the recent ones being more precisely dated than the earlier ones. In the Alps, strong correlations have been found between well-established chronologies of glacier oscillation and summer temperatures revealed by examination of tree rings using X-ray densitometry (Rothlisberger 1980). Extension of such studies to Scandinavia may well indicate the likelihood of 13th- and 15th-century advances more securely than is possible at present, and also assist in the recognition of cold periods within the last few centuries.

In the meantime, it is worth noting that while glacial retreat has predominated in both the Alps and Scandinavia since about 1850, there have been important differences in the course of events in the two regions on both longer and shorter time scales (Fig. 7.5). All the available evidence suggests that Scandinavian glaciers advanced between the late 17th and the mid-18th century to reach maximum positions which have never been regained since. In the Alps, mid-18th century maxima were in many cases not the greatest, and were

Figure 7.5 The known oscillations of representative glaciers from the Alps and Scandinavia after Faegri (1948), Grove (in preparation) and Röthlisberger (1980).

commonly exceeded in both the 17th and the 19th centuries. Between 1953 and 1980, the measured mass balances of Storglaciären in Kebnekaise and Hintereis-ferner in the eastern Alps have been in phase in some years, and in anti-phase in others (Hoinkes 1968, Schytt 1981, Fig. 7.6). This may be connected with the degree of meridional development of the cold trough in the upper westerlies over Europe, which has differed from year to year (Lamb 1968) but has not been fully explained. There are thus similarities in glacial behaviour in the two areas, but also important differences. This being so, the possibility of divergences of behaviour earlier in the historic period is reinforced.

7.4 Conclusions

The direct evidence of 13th or 14th century enlargement is tenuous for both northern and southern Scandinavia, but no counterindications have been identified. Field evidence which may indicate 16th- and early to mid-17th century advances is available for northern Scandinavia, but strong counterindi-cations exist for southern Norway. If either the glaciers in Scandinavia as a whole remained withdrawn while those in the Alps advanced, or those in southern Norway remained withdrawn while those in both northern Norway and Sweden and the Alps advanced, then meteorological or other explanations must be sought.

Gordon Manley (personal communication, December 1978) surmised that late expansion of Jostedalsbreen and other Norwegian glaciers might be accounted for in terms of low sea temperatures over the Norwegian Sea. The associated high pressure would deflect lows south-eastwards, giving more north-easterly winds over Norway. Manley (*loc. cit.*) noted that: 'Norway would be relatively dry compared with today . . . The reversal towards more

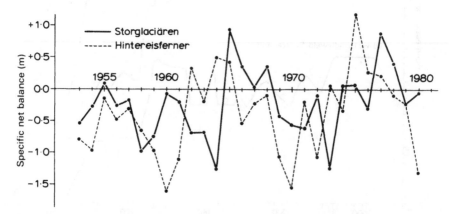

Figure 7.6 The specific net mass balances of Storglaciären and Hintereisferner (Eastern Alps) compared. The variations of Storglaciären are representative for the glaciers in Scandinavia (Schytt 1981). This in-phase-antiphase relationship has not been fully explained (Diagram after Schytt 1981.)

"cold westerly" weather after 1670 would then be assisted by the relatively cool sea surface which presumably still prevailed over the N.E. Atlantic. You'd need not only pretty cold S.W. winds in winter, but lots of instability showers with westerly winds in summer to get the snowline down on the Jostedalsbree.'

More information about temperatures over the Norwegian Sea has become available since Manley advanced this preliminary model (Lamb 1979), but details of summer temperatures, wind directions and precipitation over mainland Scandinavia must await further dendrochronological studies, together with serious analysis of documentary and phenological evidence. Even a preliminary survey of published literature revealed the possibilities of such an approach. Thus, in June 1644, Karl Olsen Svidal and Morit Berg, living in Jølster, just west of Jostedalsbreen, explained that they could not pay their taxes '. . . for there is such great misery here that very many in this little *skipreide* are starving and will soon die of hunger, some having to eat chaff mixed with bark instead of bread' (*Samlinger til det Norske Folks Sprog og Histoire V* 1838).

Crop failure evidently continued for some years, for in an open letter to the farmers of Norway dated 5 January 1648, Christian IV wrote that: 'Whereas it has become all too burdensome for Our dear subjects to continue to pay the *Six-Daler* [*Six-Daler* tax was presumably an occasional tax] because of crop failures, poor fisheries and cattle pestilence etc., it has graciously pleased Us to grant Our dear subjects the same relief granted them in previous years . . . ' (*Norske Rigsregistranter* 1645–8). The following year, Christian's successor Fredrik III wrote an open letter to the farmers of Norway which makes it clear that difficulties were widespread: 'with reference to Hannibal Schested's humble memorandum and explanation of the poverty of the common people because of crop failure, poor fisheries and cattle pestilence . . . We are pleased to grant them divers abatements and immunities . . . ' (*Norske Rigsregistranter* 1648–9). Hannibal Schested was *Stattholder* [Governor General of Norway] and his memorandum was in general terms for the whole country.

Conditions evidently continued to be very difficult during the next decade. In Dale Skipreide, in Indre Sogn, a reduction of *leidang*, one of the smaller taxes was granted to 53 farms in July 1653 (*Riksarkivet Oslo MSS*). This relief probably covered all the farms in the area. Unfortunately, the report of the assessment made in connection with it has not survived, but it is difficult to see what circumstances other than crop failure, and probably repeated crop failure, could have been responsible. We know that there was 'crop failure in Sogndal and Laerdal and great hunger in Auland' in 1661 from a diary kept by Iver Erikson Nordal (Finne-Grønn 1945), parson at Leikanger, Sogn. The following autumn 'there was very little hay or fodder'. Then, on the 8 September 1662, 'came the great floods which caused so much great damage . . . such as have not been known in the memory of man', recorded by Lensmann Nedrebø, who was born in April 1627, lived at Nedrebø in Jølster and conveniently kept a diary.

The mention of floods 'such as have not been known in the memory of man' in this one diary is certainly no clear indication of an increase in precipitation such as Manley was envisaging might have occurred in the latter part of the 17th century. However, the long record of famine and crop failure persisting over several decades does coincide with the period during which we have to presume the ice caps in southern Norway were thickening and building up before spilling over into the surrounding valleys. Unless we turn to a hypothesis that the ice caps in Norway required a century longer than the alpine glaciers before frontal advances took place, a surprisingly long lag period, we are forced to turn to meteorological explanations of the difference in timing. It must be hoped that a really thorough examination of the Norwegian records, such as Pfister (1981) has conducted in Switzerland, may shortly verify Manley's explanation or suggest a better one.

Acknowledgements

The translations from manuscript and printed sources on which this study depends were done by Arthur Battagel. A. T. Grove, Dr J. A. Matthews and Arthur Battagel read large parts of the manuscript and made helpful suggestions. The figures were drawn by Arthur Shelley, and the typing was done by Morag Meyer.

Manuscript references

Riksarkivet Oslo

Matrikkel over Nordfjord 1667, **44**
Matrikkel over Nordfjord, 1723, **147**, 89a–121a.
Rentekammerarkivet, Fogedregnskaper Sunnfjord–Nordfjord, 1702
Rentekammeret, Amtsregnskap, Bergens Stiftamt 1661, Pakke 2, folder 3.
Rentekammeret, Ordnings Avdeling Affaednings – Forretninger, Bergens Stiftamt. 1702–84, Pakke 3.
Rentekammeret Affaeldningsforretninger Hardanger og Sunnhords Fogderier, 1702–84.

Statsarkivet Bergen

(a) Indre Sogn Fogderi Tingbok No. 14a 1684, folios 38, 38b.
(b) Tingbok for Hardanger Fogderi 1667, folios 14b, 15.
(c) Nordfjord Sorenskriveri Tingbok 1734, folios 178–81.

References

Aaland, J. 1932. *Nordfjord Fran Gamle Dager til no Tid*. Einskilde Bygder. Innvik-Stryn II. E: Nemnd.
Aaland, J. 1973. *Nordfjord Fraa Gamle Dager til no Tid*. Einskilde Bygder. Innvik-Stryn. Sandane.
Andersen, J. L. and J. L. Sollid 1971. Glacial chronology and glacial geomorphology in the marginal zones of the glaciers Midtdalsbreen and Nigardsbreen, south Norway. *Norsk Geogr. Tiddskr.* **25**, 1–38.
Bell, W. T. and A. E. J. Ogilvie 1978. Weather compilations as a source of data for the reconstruction of European climate during the medieval period. *Climatic Change* **1**, 331–48.

Benedict, J. B. 1967. Recent glacial history of an alpine area in the Colorado Front Range, USA. I. Establishing a lichen growth curve. *J. Glaciol.* **6**, 817–32.

Bray, J. R. 1982. Alpine glacial advance in relation to proxy temperature index based mainly on wine harvest dates, AD 1453–1973. *Boreas* **11**, 1–10.

Damon, P. E., C. W. Ferguson, A. Long and E. I. Wallick 1974. Dendrochronological calibration of the radiocarbon timescale. *Am. Antiquity* **39**, 35–366.

Delibrias, G., M. Ladurie and E. L. Ladurie 1975. Le fossil forêt de Grindlewald: nouvelles datations. *Ann: Econ. Socs Civils.* **1**, 137–47.

Denton, G. H. and W. Karlén 1973. Lichenometry: its application to Holocene moraine studies in southern Alaska and Swedish Lapland. *Arct. Alp. Res.* **5**, 347–72.

Diplomatarium Norvegicum 1847. Vols. I–XX. Christiania.

Eide, T. O. 1955. Breden og Bygde. *Norske Tidds. Folkelivsg.* **5**, 1–42.

Elven, R. 1978. Subglacial plant remains from the Omnsbreen glacier area, south Norway. *Boreas* **7**, 83–9.

Faegri, K. 1948. On the variation of the western Norwegian glaciers during the last 200 years. *Procés-verbaux de Séances de l'Assemblé Genéral d'Oslo de l'Union Géodésique Internationale*, 293–303.

Finne-Grønn, S. E., (ed.) 1945. *Leganger'ske Optegnelser Fra Aarene 1621–1665.*

Griffey, N. J. and J. A. Matthews 1978. Major neoglacial glacier expansion episodes in southern Norway: evidences from moraine ridge stratigraphy with ^{14}C dates on buried palaeosols and moss layers. *Geogr. Ann.* **60A**, 73–90.

Grove, J. M. 1966. The Little Ice Age in the massif of Mont Blanc. *Trans Inst. Br. Geogs* **40**, 129–43.

Grove, J. M. 1972. The incidence of landslides, avalanches and floods in western Norway during the Little Ice Age. *Arct. Alp. Res.* **4**, 131–8.

Grove, J. M. and A. Battagel 1983. Tax records from western Norway, as an index of Little Ice Age environmental and economic deterioration. *Climatic Change* **5**, 265–82.

Hoel, A. and W. Werenskiold 1962. Glaciers and snowfields in Norway. *Norsk Polarinst. Skrift.* **114**, 1–291.

Hoinkes, H. 1968. Glacier variation and weather. *J. Glaciol.* **7**, 3–19.

Hovstad, H. 1971. Pest og Krise. *Den Norske Turistforeningenss Arbok*, 33–9 & 41–9.

Ingram, M. J., D. J. Underhill and G. Farmer 1981. The use of documentary sources for the study of past climates. In *Climate and history. Studies in past climates and their impact on man*, J. M. L. Wigley, M. J. Ingram and G. Farmer (eds), 180–213. Cambridge: Cambridge University Press.

International Study Group 1982. An inter-laboratory comparison of radiocarbon measurements in tree rings. *Nature* (Lond.) **298**, 619–23.

Karlén, W. 1973. Holocene glacier and climatic variations Kebnekaise Mountains, Swedish Lapland. *Geogr. Annlr.* **55A**, 29–63.

Karlén, W. 1975. Lichenometrisk datering i norra Skandinavien – metodens tillfortit-lighet och regionala tillampning. *Naturgeografiska Institutionen Forskningsrapport* **22**, 1–71.

Karlén, W. 1979. Glacier variations in the Svartisen area, northern Norway. *Geogr. Annlr.* **61A**, 1–28.

Karlén, W. 1981. A comment on John A. Matthew's article regarding ^{14}C dates of glacier variations. *Geogr. Annlr.* **63A**, 19–21.

Karlén, W. and G. H. Denton 1976. Holocene glacial variations in Sarek National Park, Northern Sweden. *Boreas* **5**, 25–56.

Kasser, P. 1967. *Fluctuations of glaciers 1960–65.* International Commission of Snow and Ice of the International Association of Scientific Hydrology and UNESCO.

Kasser, P. 1973. *Fluctuations of glaciers 1965–70.* International Commission on Snow and Ice of the International Association of Hydrological Sciences and UNESCO.

Klein, J., J. C. Lerman, P. E. Damon and E. K. Ralph 1982. Calibration of radiocarbon dates. Tables based on the consensus data of the workshop on calibrating the radiocarbon timescale. *Radiocarbon* **24**(2), 103–50.

Laberg, J. 1944. Jostedal. *Tidsskrift utgitt av Historielaget for Sogn* **11**, 5–85.

Laberg, J. 1948. Jostedal. *Tidsskrift utgitt av Historielaget for Sogn* **13**, 200–24.

Lamb, H. H. 1968. Glacier variation and weather: comments on Professor Hoinkes' paper. *J. Glaciol.* **7**, 129–30.

Lamb, H. H. 1979. Climatic variation and changes in the wind and ocean circulation: the Little Ice Age in the Northeast Atlantic. *Quatern. Res.* **11**, 1–20.

Le Roy Ladurie, E. 1971. *Times of feast, times of famine: a history of climate since the year 1000.* New York: Doubleday.

Liestøl, O. 1967. Storøbreen glacier in Jotunheimen, Norway. *Norsk Polarinst. Skrift.* **141**, 15–63.

Liestøl, O. 1973. Glaciological work in 1973. *Norsk Polarinst. Årbok* 181–92.

Liestøl, O. 1979. Svartisen. Fjell og Vidde. *Den Norske Turistforening Arbok*, 137–43.

Manley, G. 1966. The problem of the climatic optimum: the contribution of glaciology. In *World climate 8000–0 BC*, J. S. Sawyer *et al.* (eds), 34–9. London: Royal Meteorological Society.

Manley, G. 1974. Central England temperatures: monthly means, 1659–1973. *Q. J. R. met. Soc.* **100**, 389–405.

Matthews, J. A. 1974. Families of lichenometric dating curves from the Storbreen Gletschervorfeld, Jotunheimen, Norway. *Norsk. Geogr. Tidsskr.* **28**, 215–35.

Matthews, J. A. 1977. A lichenometric test of the 1750 end-moraine hypothesis: Storbreen Gletchervorfeld, southern Norway. *Norsk. Geogr. Tidsskr.* 31, 129–36.

Matthews, J. A. 1980. Some problems and implications of ^{14}C dates from a podzol buried beneath an end moraine of Hangabreen, southern Norway. *Geog. Annlr.* **62A**, 185–208.

Matthews, J. A. 1981. Natural ^{14}C age/depth gradient in a buried soil. *Naturwiss.* **68**, 472–4.

Matthews, J. A. 1982. Soil dating and glacier variations: a reply to Wibjörn Karlén. *Geog. Annlr.* **62A**, 15–20.

Messerli, B., P. Messerli, C. Pfister and H. J. Zumbühl 1978. Fluctuations of climate and glaciers in the Bernese Oberland, Switzerland, and their geoecological significance, 1600 to 1975. *Arct. Alp. Res.* **10**, 247–60.

Müller, F. 1977. *Fluctuations of glaciers 1970–75.* International Commission on Snow and Ice of the International Association of Hydrological Sciences and UNESCO.

Norske Lensrekneskapsboker 1938. *Skatten av Bergenhuslen 1563*, vol. III. Oslo: Riksarkivet.

Norske Rigsregistranter 1523–1660. Vols I–X. Christiania.

Østrem, G., N. Haakensen and O. Melander 1973. *Glacier atlas of Northern Scandinavia.* Norges Vassdrags – og Elektrisitetsvesen Meddelese 22 Fra Hydrologisk Avdelning, Oslo.

Østrem, G., O. Liestøl and B. Wold 1976. Glaciological investigations at Nigardsbreen Norway. *Norsk Geogr. Tidssk.* **30**, 187–209.

Øyen, P. A. 1899. Bidrag til vore Braeegnes Geografi. *Nyt. Mag. Naturviden.* **37**, 156–213.

Pálsson, S. 1795. Forsög til en physisk, geographisk og historisk Beskrivelse af de Islandiske Isbiærge. In *Ferdabók Sveins Pálssoner. Dagbaekur og ridabók 1791–1797*, 1945, J. Eythorsson (ed.), 425–552.

Pfister, C. 1980. The Little Ice Age: thermal and wetness indices for Central Europe. *J. Interdisc. Hist.* **10**, 665–96.

Pfister, C. 1981. An analysis of the Little Ice Age climate in Switzerland. In *Climate and* history. *Studies in past climates and their impact on Man*, J. M. L. Wigley, M. J. Ingram and G. Farmer (eds), 214–48. Cambridge: Cambridge University Press.

Porter, S. C. 1981a. Glaciological evidence of Holocene climatic change. In *Climate and history. Studies in past climates and their impact on Man*, J. M. L. Wigley, M. J. Ingram and G. Farmer (eds). 82–110. Cambridge: Cambridge University Press.

Porter, S. C. 1981b. Lichenometric studies in the Cascade Range of Washington: establishment of *Rhizorcarpon geographicum* growth curves on Mount Rainier. *Arct. Alp. Res.* **13**, 11–23.

Ralph, E. K., H. N. Michael and M. C. Han 1973. Radiocarbon dates and reality. *Masca Newsletter* **9**, 1–20.

Rekstad, J. 1901. Iagttagelse Fra Braeer i Sogn og Nordfjord. *Norges Geol. Unders. Aarbok* **34**, 1–48.

Röthlisberger, F. 1980. Tree-rings and climate – a retrospective survey and new results. World Meteorological Organisation. *WMO Bull.* **29**, 170–3.

Samlinger til det Norske Folks Sprog og Historie V. 1938. Christiania.

Schweingruber, F. H., H. C. Fritts, O. U., Bräker, L. G. Drew and E. Schär 1978. The X-ray techniques as applied to dendroclimatology. *Tree Ring Bull.* **38**, 61–91.

Schytt, V. 1981. The new mass balance of Storglaciaren, Kebnekaise, Sweden related to the height of the equilibrium line and to the height of the 500 mb surface. *Geogr. Annlr.* **63A**, 219–23.

Stuiver, M. 1978. Radiocarbon timescale tested against magnetic and other dating methods. *Nature* (Lond.) **273**, 271–4.

Thorarinsson, S. 1943. Oscillations of the Icelandic glaciers in the last 250 years. *Geogr. Annlr.* **25**, 1–54.

Tvede, A. M. and Liestøl, O. 1977. Blomsterskardbreen, Folgefonni, mass balance and recent fluctuations. *Norsk. Polarinst. Årbok 1976*, 13–21.

Webber, P. J. and J. T. Andrews 1973. Lichenometry: a commentary. *Arct. Alp. Res.* **5**, 295–302.

Wold, B. 1982. Breer i Norge. *Fjell og Vidde. Det Norske Turistforening Årbok*, 17–21.

8

Snow cover, snow lines and glaciers in Central Europe since the 16th century

CHRISTIAN PFISTER

Changes in the duration of snow cover in the lowlands of Switzerland from the 16th century are analysed from historical sources. The variability of thaw at higher altitudes from the 18th century is compared with the development of vine grapes in the lowlands, and with glacier fluctuations in the Alps. In the lowlands, positive extremes in the duration of snow cover were far more pronounced before 1840, due to low temperatures in March. The mean duration of snow cover was significantly higher in the coldest period of the Little Ice Age, from 1683 to 1700. For the alpine thaw, the long series of Saentis (2465 m) is analysed, comprising the periods 1821–51 and 1886–1980. During the climatic optimum (1920–53), Saentis became snow free around a month earlier than in the other periods considered. The late thaws in 1821–51 and in 1886–1919 are attributed to lower temperatures in May and June, whereas the delay in 1954–80 is primarily related to higher precipitation in winter and spring, i.e. more heat was required to melt the larger amounts of snow. The dates of thaw at similar altitudes are highly correlated with each other. The thaw on Saentis is also significantly correlated with the coloration of the Red Burgundy grapes in the lowlands, and with the percentage of advancing glaciers in the Alps.

8.1 Introduction

Snow cover has a considerable bearing upon a country's economy. Whether its effect is positive or negative depends upon the economic regime and the altitude of the area under study. The impact may change from one season to another.

It is important to distinguish between lowland and upland areas. Below altitudes of 800–1000 m above sea level, the snow cover usually melts away completely several times during the winter months. However, at higher altitudes, it remains from the first heavy snowfalls in late autumn or early winter until the thaw in spring or early summer (Gensler 1966, Primault 1969, Witmer 1983).

In the lowlands, the impact of snow cover is mainly negative. When it is deep, it blocks the highways, interrupts railway traffic, and brings building activity to a standstill. Considerable amounts of manpower and salt are needed to clear the roads. However, snow cover may protect seeds from severe frosts. In the uplands, the main source of income in winter until Easter is ski tourism, which relies on a snow cover of sufficient depth. After Easter the value of the snow cover changes, because summer tourism and farming become the dominant activities. A delay in snow-melt delays grass growth, which affects the supplies of hay and reduces the grazing time in the alpine meadows.

The impact of climatic fluctuations on the duration and variability of snow cover in the lowlands and on the time of thaw in the uplands will be considered – a theme pursued by Professor Gordon Manley in Britain. For the lowlands, several records at different places from 1683 will be examined. For the uplands, the long series at Saentis from 1886 to the present is compared with a similar set of observations from 1821 to 1851. An analysis is made of the temporal variability of the climatic factors that affect the melting date. The times of flowering and maturity of crops, particularly vines, in relation to the melting dates in higher altitudes are also investigated. In Switzerland, the first regular phenological observations date from the early 18th century; the earliest vine harvests date from the end of the 15th century (Pfister 1984).

Finally, the Saentis series is compared with the records of the advance and retreat of Swiss glaciers from 1880 to the present.

8.2   Dates of snow cover

The existence of a snow cover is essentially determined by visual observation, thus limiting recording to the area around the climatic stations. Only recently have satellites made it possible to note the changing distribution of snow cover on a global scale. The definition of snow cover, as laid down by the Swiss meteorological headquarters, allows for some variation of interpretation by the observer. A day with snow cover should be recorded if, at the time of the 07.30 h observation, the countryside surrounding the station at the same level is half covered with snow, a criterion which is not always easy to assess.

A fresh snow layer is an eye-catching meteorological element that will not be overlooked by an observer. It is therefore not surprising that snow cover is among the elements which are most frequently reported in the Swiss chronicles. But, according to the nature of this type of source, these records are limited to outstanding events (Pfister 1978).

The earliest diary in which snow cover is regularly reported was kept by Johann Heinrich Fries, who was a parson in Zürich. It began in July 1683 and ended in October 1718. Another reliable diary was kept in Winterthur (a town situated some 20 km north-east of Zürich) by Hans Rudolf Rieter, who was a baker and a member of the local government. It began in 1721, two years after the Fries diary ended, and covered the period up to 1738 (Pfister 1977).

Johann Jakob Sprüngli, a Bernese parson, was probably the first observer to record the formation and disappearance of snow cover at higher elevations. From 1766 to 1784, for instance, he noted the date on which the last snow patch melted on the mountains that he could observe: Gurnigel (1541 m, 35 km south of Berne) and Stockhorn (2100 m, 45 km south of Berne). Sprüngli also described the lower limit of every fresh-snow layer during the summer months. He regularly observed the flowering and ripening of well over 100 plants (Pfister 1975). The daily meteorological observations of Pater Haslinger (1796–1833) also contain detailed indications of the pattern of snow cover in the area of Linz, in Upper Austria (Mueller 1977).

The first statistics on snow cover may have been published by Johann Rudolf von Salis, who worked on the estate of his family in Marschlins near Chur (Canton of Grisons) (Salis 1811).

Unique data on snow limits were produced by Johannes Zuber, who was an engineer and cartographer in St Gallen. Zuber noted the daily altitude of the snowline between Lake Constance (386 m) and the top of Saentis (2465 m) from 1821 to 1851. Several frequently recurring altitudes in his data may refer to specific villages, forests, alpine meadows and summits. To monitor the entire area, Zuber collected data from a vantage point near St Gallen every day and obtained information from people living closer to Lake Constance and to the Saentis region (Denzler 1855). Monthly estimates of the duration of snow cover for various levels have been produced from his data that are all significantly correlated with the corresponding monthly mean temperatures in Basel (Pfister 1984).

In 1864, the *Schweizerische Naturforschende Gesellschaft* (forerunner of the Swiss Academy of Science) created a national network of weather stations that passed under the surveillance of the newly established National Weather Service in 1881. Strangely enough the data on snow cover were not published in the yearly annals until 1959. Schuepp *et al.* (1980) have recently put together some long series from the original observations, among which the secular series at Zürich (1881 to the present) deserves mention (Uttinger 1962). On the top of Saentis (2465 m), where a station was established in 1882, snow has always been observed and measured, but the record has not been published except for the last decades. The data from the manuscript sources have been kindly provided by Dr Ernst Ambühl, Bern. The station was, at the time of establishment, one of the highest in the world. Unfortunately, the places of observation and measurement have been transferred several times during the last 100 years. Up to 1970, the snow was measured in a place 150 m below the summit that was exposed to the south-east. Since 1971, the observations have been made at about the same altitude but on a north-west exposure (personal communication, Dr Gensler, Zürich). This shift might have somewhat delayed the recorded time of thaw compared to that noted in the earlier observations. On the other hand, the trend of the Saentis series over the past 10 years is in good agreement with other alpine stations at lower altitudes (see Section 8.3).

During World War II, an avalanche-monitoring service was created in the Swiss armed forces. In 1946/7, it passed to the direction of the Federal Institute of Snow and Avalanche Research (Davos). The snow-monitoring network of this service is mainly situated in the Alps, and at present comprises some 60 stations. A part of the huge body of unpublished data has recently been analysed by Witmer (1983), and data for a sample of stations are regularly published (EISLF).

8.3 Formation and melting of snow cover as a result of accumulation and ablation

In most studies, snow-melt has been analysed with reference to microclimatic factors and landforms (Kronfuss 1967, Meier & Schaedler 1979). Fine summaries of recent work are provided by Foijt (1974) and Witmer (1983).

In Switzerland, the number of days with snow cover increases with altitude in an almost linear way, according to the formula

$$N = 0.12h - 12$$

where N is the number of days with snow cover and h the altitude of the station above sea level (Schuepp et al. 1980). For specific locations, orographic factors (such as windward or leeward slope) and local climatic parameters (such as windiness, aspect and slope angle) must also be considered (Witmer 1983).

Following the formation of a snow cover, its persistence is related to the sequence and duration of weather situations favouring accumulation or ablation. Whether a particular situation has the characteristics of accumulation or ablation is mainly a function of altitude.

In *winter*, westerly and southwesterly weather situations generally produce rainfall in the lowlands (where they promote a melting of the snow cover), while at higher altitudes, where the precipitation falls in the form of snow, they give increased accumulation. In contrast, anticyclonic situations tend to preserve an existing snow cover in the lowlands, particularly when the sun is shielded by a layer of fog during inversions. In late winter and early spring, such situations often have a melting effect above the inversion, especially on steep slopes exposed to the south (Fliri 1975).

In *spring*, the limit between accumulation and ablation effects in westerly and southwesterly situations shifts upwards with the rise in temperature. The anticyclonic situations tend to have an ablative effect at every level. On the other hand, mean precipitation increases from March to May in the order of 10% to 60%. At higher altitudes, a considerable portion of this total falls in the form of snow, e.g. in April, 50% to 60% at altitudes of 1200 m, 86% at Rigi-Kulm (1737 m), 98% at Saentis (2465 m) (Uttinger 1933).

This helps to explain why, at altitudes above 1800–2000 m, the snow accumulation in spring is far more important than that in winter. In extreme cases, the fresh-snow layers in April and May may increase by 120 cm within 24 hours (Gensler 1966). The average maximum depth of snow from 1945/6 to 1969/70 at Sonnblick (3080 m) was recorded on 29 April, according to Lauscher and Lauscher (1971).

May is the most critical month for snow melt above 1800 m. On Saentis, snow still accounts for 80% of total precipitation (Uttinger 1933), and at

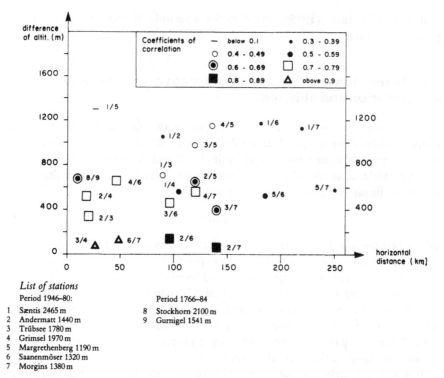

Figure 8.1   The covariation of the snow-melt between stations in the Swiss Alps above 1200 m as a function of their vertical and horizontal distances from one another.

Sonnblick Observatory (3080 m) for almost 100% (Lauscher & Lauscher 1971). However, the amount of snow that is melted in anticyclonic warm and sunny periods may exceed 60 cm within 24 hours (Gensler 1966). In June and July, warm thunderstorm rains, together with radiation and high temperatures, promote thaw up to fairly high levels. In contrast, an influx of subpolar air masses often brings fresh snowfall down to altitudes of 1500–1200 m. On the northern slope of the Swiss Alps, the temporary snowline rises (on average) 100 m every 8 days, while in the inner valleys (Engadin, Wallis) the shift is 100 m every 6 days (Gensler 1966). In the Austrian Alps, the rise of the temporary snowline is more gradual at altitudes above 1500 m than below (Lauscher 1975).

Figure 8.1 shows the correlation matrix of the melting dates of seven 'contemporary' and two 'historical' stations as a function of their horizontal and vertical distances from one another. Obviously, the agreement of the melting dates for two stations depends primarily on their difference of altitude. The dates for stations at nearly the same elevation, such as Andermatt and Morgins, are highly correlated, although their horizontal distance apart is almost 150 km. On the other hand, the correlation between Margrethenberg (1190 m) and Saentis (2465 m) is near zero, although the two places are within a horizontal distance of less than 30 km.

8.4 The duration of snow cover in the lowlands since the 16th century

Mueller (1977) found a remarkably similar pattern of snow cover when he compared the figures computed from the Haslinger diary (1796–1833) with the snow-cover conditions of today. The Zurich series from 1881 to 1960 (Uttinger 1962) and the Munich series from 1887 to 1977 (Baumgartner & Mayer 1978) do not disclose any significant trend. Can it be concluded, therefore, that long-term fluctuations in the duration of snow cover do not correspond to changing climate?

Figure 8.2 displays the positive and negative extremes of snow duration at altitudes between 400 m and 500 m that have been observed in the last 450 years, in conjunction with indications of the freezing of lakes in the alpine borderland (Pfister 1984). The bars pointing to the right refer to snow duration, those pointing to the left to the freezing of lakes and rivers.

The maximum figures for the duration of snow cover are between 80 and 90 days from 1840 to the present. However, from 1560 to 1840, this level was exceeded in at least 25 winters. In the two longest winters – 1613/14 and 1784/5 – the snow cover lasted (without interruption) for about five months, and did not melt before 20 April. An extended duration of snow cover points to very low temperatures in March and April. The most extreme example documented with evidence from thermometers is March 1785, when the mean temperature was 8°C below the 1901–60 average in Basel (Bider *et al.* 1958). A 1.2°C long-term warming of March has been measured from the end of the 19th century. This may help to explain why a permanent snow cover in March has not been recorded at altitudes below 500 m in the present century.

During very snowy winters, the lakes in the alpine borderland are often covered with ice. It has been established that the lakes freeze in a specific rank order (Pfister 1984) according to their surface area, depth, and individual characteristics. The freezing of lakes is primarily a function of the sum of below-freezing-point daily mean temperatures and other factors such as wind speed. The record of the freezing of the Zurichsee is best known. The bars (in Fig. 8.2) of medium length give the years in which the surface of this lake was covered with ice. The longest bars stand for the freezing of great or deep lakes, such as Lake Constance. In those years, heavy cargoes of ice could be carried on the smaller lakes, and between 1680 and 1700, freezing and snowy winters occurred very often (Pfister 1984). This period of the coldest winters shows up also in the freezing figures for the Dutch canals (De Vries 1977) and in the central England temperature series (Manley 1974). The smallest bars indicate the winters in which only small lakes were covered with ice.

A very small number of days with snow cover may accompany warm winters, but may also coincide with extended dry and cold anticyclonic situations. For the pre-industrial periods, the hypothesis of warmth needs, therefore, to be supported with observed signs of vegetation activity. Between 1590 and 1615, the rapid oscillations between short and long winters indicate a high degree of variability.

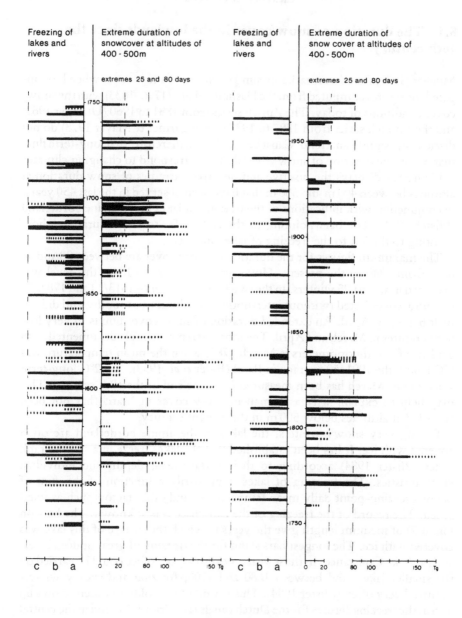

Figure 8.2 Extreme duration of snow cover in the lowlands, and freezing of lakes in the northern borderland 1525–1970. $Tg = Tag$ (day).

Table 8.1 Variations in the average number of days with snow cover for selected periods.

Place	Years	Snow cover (days)	Source
Zürich (430 m)	1683/4–1699/1700	70	Pfister (1977)
Winterthur (445 m)	1721/2–1737/8	54	Pfister (1977)
Zürich (430 m)	1880/1–1899/1900	52	Uttinger (1962)
Zürich (570 m) *	1963/4–1982/3	45	*Annalen SMA*

* Location of the National Weather Service. The values are reduced to the old location.

Mean figures for the duration of snow cover before the creation of the National Weather Service can be estimated from detailed weather diaries for Zürich and Winterthur (Table 8.1). The mean value of 70 days for the period 1683–1700 coincides with the coldest winter decades of the Little Ice Age; a duration of somewhat more than 50 days can be taken as proxy for the average winter and early spring in the Little Ice Age, while 45 days marks the level of the last two decades. The estimated winter temperature for the 1683–1700 period is 3.5°C below the average for 1963/4–1982/3 (Pfister 1984). The significant warming in winters in the 20th century can also be derived from the falling trend in the number of days with snow cover in Zürich (Fig. 8.3), although the picture is somewhat masked by the inclusion of snow cover in November, March and April. A significant increase in precipitation accompanied the warming of the winter months, and this has promoted a recurrent melting of the snow cover in the lowlands.

8.5 Changes in the seasonality of snowlines

In Table 8.2 the average date of alpine thaw is shown for three 'historical', and four recent series. Up to about 1400 m the snow melted later in the Little Ice Age Period than in the 20th century, which can be attributed to lower temperatures in March and April. For the upper levels a more detailed analysis is needed. Figure 8.4 shows a comparison between the mean seasonal rise of the temporary snowline at stations on the northern slope of the Alps from 1913 to 1922 (Lugeon 1928) and the average Saentis profile over the period 1821–51. The two lines are roughly parallel up to a level of 2000 m. In the higher regions, the averages shift considerably over time.

The recent Saentis series (Fig. 8.5) can be divided into three subperiods: 1886–1919, 1920–53 and 1954–80. In 1920–53, Saentis became snow free about a month earlier than in 1886–1919 and in 1954–80. The differences are statistically highly significant (<0.001). Correspondingly, the 1886–1980 Saentis series comprises a rising trend from 1886 to 1919 that becomes U-shaped in the next six decades. The averages of the period 1920–53, i.e. at the bottom of the U, are significantly (<0.001) different from those for 1886–1919 and 1954–80. Com-

Table 8.2 Average dates of the alpine thaw during the Little Ice Age period and in the 20th century.

Place	Altitude (m)	Period	Mean	Number of observations	Standard deviation	Source
Margrethenberg (Saentis area)	1190	1954–80	28 March	26	25	EISLF
	1246	1821–51	2 May	29	17	Denzler (1855)
						Pfister (1984)
Morgins (W. Alps) (Saentis area)	1380	1958–80	25 April	19	18	EISLF
	1500	1821–51	23 May	29	19	Denzler (1855)
						Pfister (1984)
Gurnigel* (Saentis area)	1541	1766–84	6 June	19	18	Pfister (1975)
	1800	1821–51	9 June	29	19	Denzler (1855)
						Pfister (1984)
Trübsee (Central Switzerland) (Saentis area)	1780	1958–80	11 June	27	15	EISLF
	2100	1821–51	28 June	29	22	Denzler (1855)
						Pfister (1984)
Pilatus (Central Switzerland)	2121	1930–41	27 May	12	—	Roshardt (1946)
Stockhorn* Saentis	2100	1766–84	23 August	19	19	Pfister (1975)
	2465	1821–51	2 July	29	22	Denzler (1855)
						Pfister (1984)
		1886–1919	13 July	34	23 ⎫	National Weather Service,
		1920–53	14 June	34	16 ⎬	original
		1954–80	7 July	26	24 ⎭	observations

*Melting of last snow patch.

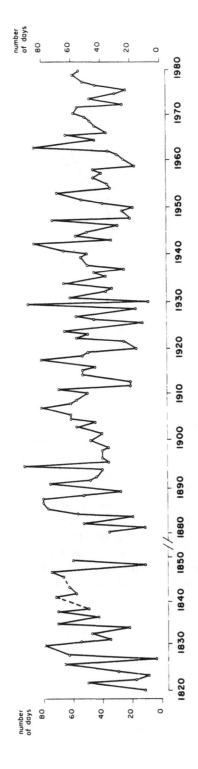

Figure 8.3 Number of days with snow cover in north-east Switzerland (Saentis Region) at 430 m, 1820–51 and in Zürich 1881–1981.

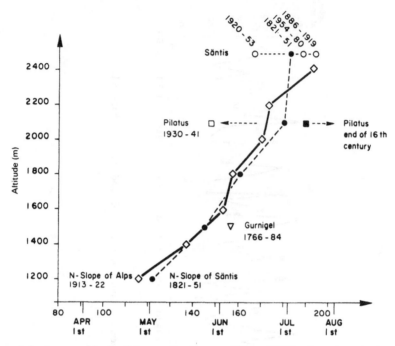

Figure 8.4 Average altitude of the snowline between 1200 and 2500 m from the end of the 16th century to 1980.

pared to the period 1886–1919, Saentis became snow free on average around a month earlier in 1920–53. The average snow-melting on the summit of Pilatus (near Lucerne) from 1930 to 1941 was advanced by a month compared to the corresponding level in the Saentis region from 1821 to 1851. On the other hand, the thaw occurred 10 days or more later in the final decades of the 16th century than it did in 1821–51. The scientist and statesman Renward Cysat (1545–1613), who used to climb the mountains around his native Lucerne several times every year, reported at the end of the summer of 1608 (one of the poorest in the last 500 years) that the top of Pilatus did not become snow free before 9 September, whereas this usually occurred in July. The data of this reliable observer suggest that, around 1600, the snow cover above 1800–2000 m melted more than 35 days later than it has done during the climatic optimum of the 20th century.

Which are the climatic factors that can account for the observed differences in the time of thaw? In order to deal with this problem, climatic series have been analysed using stepwise multiple regressions. For the monthly mean temperatures, the Basel series was again taken, whereas precipitation was documented from wetness indices for the Swiss Plateau (Pfister 1984). The results are shown in Table 8.3.

In the period 1821–51, the time of thaw was mainly a function of temperature. At 1500 m, the temperatures in May and April were dominant; on Saentis the

Table 8.3 Climatic factors affecting the snow-melt above 1300 m.

Place	Altitude (m)	Period	Average	April	May	June	July	Winter	Spring	r^2
Morgins (W. Alps)	1380	1959–80	25 April	2 (−)					1 (+)	0.44
Saentis area	1500	1821–51	23 May	2 (−)	1 (−)					0.35
Trübsee (central Alps)	1780	1958–80	11 June		1 (−)				2 (+)	0.38
Saentis	2465	1821–51	2 July		1 (−)	2 (−)		3 (−)		0.36
		1886–1919	13 July		2 (−)	1 (−)	3 (−)			0.42
		1920–53	14 July		2 (−)	1 (−)			3 (+)	0.36
		1954–80	7 July		1 (−)			2 (+)	3 (+)	0.35

The numbers indicate the rank order of the relevant factors in the regression; the signs give the direction of the relationship.
r^2 = total of variance explained.

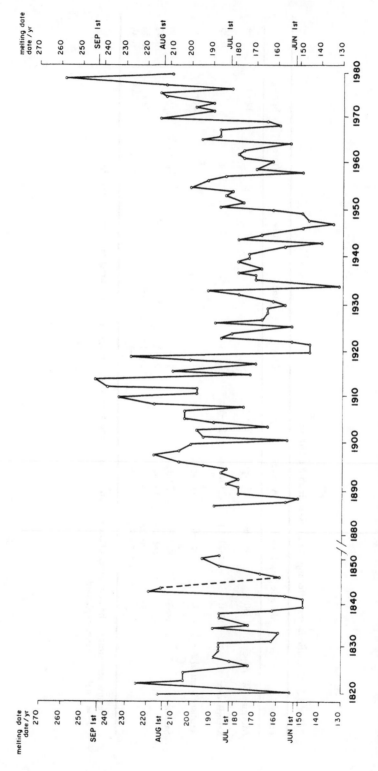

Figure 8.5 The date of the thaw on Saentis (north-east Switzerland) at 2465 m, 1820–51 and 1886–1981.

dominant temperatures were those of May and June. The negative correlation with winter precipitation might be a spurious result.

In 1886–1919, the date of thaw is significantly related to temperatures from May to July, and in 1920–53, primarily to temperatures in May and June. However, in the last period, the importance of accumulation in winter and spring is more important than the temperatures in June and July. The following pattern emerges when the climatic changes that have occurred over the four periods are considered.

(a) In 1886–1919, the snow cover on Saentis disappeared 12 days later on average than during the period 1821–51. This may mainly be explained by the significant 8.5% increase in spring precipitation in 1886–1919.

(b) In the second case, the delay in thaw may be connected with a slight lowering of temperatures in April and July. Obviously, the effects of the increased accumulation in the spring period were enhanced by a less effective ablation in midsummer.

(c) A significant warming (at the 0.5 level) of the critical months from April to July promoted an early and intensive ablation, in the period 1920–53, that may have been the climatic optimum of the last 500 years.

(d) In 1954–80, the average date of thaw was again 23 days later than in the previous period. This setback was accompanied by a significant (at the 0.5 level) 5% increase in winter precipitation, and a slight cooling in April and July.

The snow cover disappeared six days later than in the Little Ice Age (i.e. 1821–51), although April, May and June were somewhat warmer. The thermal gradient was not sufficient to offset the large positive differences of 27% in winter and 15% in spring between the two periods. The deficit in winter and spring precipitation before 1900 is not limited to the period 1821–51. It is, in conjunction with lower temperatures, the most typical feature of the Little Ice Age climate since 1525 (Pfister 1984). Therefore, the average date of thaw may not differ very much between the Little Ice Age period and the last three decades. However, the underlying patterns of ablation and accumulation are somewhat different. In the Little Ice Age period, accumulation in winter and spring was smaller on average than in the 1901–60 period. Hence, less heat and radiation were required to melt the snow cover at altitudes above 1300–1500 m. However, the considerable delay of the thaw in the last 25 years is far less related to a cooling of summers than to marked increases of precipitation in spring and winter.

8.6 Patterns in plant development and dynamics of the temporary snowline

Reliable series of the date of thaw are quite rare for the time before 1945. For this reason it may be that historical proxy data such as phenological observations, that go further back in time, can be used. During the disappearance of the snow cover in the mountains, the same pattern of snow patches appears every year. Some patches may take the form of animals and human faces, and rural communities noticed that the appearance of such a sign often preceded the maturity of a certain crop (Kronfuss 1967). The same author has demonstrated close relationships between phenophases and the rise of the snowline on mountains.

The different stages in the development of vine grapes are particularly appropriate for comparison because they are well documented in the historical past. The beginning of the vine bloom occurs largely in response to temperature conditions in May, while the start of coloration of the Red Burgundy grapes is linked to temperatures for May to July (Pfister 1984).

In Table 8.4, the average dates of phenophases and the altitudes of snowlines are compared, and coefficients of correlation are given. The average beginning of the vine bloom in the vineyard of Hallau (Canton of Schaffhausen, northern Switzerland) coincides roughly with the average thaw at 1800 m. The coloration of the Red Burgundy grapes in the same vineyard is taken for comparison with the melting date on Saentis for the period from 1886. Similar data from a vineyard at the shore of the Zurichsee are compared with the disappearance of the last snow patch on the mountains near Thun for the 18th century.

According to von Rudloff (1967), the time of melting of snow patches provides a good yardstick for the analysis of climatic changes. All phenophases are significantly correlated with the corresponding melting dates, which confirms the graphical comparison in Fig. 8.6. From the regression models, it can be concluded that the beginning of coloration of Red Burgundy grapes has a higher predictive power for the melting date on Saentis than does any individual monthly temperature mean.

In order to focus more closely upon the relationship, those years in which the thaw was far delayed or advanced with respect to the blossoming and the maturation of crops need to be given special attention. In 1770, the thaw was so far delayed that the snow cover was not completely melted on the Stockhorn range (2100 m) before the first fresh-snow layer in autumn (cf. Fig. 8.4). Contemporary reports point to the conclusion that this extreme delay must be associated with an extremely high amount of snow accumulated at higher altitudes rather than with a pronounced thermal deficit during the summer months (Pfister 1975). These large amounts of snow were accumulated during an almost uninterrupted cold and wet period that began in February and ended in the first days of May. In June and July, temperatures were somewhat below the 1901–60 average, and the melting was hindered by recurrent fresh-snow layers. Obviously, the ablation-producing weather situations were too short and not pronounced enough to melt the huge masses of snow. In this year, the temporary snowline may have remained 300–500 m below its present average. In 1816, (the 'year without a summer'), the lowering of the snowline was even more pronounced: on slopes exposed to the south above 2300 m, and on those exposed to the north above 1800–2000 m, the snow cover did not melt at all, and summer temperatures were extremely low (Messerli et al. 1978).

Table 8.4 Phenophases and the temporary snowline.

Period	Phenophase	Place	Average	Place (snowline)	Average	r
1958–80	wine bloom begins	Hallau SH	12 June	Trübsee (1780 m)	11 June	0.62
1766–84		Zollikon ZH	9 June	Gurnigel* (1541 m)	6 June	0.78
1886–1980†	coloration of Red Burgundy grapes	Hallau SH	12 August	Saentis (2465 m)	2 July	0.53
1766–84		Zollikon ZH	14 August	Stockhorn* (2100 m)	23 August	0.64
1766–84	average harvest date for barley, rye and spelt	Gurzelen BE	14 July	Stockhorn* (2100 m)	23 August	0.84

* Melting of last snow patch.
† The series includes several gaps.
r = coefficient of correlation. Significance: <0.001 in all cases.
Sources: Pfister (1975, 1984).

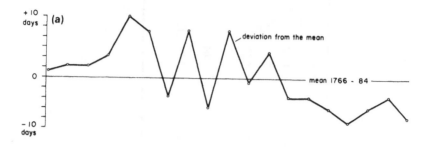

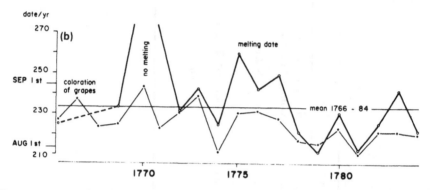

Figure 8.6 Plant development in the lowlands and snow-melt in the Alps 1766–84. (a) Mean harvest date of barley, rye and spelt at Gurzelen BE (591 m), showing deviations from the mean. (b) Coloration of the Red Burgundy grapes at Zollikon ZH (408 m), and the melting of the last snow patch in the Stockhorn range (2000 m).

The year 1779 was the only one in which the snow-melt on the Stockhorn range preceded the coloration of the grapes in Zolliken (see Fig. 8.4). Winter and spring had been exceptionally dry; total rainfall in Bern from January to May was only 54% of the 1901–60 average, and the level of the lakes in the alpine borderland was extremely low (Pfister 1975). The snow layer on higher levels was probably very thin, and it had therefore vanished in that year before the grapes began to mature. It may be hypothesised, that snow–melting and plant development are roughly parallel in warm and dry situations. However the threshold, above which grape development begins, may not be identical with the critical values of heat and radiation required for ablation. Also, the amount of heat required to melt a snow cover varies according to the initial depth of the snow, while it remains nearly constant for promoting a given phenophase.

In order to focus more closely upon the process of ablation and plant development, the lag between the snow melt on Saentis and the coloration of the Red Burgundy grapes in Hallau (Table 8.4) has been examined for the years with a snow melt after 10 July and for those with a snow melt before 20 June. The lag was 27 days in late years, but 69 days in early years. The differences between the means are highly significant. The later Saentis becomes snow-free,

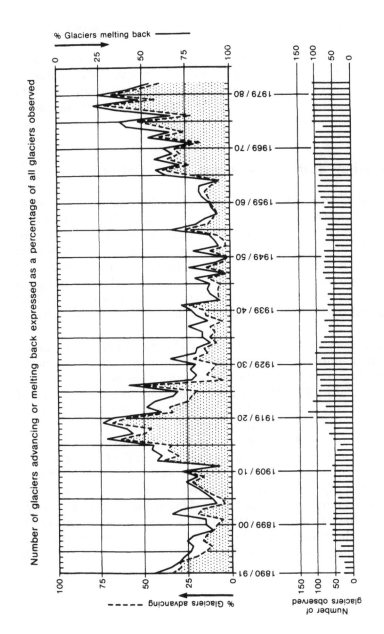

Figure 8.7 Fluctuations of the glacier snouts in the Swiss Alps, 1890–1 to 1981–2.

the smaller the delay in the colouring of grapes. The slow down of ablation with respect to the development of grapes in cool summers could be attributed to the different effect of cold and wet situations. While plant growth is only halted, the snow-cover is increased by accumulation, which is equivalent to a setback in the melting process. This may help to explain why the beginning of coloration of Red Burgundy grapes can only account for 28% of the total variance in the melting date on Saentis. Perhaps a more refined model could provide better estimates.

8.7 Fluctuations of snowlines and glacier variations

The relationship between glacier fluctuations and melting dates can be investigated from the Saentis series. Considering the percentage of glaciers retreating or advancing from 1886 to 1980 (Gletscherbericht 1983, Fig. 8.7), the similarities with the Saentis curve (see Fig. 8.5) are striking. The coefficient of correlation is 0.55 if a time lag of 3 years is included to allow for the reaction of the glacier snouts (Gamper & Suter 1978). Also, Janin (1968) has found good agreement between the total snow accumulation on the Grand St Bernhard from 1851 to 1961 and the fluctuations of the snout of the Glacier du Lys (French Alps).

Glacier fluctuations are explained mainly as a function of temperatures and weather conditions during the potential period of ablation (i.e. May to September). Ablation begins as soon as the glacier surface has become snow free (Hoinkes 1967, Vivian 1975). During the period of ablation, the frequency of fresh-snow layers is decisive (Fliri 1975, Messerli et al. 1978). Earlier, it was demonstrated, that the snow-melt at two stations separated by a small vertical distance is highly correlated over considerable horizontal distances. Therefore, the date of thaw on Saentis may roughly coincide with the beginning of the ablation period on glacier surfaces at altitudes between 2300 m and 2700 m.

In conclusion, at higher altitudes the increase in winter and spring precipitation in the 20th century has not only retarded the melting of the snow cover, but has also contributed to promoting the glacier advances in the last decade. On the other hand, the warming of the winters, together with the increased precipitation, accompanied a dwindling duration of snow cover in the lowlands.

Acknowledgements

Acknowledgements are due to Dr Ernst Ambühl, Liebefeld-Bern, and Dr Markus Aellen, Institute for Glaciology, ETH, Zurich, for providing data: to Professor Roger Barry, Boulder (Colorado, USA), for reading the manuscript; to Dr G. Gensler, Schweizerische Meteorologische Anstalt, Zurich, for making helpful suggestions; and to Ralph Rickli, Geographisches Institut der Universität Bern, for drawing the figures.

References

Annalen SMA. 1961–79. *Annalen der Schweizerischen Meteorologischen Zentralanstalt.* Zürich.

Bider, M., M. Schuepp and H. von Rudloff 1958. Die Reduktion der 200 jährigen Basler Temperaturreihe. *Arch. Met. Geophys. Bioklim.* **B 9**, 360–412.

Baumgartner, A. and H. Mayer 1978. Die Schneedecke in München von Oktober 1887 bis April 1977. *Met. Rundschau* **31/4**, 6–16.

Cysat, R. 1969. *Collectanea pro Chronica Lucernensi et Helvetiae.* Luzern: J. Schmid.

De Vries, J. 1977. Histoire du climat et économie. Des faits nouveau une interprétation différente. *Annales: économies, sociétés, civilisation* **32**, 198–226.

Denzler, H. 1855. Die untere Schneegrenze während des Jahres vom Bodensee bis zur Säntisspitze. *Neue Denkschriften der allg. Schweiz Gesellschaft für die gesammten Naturwissenschaften* **14**, 1–59.

EISLF 1936/7–1979/80. *Schnee und Lawinen in den Schweizer Alpen. Winterberichte des Eidgenöss.* Instituts für Schnee-und Lawinenforschung, Weissfluhjoch. Davos. nos. 1–44.

Fliri, F. 1975. Das Klima der Alpen im Raume von Tirol. VIII. Die Schneedecke. *Monographien zur Landeskunde Tirols, Innsbruck–Munchen*, 247–74.

Foijt, W. 1974. Die Schneedecke im Erzgebirge. *Abh. des Met. Dienstes der DDR*, 14/111. Berlin.

Gamper, M., and J. Suter 1978. Der Einfluss con Temperaturänderungen auf die Lange von Gletscherzungen. *Geograph. Helvet.* **43**, 183–9.

Gensler, G. 1966. Typische Schnee- und Witterungsverhältnisse in Frühling in den Alpen: Typische Schnee- und Witterungsverhältnisse in Sommer in den Alpen. *Die Alpen* **62**, 50–54, 81–5.

Gletscherbericht 1983. Die Gletscher der Schweizer Alpen 1975/76 und 1976/77. *Glaziologisches Jarhbuch der Gletscherkommission der Schweiz. Naturforschenden Gesellschaft SNG*, hg, durch die Versuchsanstalt für Wasserbau, Hydrologie und Glasiologie/ VAW der Edig. Teschn. Hochschule Zürich ETHZ.

Hoinkes, H. 1967. Gletscherschwankung und Wetter in den Alpen. *9. Internat. Tagung f. alpine Meteorologie 14–17.9.1966*, Schweizerischen Meteorologischen Zentralanstalt (ed.), 9–24. Zürich.

Janin, B. 1968. *Le col du Grand Saint Bernard. Climat et variations climatiques.* Thèse complémentaire pour le Doctorat ès Lettres. Aosta.

Kasser, P. 1981. Rezente Gletscherveränderungen in den Schweizer Alpen. *Gletscher un Klima Jahrb. der Schweiz. Naturf. Gesellschaft* (1978), 106–38.

Kronfuss, H. 1967. Schneelage und Ausaperung an der Waldgrenze. *Mitt. der forstl. Bundesversuchsanst.* **75**, 207–41.

Lauscher, A. and F. Lauscher 1971. Der Aufbau und Abbau der Schneedecke auf dem Sonnblick im Wechselspiel der Wetterlagen. *Jahresber. des Sonnblick-Vereins 1970–71*, 1–30.

Lauscher, F. 1975. Die Zeitpunkte grösster Schneehöhen in den Ostalpenländern. *Wetter und Leben* **27**, 26–30.

Lugeon, J. 1928. *Le cycle des précipitations atmosphériques.* Neuchâtel: Borel.

Manley, G. 1974. Central England temperatures: monthly means 1659 to 1973. *Q. J. R. met. Soc.* **100**, 389–405.

Meier, R. and B. Schaedler 1979. Die Ausaperung der Schneedecke in Abhängigkeit von Strahlung und Relief. *Arch. Met. Geophy. Bioklim.* **B 27**, 151–8.

Messerli, B. *et al.* 1978. Fluctuations of climate and glaciers in the Bernese Oberland, Switzerland, and their geoecological significance, 1600 to 1975. *Arct. & Alp. Res.* **10/2**, 247–60.

Mueller, W. 1977. Gab es im fruhen 19 Jahrhundert in Oberösterreich häufiger Schnee als in der Gegenwart? *Wetter und Leben* **29**(2), 75–82.

Müller, G. 1982. Die Wetterstation auf dem Säntis. In *Hudert Jahr Wetterwarte Säntis*, Schweiz. Meteorolog. Anstalt (ed.), 5–15. St Gallen.

Pfister, Ch. 1975. Agrarkonjunktur und Witterungsverlauf im westlichen Schweizer Mittelland. *Geogr. Bernensis* **G** 2, Bern.

Pfister, Ch. 1977. Zum Klima des Raumes Zürich im späten 17. und frühen 18. Jahrhundert. *Vierteljschr. natf. Ges. Zürich* **122**, 447–71.

Pfister, Ch. 1978. Fluctuations in the duration of snow cover in Switzerland since the late seventeenth century. *Proceedings of the Nordic Symposium on Climatic Changes and Related Problems*. Danish Meteorological Institute, Climatological Papers **4**, 1–6. Kobenhavn.

Pfister, Ch. 1984. *Das Klima der Schweiz von 1525 bis 1863 und seine Bedeutung in der Geschichte von Bevölkerung und Landwirtschaft*. Habil. Schrift, University of Bern.

Primault, B. 1969. *La durée de l'enneigement dans le Canton de Vaud*. Documents de l'Aménagement Régional. Office Cantonal Vaudois de l'Urbanisme, 6.

Roshardt, A. 1946. Der Winter in der Innerschweiz. Eine vergleichende Studie auf Grund zwanzigjähriger Beobachtungen. *Mitt. der naturf. Geselleschaft Luzern* **15**, 107–247.

Rudloff, H. von 1967. *Die Schwankungen und Pendelungen des Klimas in Europe seit dem Beginn der regelmässigen Instrumenten-Beobachtungen (1670)*. Braunschweig.

Salis, J. R. 1811. Einige Resultate aus sechs- und zwanzig-jährigen Witterungsbeobachtungen in Marschlins. *Der neue Sammler. Ein gemeinnütziges Archiv für Bünden*, hg. von der Oekonomischen Gesellschaft (Chur), 193–211.

Schuepp, M., G. Gensler, and M. Bouet 1980. Schneedecke und Neuschnee. *Klimat. der Schweiz* 24, Zürich.

SMA 1887–1979. *Annalen der Schweizerischen Meteorologischen Zentralanstalt*. Zürich.

Uttinger, H. 1933. Die Schneehäufigkeit in der Schweiz. *Ann. Schweiz Met. Zentr. Anst.* **70**, 1–8.

Uttinger, B. 1962. Die Dauer der Schneedecke in Zürich. *Arch. Met. Geophys. Bioklim.* **B 12**, 404–21.

Vivian, R. 1975. *Les glaciers des Alpes Occidentales*. Grenoble: Allier.

Witmer, U. 1983. *Eine Methode zur flächendeckenden Kartierung von Schneehöhen unter Berücksichtigung von relief-bedingten Einflüssen, Bern*: Diss. phil. nat. Geogr. Institut der Universität Bern.

9

Peat stratigraphy and climatic change: some speculations

KEITH E. BARBER

Questions about the relationship between climatic change and variations in peat stratigraphy fall into two groups. Temporal questions concern the length of the climatic record available from peat samples, the magnitude of climatic variations that can be detected and the problem of recurrence surfaces. Behavioural questions are related to the mechanisms by which peat responds to changes in exposure, temperature and precipitation. The possible reasons why some bogs are climatically more sensitive than others are investigated.

9.1 Introduction

Peat bogs are unique plant communities. Not only do they exhibit a special relationship with climate and its variations, but, alone amongst terrestrial ecosystems, they also form a record of that relationship in the peat itself. A single shoot of the bog moss, *Sphagnum*, may not look like a potent agent of environmental change, but when packed together in their millions on the surface of a wet bog or mire, they have the capacity to change the local environment completely. They do this by means of special leaf cells which absorb water; by extracting mineral nutrients such as calcium from the water and giving up hydrogen ions in exchange, they acidify the water around them. Walker (1970) showed how the succession of plant communities in small lake basins in Britain could be influenced by the ingress of *Sphagnum* species. The climax stage of such hydroseres was thought to be deciduous woodland of oak, beech and lime, at least in the sub-oceanic climate of lowland England. By examining the stratigraphic record in such basins, Walker was able to demonstrate that once *Sphagnum* mosses (of which there are some 40 species in north-west Europe) invaded the hydrosere, they changed it rapidly into a raised bog ecosystem. Such ecosystems appear to have been fairly stable end-points of successions for the past 7000 years; that is, for well over half of the period covered by postglacial time. In accumulating peat above the groundwater table, such bogs become dependent upon rainfall, or, more precisely, upon the ratio of precipitation to evaporation, to sustain their growth, in contrast to valley bogs and some parts of blanket bogs. These last two types of peatland are also dominated by *Sphagnum* mosses – except where man's activities in polluting the atmosphere and draining or burning the peatlands have made conditions unsuitable – and though there must be some kind of climatic 'signal' in their peat, the 'noise' of groundwater inputs and mineral flushes in such systems obscures the evidence. It may still be possible to discern moister and drier episodes of peat formation in such mires by other means, including the use of microfossil indicators (van Geel et al. 1981), but almost all the work in reconstucting past climate from peat stratigraphy has been undertaken on raised or ombrotrophic bogs.

The relationship of bog growth and climate change, based on peat stratigraphic researches at Bolton Fell Moss, Cumbria, has already been considered (Barber 1981a), and the usefulness of the dual record of peat and pollen reviewed (Barber 1981b), as well as the use of peat stratigraphic researches as a proxy climatic record for the later prehistoric period (Barber 1982). This contribution will be rather more speculative – in the tradition of Gordon Manley himself. Many of the basic questions concerning climatic change, bog growth and past accumulation are still not answered with any accuracy, even if the limits within which an answer must fall are known. The questions dealt with in this chapter may be listed as:

(a) How long a climatic record can we get from peat?
(b) How short a climatic event can we resolve?
(c) Do recurrence surfaces represent breaks in accumulation?
(d) What kind of climatic elements are recorded?
(e) Why are some bogs more climatically sensitive than others?
(f) Are there limits to the growth of peat bogs?

These six questions may be grouped into two sets which, for convenience, have been labelled *temporal* and *behavioural* – though the latter term is not to be taken as an attempt to re-open the argument of earlier times as to whether bogs were 'quick' or 'dead'!

9.2 Temporal questions

In north-west Europe, ombrotrophic peat formation began generally at around the time of the Boreal–Atlantic Transition, c. 5000 BC. The submergence of a land connection across the southern North Sea restored the full circulation to the waters of the North Sea, and ushered in a moister and warmer climatic period, still conveniently referred to by the title given to it by Blytt and Sernander in the last century – the Atlantic period. Pollen-analytically, this is marked by the massive expansion of alder, invading not only the wet valley bottoms, but also becoming a component of the damp mixed oak forest (Godwin 1975), though Magny (1982) has cast doubt on this. It is well to remember that most pollen diagrams of the period come from lake, coastal or riparian situations which would be expected to show the expansion of such a hygrophilous species.

However, most workers agree with Lamb (1977, 1982) on the essentially 'warmer but wetter' view of the Atlantic period. Allied to this, the successional factor must be stressed: raised bogs represent the culminating stage of the hydrosere, and adequate time must be allowed for the water to be colonised by aquatic plants, reedswamp and fen herbs and trees, before ombrotrophic *Sphagnum* communities follow. Walker (1970), in his classic study, has examined the rates of change and the directions taken by such hydroseres in Britain, and it is notable that no instance of bog peat forming is recorded between 10 000 and 7800 BC; only one instance between 7800 and 7000 BC; two between 7000 and 5400 BC; five between 5400 and 3000 BC; and 14 between 3000 and 500 BC. This is not to say that ombrotrophic peat formation was impossible before the Atlantic period, merely that it was rare. The two pre-Atlantic examples used by Walker are Cranes Moor in Hampshire and Scaleby Moss in Cumbria. The latter developed in a shallow basin which passed rapidly through the early hydroseral stages (Walker 1966), and the former site is a peculiar one (at present under investigation), where it seems that *Sphagna* directly colonised pools in the impoverished Eocene sand during the early Holocene. Elsewhere in Europe, Overbeck (1975) and Tolonen (1966) have reviewed the age of ombrotrophic peat initiation, showing clearly the increased incidence after the Boreal–Atlantic Transition.

Walker (1970) also reviewed the rapidity with which ombrotrophic bog could become established, and concluded that it could be as rapid as 1000 years from the time of *Sphagnum* invasion of the previous stages of floating-leaved macrophytes or reedswamp, but that if fen intervened (especially if carr woodland accompanied it), it could take between 2500 and 4000 years. Recently, workers in the Netherlands have shown, by means of very detailed micro- and macrofossil diagrams, just how quickly such ombrotrophication can occur. van Geel *et al.* (1981) analysed a section of peat from De Borchet, near Denekamp in the eastern Netherlands, and found that a few centimetres of *Drepanocladus* peat had succeeded late-glacial layers of sandy gyttja with birch leaf remains (dating from 10 150 – 9800 BP), and that these were succeeded directly by *Sphagnum* peat of low humification (dating from 9730 BP). *Sphagnum* remains are then present in this oligotrophic (but not ombrotrophic) bog until 5000 BP. This must be one of the oldest Holocene peat profiles from which climatic evidence has been adduced, not only from macrofossils such as *Sphagnum* remains, but also from the occurrence of various microfossils which have been studied in detail. Five relatively wet and five relatively dry phases were discerned between 9700 BP and 6200 BP.

Bakker and van Smeerdijk (1982) studied a Late Holocene section of marine clay and peat deposits in the western Netherlands, and were able to establish the durations of the various vegetational stages. They comment that 'the relatively short time-span between the end of the salt-marsh phase and the beginning of bog growth is rather striking'. The vegetation passed through phases of reedswamp (lasting 70 years), transitional fen (40 years), open moss fen (70 years) and open fen carr (240 years) – a total of only 420 years – before beginning ombrotrophic bog growth. This lasted for about 3000 years before being disturbed by man. In the absence of such disturbance, ombrotrophic peat suitable for reconstructing past climates may be encountered from virtually any period of the Holocene (the past 10000 years), but particularly during the last 7000 years. In many areas, particularly rapid vertical growth and lateral transgression occurred after about 2500 years ago, the beginning of the Sub-atlantic period, which is generally accepted as being cooler and wetter.

Whether it is possible to follow the record up to the present day depends primarily on whether or not the bog has been disturbed, especially by drainage and peat cutting, though there is the suggestion of irrecoverable drying-out of the peat bogs in the later Sub-boreal such as is suggested at Shapwick Heath in Somerset at a later date (Barber 1982). Man's impact is especially severe in the Netherlands, where it is unlikely that any peat remains undisturbed beyond a level dating to about AD 300 due to the cultivation of buckwheat (*Fagopyrum esculentum*), which was preceded by drainage and burning (W. A. Casparie, personal communication). In all the heavily populated areas of north-west Europe, over 90% of raised bogs have been badly disturbed, if not completely destroyed. Ironically, however, during the early part of the destruction of a bog, the stratigraphy can be recorded in detail (Barber 1981a), but there is no substitute for a complete and undisturbed profile from a living bog surface.

The question of the resolution of events is an important one. It is usual to take a 1-cm thick sample for micro- and macrofossil analysis (though possible to take much less, e.g. Garbett 1981), and with a very rough average peat accumulation rate during the Sub-atlantic of 1 cm every 17 years, this allows resolution of events of about 50 years duration, on the assumption that an 'event' must be defined by three consecutive samples. While this is of use in recording climatic variation over periods of millennia, it points up the fact that bogs with higher rates of accumulation are potentially very valuable indeed. To the figures given elsewhere (Barber 1981a, pp. 202–204, Barber 1982), the extraordinary example of Rimsmoor, Dorset (Waton 1982, 1983), may be added; here, peat rich in oligotrophic *Sphagna* has accumulated at a more or less constant rate of about 4 years/cm for the past 7000 years, due to the slow subsidence of the base of the doline, or sink-hole. The Boreal peats of Cranes Moor in the New Forest, almost 3 m thick in parts, and almost exclusively of oligotrophic *Sphagna*, are also potentially of great value. It is known that peat-cutting has taken place here (Barber 1981c), so that perhaps the last 5000 years of record are missing, but the type of peat in an area of the bog shielded from water inflows, backed up by detailed levelled stratigraphic boreholes (A. R. Tilley, personal communication), suggests that the bog was once ombrotrophic and so could provide a high-resolution surface wetness curve for the whole Boreal period. This will be especially interesting in the light of the dearth of records from this period, for the reasons discussed above.

Together with the rapid rates of accumulation quoted by Turner (1981), these examples prompt the question of what is the 'ultimate resolution' of such data. Here a cautionary note must be entered: peat accumulation may be rapid, or appear to be so, between two radiocarbon-dated horizons, but it may also have fluctuated markedly over a few centimetres. Donner *et al.* (1978) showed that pollen influx diagrams from a peat bog exhibited marked variations compared with an adjacent lake, and Middledorp (1982) confirmed this in a detailed study of a single section. The influence of such fluctuations in the detailed growth of a bog on a derived climatic curve may not be too important. The fluctuations can be minimised by choosing to analyse a section which does not display marked vegetational changes (e.g. from hummock to pool), and by considering multiple analyses (pollen, rhizopods, *Sphagnum* species, non-pollen microfossils), and/or multiple cores from the same bog. Many of these analyses are very time consuming, and detailed analyses of virtually all the biological constituents of a single core takes several months, if not years. We have to add to this realisation the fact that the general course of postglacial climatic change is now fairly well known, and enquire the purpose of detailed, fine-resolution work. It could be suggested that if, as now seems possible, we can almost arrive at a proxy version of the decadal moving averages of the instrumental climatologist, then the effort will be worthwhile in terms of our scientific understanding of how the climatic systems works, and the importance of such results to palaeoecology and archaeology.

Recurrence surfaces represent big changes in the surface wetness of a peat bog. In the stratigraphy they are seen as levels at which dark humified peat is succeeded by lighter unhumified peat, often with pool muds containing semi-aquatic species immediately on top of the dark peat. They must represent a gross change in the surface wetness condition of the bog, though whether this was brought about by the crossing of a threshold in the bog's hydrology while climate slowly changed to a cooler and/or wetter state, or whether the climatic change itself was also sudden, is open to question. Arguments about the number and nature of recurrence surfaces have been legion since Weber (1900) first accurately described the most marked recurrence surface (Grenzhorizont) in Germany, and they have been reviewed by Turner (1981), Barber (1982), and Godwin (1981). The only question that need concern us here, therefore, is the possibility of gaps in the record if, as was indeed supposed by Weber, bog growth ceased completely for some decades or centuries before the laying down of the unhumified peat. This cessation does seem to have happened in a number of well dated instances, with a possibility of greater likelihood in the more continental climates of Sweden and north Germany, though the evidence is not always as clear cut as one would like (Turner 1981, Barber 1982). But there are more examples of both synchroneity of change and of continuity of peat accumulation across the stratigraphic boundary (Aaby 1976). The rate of peat accumulation may, of course, change quite drastically as one passes from humified to unhumified peat – figures as low as 80 years/cm to 25 years/cm for the humified peat, and as varied as 20 years/cm to 2 years/cm for the unhumified peat, have been recorded.

For the purposes of constructing any kind of continuous sequence of climatic variation from peat stratigraphy, a bog without any obvious hiatus must be chosen. Careful macrofossil analysis of the top of the humified peat will reveal the past state of the surface, and any appreciable period of non-accumulation should show up as totally disintegrated plant materials of a purplish-black colour, and there are often other indicators such as an old cracked, oxidised peat surface. Radiocarbon datings on either side of any pronounced recurrence surface should confirm the diagnosis.

9.3   Behavioural questions

While there is no doubt that a relationship (often a close one) exists between climatic change and changes in peat stratigraphy, the exact nature of the relationship is neither as clear nor as well defined as it may appear. Walker (1970), Moore and Bellamy (1974) and Godwin (1975) are all vague on the question of the exact precipitation/evaporation conditions for bog growth. Obviously summer water deficits must be something of a determining factor, but how does a long cold winter, with water from snow-melt and low evaporation for much of the year, 'compensate' for a short hot continental summer? What of other factors such as average cloudiness and exposure to high winds? What of the great inertia represented by the sodden mass of peat itself, and the self-regulatory tendencies of plants? *Sphagnum* mosses have no stomata or other specialised control devices to reduce evapotranspiration, and apart from the cover afforded by vascular plants such as heather and cotton grass, they are at the mercy of climate. Drier, hotter summers would lead to a reduction of surface variance due to the inhibition of *Sphagnum* growth on higher, more exposed, microhabitats (Clymo 1973). The exploration of this climate/stratigraphy relationship (Barber 1981a) at Bolton Fell Moss, Cumbria, led to the rejection of any idea of autonomous cyclic processes at work in peat growth, and instead an emphatic vindication of the controlling influence of climate on the relative abundances of the various bog species and hence the composition of the peat, and on the degree of humification. The graph of changing surface wetness on the bog (Fig. 9.1) could be closely related in a qualitative sense with Lamb's indices of high summer wetness and winter severity (Lamb 1977). In this way, it is perhaps not possible to unravel the effects of the different climatic elements by analyses of the major plant remains alone – it is rather like attempting to say exactly which element and which part of the year's weather were responsible for a good vintage: there are indications, but they defy precise quantitative analysis.

It may be, however, that two new lines of research will allow us to come nearer the goal. The first is the demonstration by van Geel and his colleagues that several hundred types of non-pollen microfossil (fungal and algal remains) previously ignored by palynologists, can be identified or at least catalogued, and associated with particular moisture regimes. Zygospores of *Mougeotia*, for example, are associated with shallow pools in spring which then dry up in summer, ascospores of *Neurospora* with fires, and Type 10 fungal remains with

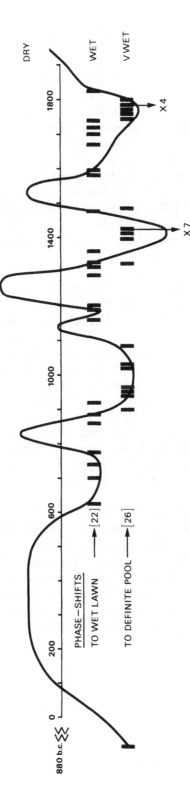

Figure 9.1 Surface wetness curve from Bolton Fell Moss. *NB*. The curve is generalised, and includes data other than the wet phase-shifts shown (see Barber 1981a for details). The relative height of the 'dry' peaks reflects the area and vegetation/humification of hummocks.

drier peat associated with *Erica tetralix* and *Calluna vulgaris* (van Geel 1978, Bakker & van Smeerdijk 1982). Analyses of such microfossils from two profiles from Cranes Moor, New Forest, have shed new light on the comparative development of two parts of the bog (K. E. Barber, unpublished data), and there is no doubt that, allied with rhizopod and macrofossil analysis, they can help us in determining the relative dryness of past summers.

The second line of new work is that of stable isotope analysis of peat. Variations in the ratio of $^{18}O:^{16}O$ and of D/H are known from work on ice and ocean cores to be related to temperature and humidity. Following earlier work by Schiegl (1972) and Gray and Thompson (1976), Brenninkmeijer has been refining these methods for use on peat. Preliminary results (Brenninkmeijer *et al.* 1982) open the prospect of being able to resolve the 'cooler and/or wetter' conundrum.

They may also help in determining why some bogs appear to be more sensitive to climatic change than others. Manley (1965, p. 374) himself commented on this, ascribing the better registrations of climatic change of bogs of eastern Sweden compared with those of England to the greater range of temperature fluctuations in summer. More research is certainly needed on the registration of climatic change. A transect study of bogs from western Ireland across to southern Finland could prove rewarding. Bolton Fell Moss (Barber 1981a) was certainly sensitive to climatic change of quite a small magnitude, such as the amelioration of climate between AD 1500 and 1550 (Lamb 1982), as were the bogs in Denmark studied by Aaby (1976). In western Ireland, however, though no really detailed work is available, one may surmise that in such an extremely oceanic climate, the ombrotrophic blanket bog covering much of the landscape would hardly have registered very small changes in climate such as the amelioration of the early 1500s: in other words, it is a question of marginality.

Once ombrotrophic bog growth is well under way, the question arises as to whether there are limits to such growth, both vertical and lateral. Lateral spread of bogs in the Harz Mountains of Germany has been documented by Beug (1982), who found that at altitudes of 700–900 m topographic factors were more important than climatic factors. Where such topographic controls are of negligible importance, ombrotrophic bogs can transgress over adjacent land to cover huge areas. Examples of large bog systems in Britain, bounded only by large rivers or hillsides, include the bogs at Tregaron (Godwin & Conway 1939) and at the Silver Flowe (Ratcliffe & Walker 1958), besides the numerous bogs in the Cumbrian lowlands, the central plain of Ireland, and the midland valley of Scotland. The Bourtanger Moor area on the border of the Netherlands and Germany was once a complex of bogs about 50 km long and up to 20 km wide. Due to the coalescent growth of several raised bogs in this area, lakes (or *Hochmoorsee*) have been ponded up in the past, occasionally overflowing with catastrophic results (Casparie 1972). At one time, about 35% of the adjacent German province of Ostfriesland was covered with peat bogs (K.-E. Behre, personal communication). It is clear, therefore, that lateral bog growth may extend over many kilometres or even hundreds of kilometres as in northern Canada and Siberia. With such a horizontal scale, the vertical scale point so far reached in postglacial times, of 10 or more metres, may seem insignificant. It is usual to draw up stratigraphic cross-sections of peat bogs at a greatly exaggerated vertical scale; this may have contributed to the concern in the earlier literature about limiting heights and the possibilities of wholesale drainage, and of height above the watertable being a contributory cause of recurrence surfaces. As an antidote, it is instructive to look at true-scale diagrams such as those in Osvald (1923). However, lateral water movement does occur on raised bogs, and the convexity of the surface, though slight, is visible. Whether such lateral movement, in a natural bog system, is ever great enough to drain a bog and limit its height is open to doubt. One may point out that the most marked recurrence surface (RYIII at about 500 BC) is followed by the vigorous accumulation of very unhumified peat on an increasingly convex bog. Where a raised bog forms in a confined area, such as at Tregaron, there must eventually be some limiting height, but there is no doubt that the calculation of such a limit is a very complex affair. These points and the questions of water movement in peat in general may be explored by reference to the works of Ingram (1967) Rycroft et al. (1975a, 1975b) and Ivanov (1981).

In ending his 1965 paper on 'Possible climatic agencies in the development of post-glacial habitats', Manley made a plea for ecological work on peat stratigraphy in an effort to resolve whether local meteorology or factors internal to the bog were responsible for changes in bog growth. With the benefit of the sort of research work mentioned in this contribution, it is possible to conclude that there is a close, if variable, relationship between climatic change and variations in peat stratigraphy. The proxy record for climate can be extended back for more or less the whole postglacial period. There is, however, still a need for further studies. Manley (1965) himself suggested that 'the balance of evaporation, temperature and rainfall in north-east Scotland would suggest that a filled-up kettle-hole in the Black Isle might prove fruitful'. It is to be hoped that the speculations in this contribution might prompt research along the lines of this suggestion.

Acknowledgement

A. A. Balkema (Publishers) of Rotterdam have given permission for the reproduction of Figure 9.1 from Barber (1981a).

References

Aaby, B. 1976. Cyclic climatic variations in climate over the past 5,500 years reflected in raised bogs. *Nature (Lond.)* **263**, 281–4.

Bakker, M. and D. G. van Smeerdijk 1982. A palaeoecological study of Late Holocene section from 'Het Ilperveld', W. Netherlands. *Rev. Palaeobot. & Palynol.* **34**, 95–163.

Barber, K. E. 1981a. *Peat stratigraphy and climatic change: a palaeoecological test of the theory of cyclic peat bog regeneration*. Rotterdam: Balkema.

Barber, K. E. 1981b. The stratigraphy and palynology of ombrotrophic peat: a source for the reconstuction of agricultural and climatic change. In *Consequences of climatic change*, C. Delano-Smith and M. Parry (eds), 124–34. Nottingham: University Geography Department.

Barber, K. E. 1981c. Pollen-analytical palaeoecology in Hampshire: problems and potential. In *The archaeology of Hampshire: from the Palaeolithic to the Industrial Revolution*, S. J. Shennan and R. T. Schadla-Hall (eds), 91–4. Monograph No. 1. Hampshire Field Club and Archaeological Society.

Barber, K. E. 1982. Peat-bog stratigraphy as a proxy climate record. In *Climatic change in later Prehistory*. A. F. Harding (ed.), 103–13. Edinburgh: University Press.

Beug, H. J. 1982. Vegetation history and climatic changes in central and southern Europe. In *Climatic change in later Prehistory*, A. F. Harding (ed.), 85–102. Edinburgh: University Press.

Brenninkmeijer, C. A. M., B. van Geel and W. G. Mook 1982. Variations in the D/H and $^{18}O/^{16}O$ ratios in cellulose extracted from a peat bog core. *Earth & Planet. Sci. Lett.* **61**, 283–90.

Casparie, W. A. 1972. *Bog development in southeastern Drenthe (the Netherlands)*. The Hague: Junk.

Clymo, R. S. 1973. The growth of *Sphagnum*: some effects of environment. *J. Ecol.* **61**, 849–70.

Donner, J. J., P. Alhonen, M. Eronen, H. Jungner and I. Vuorela 1978. Biostratigraphy and radiocarbon dating of the Holocene lake sediments of Tyotjarvi and the peats in the adjoining bog Varrasuo west of Lahti in southern Finland. *Ann. Bot. Fennica* **15**, 258–80.

Garbett, G. G. 1981. The elm decline: the depletion of a resource *New Phytol.* **88**, 573–85.

van Geel, B. 1978. A palaeoecological study of Holocene peat bog sections in Germany and the Netherlands. *Rev. Palaeobot. & Palynol.* **25**, 1–120.

van Geel, B., S. J. P. Bohncke and H. Dee 1981. A palaeoecological study of an upper Late Glacial and Holocene sequence from 'De Borchert', the Netherlands. *Rev. Palaeobot. & Palynol.* **31**, 367–448.

Godwin, H. 1975. *History of the British flora*, 2nd edn. Cambridge: University Press.

Godwin, H. 1981. *The archives of the peat bogs*. Cambridge: University Press.

Godwin, H. and V. M. Conway 1939. The ecology of a raised bog near Tregaron, Cardiganshire. *J. Ecol.* **27**, 315–59.

Gray, J. and P. Thompson 1976. Climatic information from $^{18}O/^{16}O$ ratios of cellulose in tree rings. *Nature (Lond.)* **262**, 481–2.

Ivanov, K. E. 1981. *Water movement in Mirelands*. London: Academic Press. Translated from Russian by A. Thompson and H. A. P. Ingram.

Ingram, H. A. P. 1967. Problems of hydrology and plant distribution in mires. *J. Ecol.* **55**, 711–25.

Lamb, H. H. 1977. vol. 1: *Climate: present, past and future*, vol. 2: *Climatic history and the future*. London: Methuen.

Lamb, H. H. 1982. *Climate, history and the modern world*. London: Methuen.

Magny, M. 1982. Atlantic and Sub-boreal: dampness and dryness? In *Climatic change in later Prehistory*, A. F. Harding (ed.), 33–43. Edinburgh: University Press.

Manley, G. 1965. Possible climatic agencies in the development of post-Glacial habitats. *Proc. R. Soc. Series* **B 161**, 363–75.

Middledorp, A. A. 1982. Pollen concentration as a basis for indirect dating and quantifying net organic and fungal production in a peat bog ecosystem. *Rev. Palaeobot. & Palynol.* **37**, 225–82.

Moore, P. D. and D. J. Bellamy 1974. *Peatlands*. London: Elek.

Osvald, H. 1923. Die Vegetation des Hochmoores Komoose. *Svensk. Växtsoc. Sällsk, Handl.* 1.

Overbeck, F. 1975. *Botanisch-geologische Moorkunde*. Neumunster: Karl Wachholts.

Ratcliffe, D. A. and D. Walker 1958. The Silver Flowe, Galloway, Scotland. *J. Ecol.* **46**, 407–45.

Rycroft, D. W., D. J. A. Williams and H. A. P. Ingram 1975a. The transmission of water through peat. I. Review. *J. Ecol.* **63**, 535–56.

Rycroft, D. W., D. J. A. Williams and H. A. P. Ingram 1975b. The transmission of water through peat. II. Field experiments. *J. Ecol.* **63**, 557–68.

Schiegl, W. E. 1972. Deuterium content of peat as a palaeoclimatic recorder. *Science* **175**, 512–13.

Tolonen, K. 1966. Stratigraphic and rhizopod analyses on an old raised bog, Varrasuo in Hollola, South Finland. *Ann. Bot. Fenn.* **3**, 147–66.

Turner, J. 1981. The Iron Age. In *The environment in British Prehistory*, I. G. Simmons and M. J. Tooley (eds), 250–81. London: Duckworth.

Walker, D. 1966. The Late Quaternary history of the Cumberland Lowland. *Phil Trans R. Soc. Series B* **251**, 1–210.

Walker, D. 1970. Direction and rate in some British Post-Glacial hydroseres. In *Studies in the vegetational history of the British Isles*, D. Walker and R. G. West (eds.), 117–39. Cambridge: Cambridge University Press.

Waton, P. V. 1982. Man's impact on the chalklands: some new pollen evidence. In *Archaeological aspects of woodland ecology*, M. Bell and S. Limbrey (eds), 75–91. Oxford: BAR International Series, 146.

Waton, P. V. 1983. *A palynological study of the impact of man on the landscape of Central Southern England with special reference to the chalklands*. PhD thesis. Southampton: University of Southampton.

Weber, C. A. 1900. Über die Moore mit besonderer Berücksichtigung der zwischen Unterweser und Unterelbe liegenden. *Jahres-Bericht der Männer von Morgenstern* **3**, 3–23.

10

Geomagnetism and palaeoclimate

FRANK OLDFIELD AND SIMON G. ROBINSON

Links between magnetic measurement and the record of climatic change in Pleistocene and Late Tertiary sediments are examined. The possible reason for variations in mineral magnetism in lake sediments and deep sea cores are discussed. The supply of magnetic minerals to most sediments is dominated by continental sources, and the processes of weathering and erosion which influence the supply are obviously under direct climatic control. Apparent geomagnetic–palaeoclimatic linkages may therefore be generated by climatically modulated changes in the supply of magnetic mineral assemblages.

10.1 Introduction

In expressing his major concern with the nature and course of climatic change, Gordon Manley spoke from the standpoint of an empiricist with a deep awareness of the often conflicting nature of environmental evidence. This awareness grew not only from his critical appraisal of data from a wide range of disciplines, but also from his own direct experience working with early instrumental records. In consequence, he was, with regard to theories of climatic change, a sceptic, always ready to challenge or qualify oversimplification from his own great store of knowledge and experience. This contribution is an empirically based evaluation of one area of palaeoclimatic theorising, and it is offered in recognition of the great debt owed to Gordon Manley's unique and marvellously idiosyncratic blend of stimulus and critique.

In a recent letter to *Nature*, Kent (1982) states: 'A fundamental relationship between geomagnetic field behaviour and climate has been suggested, largely on the basis of an apparent correlation in geomagnetic and climatic indices over timescales ranging from historical observatory records of geomagnetic intensity and air temperature, to the Pleistocene record of palaeomagnetic and palaeoclimatic data in deep-sea sediments.' This chapter is concerned solely with the long-term record, and with the possible basis for links between magnetic measurements and the record of climatic change in Pleistocene and Late Tertiary sediments. Most contributors in this field have inferred some linkage between climatic and geomagnetic behaviour on a global scale. Harrison and Prospero (1977) postulated a direct effect of the magnetic field on the atmosphere, leading to a causal link between low geomagnetic field intensities and warmer climates. Other authors have suggested that the distribution of ice may have affected the geomagnetic field by influencing the nature of fluid motions within the core (Doake 1977), or that astronomical variables, such as the eccentricity of the Earth's orbit, may be controlling both climatic and geomagnetic intensity variations (e.g. Wollin *et al.* 1977).

The starting point for the development of these theories in relation to the long-term record of climatic change is a demonstrated parellelism in deep-sea sediment cores between palaeomagnetic measurements and indices of climatic change, for example oxygen isotope composition (^{18}O), foraminiferal assemblages, and calcium carbonate content (Wollin *et al.* 1971). Figure 10.1 illustrates this parallelism. The most striking apparent 'magnetic indicator' of palaeoclimate is the intensity of natural remanent magnetisation (NRM). Cool temperature episodes correspond with high NRM intensities and *vice versa*. The relationships are too close and too widespread to be merely coincidental, and they lend great plausibility to inferences of a direct link between past variations in the intensity of the geomagnetic field and palaeoclimate. In order to evaluate this apparent relationship, it is necessary to consider the magnetic properties of sediments rather more fully than is done in the papers which purport to establish the geomagnetic–palaeoclimatic link.

10.2   Some critical distinctions

Two distinctions are essential at the outset. In the first instance, it is necessary to differentiate between those magnetic properties of sediments which reflect the nature of the Earth's magnetic field at some period in the past (NRM) and those magnetic properties which are entirely independent of the ancient field but are a function of the magnetic mineralogy of the sample. These latter properties, termed *mineral magnetic* (as against *palaeomagnetic*), are measured by using artificially generated magnetic fields in ways which illuminate variations in the magnetic mineralogy and grain size within the sediments. Some of the mineral magnetic parameters used to describe sediments, identify sources and establish intercore correlations are defined briefly in Thompson *et al.* (1975) and Oldfield, Dearing, Thompson and Garret-Jones (1978). The parameters used include some which are strongly or partially concentration dependent, e.g. magnetic susceptibility (χ), saturation isothermal remanent magnetisation (SIRM), and anhysteretic remanent magnetisation (ARM), and others which are normalised and completely independent of concentration, for example the coercivity of SIRM ($(Bo)_{CR}$) and the ratios between χ, SIRM and ARM. It is worth re-emphasising that all these mineral magnetic parameters, whether concentration dependent or independent, are an expression of magnetic grain size and mineralogy; they are completely independent of the Earth's field, past or present.

A second distinction is necessary (this time within the ambit of NRM measurements) between the intensity of NRM as measured in sediment cores, and the intensity of the geomagnetic field at the time of deposition of the

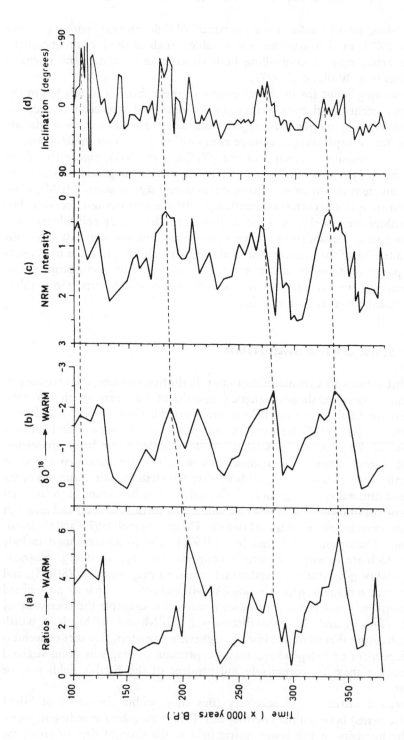

Figure 10.1 Magnetic parameters (intensity of NRM and inclination) plotted against ^{18}O, and foraminiferal evidence for climatic change from Caribbean core V12–122 (Wollin et al. 1971). (a) Foraminiferal ratios. (b) Oxygen isotope curve. (c) Intensity of NRM. (d) Palaeomagnetic inclination.

sediments. Although archaeomagnetists, by means of carefully controlled heating and cooling experiments, have developed ways of using the NRM of potsherds and baked hearths, for example, to reconstruct 'palaeointensities', the relationship between the intensity of NRM in *sediments*, and the palaeointensity of the geomagnetic field remains undefined, despite several alternative appoaches to correcting and normalising the sedimentary signal. Sedimentary NRM is acquired during or soon after deposition, and it is predominantly detrital in origin. That is to say, alignment in the magnetic field is a physical process taking place either during deposition or shortly after within the water-filled interstices of the near-surface sediment. The alignment is subsequently 'locked-in' as the sediment becomes more compact as a result of continuing deposition. Although natural remanence can be acquired chemically as a result of authigenic or diagenetic changes, *primary* NRM is largely detrital. From the locked-in orientations, the inclination (dip) and declination (horizontal component) of the ancient field can be reconstructed, though the directions will often be modified by compaction and bioturbation. Unfortunately, the intensity of NRM (i.e. the strength of the recorded signal) is not so directly linked to the nature of the ambient field. It reflects, above all, sedimentary rather than geomagnetic variables. Usually, only a narrow band in the total spectrum of magnetic minerals present will respond to and record the Earth's geomagnetic signature. The narrow band is defined by factors such as magnetic crystal size, and bulk sediment structure and granulometry. Moreover, the number of magnetic grains within the NRM 'bandwidth' will depend on the concentration of magnetic minerals within a given layer of sediment. The intensity of NRM is thus strongly dependent on variations in both quality and concentration within the sediment supply system. Generally speaking, these variations are much (often orders of magnitude) greater than are those in the intensity of the geomagnetic field; they thus overprint and obscure the latter in the sedimentary record of NRM. Therefore it is not possible to accept the intensity of NRM as a record of geomagnetic palaeointensity until ways have been found to normalise the signature for both qualitative and concentration-dependent variations. Normalisation has been attempted using additional mineral magnetic parameters and ratios (Levi & Banerjee 1976), and also by means of redeposition experiments in the Earth's field at the present day (e.g. Barton & McElhinny 1976). Up to the present time, no generally satisfactory method has been devised, and such successes as are claimed are both specific to particular sediments and locations, and unconfirmed by rigorous comparison with independent evidence. Variations in NRM *intensity* must therefore, for the present, be regarded as strongly sedimentologically modulated, and not as direct expressions of geomagnetic palaeointensity.

Two important conclusions emerge:

(a) Parallelism between palaeoclimatic indices and intensity of NRM may be sedimentologically rather than geomagnetically controlled.
(b) Parallelism between palaeoclimatic indices and *mineral* magnetic para-

meters must express linkages which are completely independent of geometric changes.

10.3   Mineral magnetism and palaeoclimate in lake sediments

Figure 10.2 shows a very simple, direct relationship between mineral magnetic and pollen-analytical evidence for climatic change. The site at High Furlong, Blackpool, spans the late-Devensian and early Flandrian period (Hallam *et al.* 1973). The pollen-analytical subzones defined record a succession from grass–herb tundra through to birch woodland during the Windermere (Allerød) Interstadial, the subsequent spread of dwarf shrub and herb communities during the Loch Lomond Stadial, and the spread of birch woodland at the opening of the Flandrian. Magnetic susceptibility (χ) measurements are highest during the cold stadial episodes, decline to a minimum during the Interstadial, and rapidly fall to around zero at the opening of the Flandrian. Magnetic susceptibility is largely a function of the concentration of magnetic minerals in the sample, and the variations shown are most readily interpreted as an expression of the changing input of unweathered mineral substrate into the sediments. Episodes marked by poorly developed vegetation cover, soil instability and active

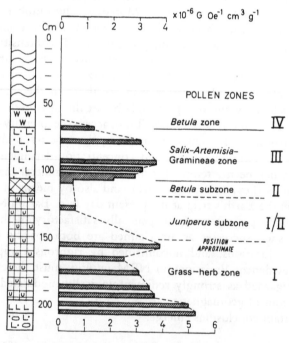

Figure 10.2   Magnetic susceptibility of samples from the Late-Devensian sediments of the site at High Furlong, Lancashire. The stratigraphy and the pollen assemblage zones are those identified in Hallam *et al.* (1973).

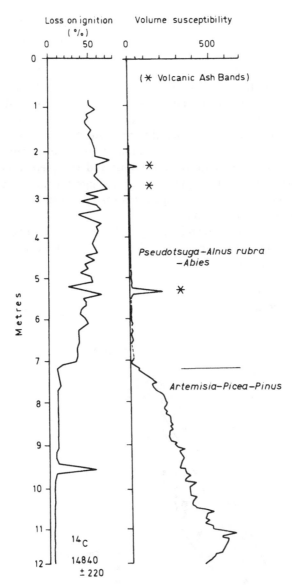

Figure 10.3 Volume magnetic susceptibility scans of a sediment core from Battleground Lake, Washington State, USA. The methods used and the palaeoecological context are noted in Oldfield *et al.* 1983.

solifluction give rise to peaks in χ; episodes of developing plant cover and soil maturation under milder climatic conditions give rise to minimum χ values.

Figure 10.3 plots volume susceptibility for a series of stratigraphically consecutive piston cores from Battleground Lake in southern Washington State (Oldfield *et al.* 1983). The 12 m of sediment span Late-Wisconsin and Holocene time, with a basal ^{14}C date of 14 840 ± 220 (QL–1539). The lake is in a closed crater basin with a small low-rimmed catchment in basalt. Below 7 m, the high susceptibility values reflect a substantial detrital input of ferrimagnetic basalt from the catchment. Above 7 m, the values are consistently low, save where volcanic ash bands are recorded. The major change from *c.* 7.5 to 7 m corresponds with the lithostratigraphic boundary, the sudden increase in loss-on-ignition values, and the change in pollen assemblage from *Artemisia–Picea–Pinus* to *Pseudotsuga–Alnus rubra–Abies* dominance. The change in magnetic mineral concentrations clearly identifies the Late Wisconsin/Holocene boundary, and reflects a diminution in allocthonous detrital input from the catchment, with the development of more stable soils and complete vegetation cover. Other associated mineral magnetic changes suggest a shift to proportionally more secondary, soil-derived, as against primary, substrate-derived, minerals above the decline in susceptibility.

In both cases, the magnetic mineralogy reflects the course of climatic changes by recording evidence of the associated changes in sedimentation, weathering and pedogenic regimes.

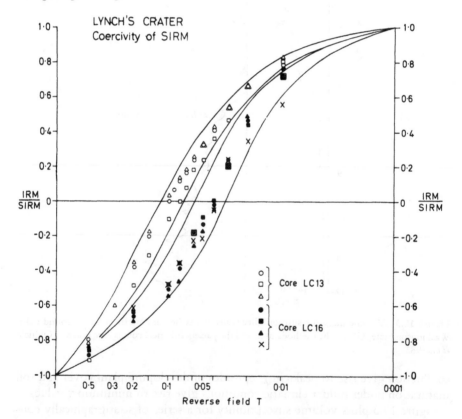

Figure 10.4 Coercivity of SIRM profiles for samples from Lynch's Crater, Queensland. Kershaw (1978) dates Core Section LC section 13 to *c.* 80 000 to 86 000 BP, and Core Section LC section 16 to *c.* 100 000 to 105 000 BP on his estimated timescale for the 40 m core.

Figure 10.4 shows plots of the mineral magnetic measurements for sets of samples from the sediments of Lynch's Crater, north-eastern Queensland, Australia. Whereas in Figures 10.2 and 10.3, the parameter plotted is largely concentration dependent, here the values are normalised, and hence completely independent of concentration. Virtually all the points lie within two quite separate envelopes of values, one for samples for Core Section LC13, the other for samples from Core Section LC16. The shape of the envelopes reflects the behaviour of each sample when it has been first placed in a 'saturating' forward magnetic field and measured to give the SIRM (here normalised to 1.0), then placed in a series of increasingly strong reverse fields, and the remanence (IRM) measured after each step. The IRM/SIRM is plotted for each sample, against each of the reverse fields used. With each stepwise reversed field increase, the initial SIRM is reduced, and values decline to IRM/SIRM = 0. The field at which this happens gives the coercivity of SIRM–$((Bo)_{CR})$. For the LC13 samples, this is just under 0.04 Tesla, but for the LC16 samples it lies between 0.07 and 0.11 Tesla. Beyond the coercive field, the samples begin to acquire a net reverse remanence, which increases as the fields used increase, and in all samples it exceeds 80% of the initial 'forward' SIRM once the samples have been subjected to a reverse field of 0.5 Tesla. Core Section 13 comes from Kershaw's (1978) pollen Zone E1, Core Section 16 from Zone E2. In E1, pollen of complex vine forest taxa dominates the palynological record, and the mean annual rainfall suggested by the reconstructed vegetation type is a little in excess of contemporary rainfall at the site. In E2, higher pollen representation of sclerophyll and gymnosperm elements suggests drier conditions than those prevailing at present. Here the *normalised* mineral magnetic parameters are related to palaeoclimate in ways which are not immediately apparent, though the linkage between the two persists for most of the very long sediment record at the site (Thompson, personal communication). These mineral magnetic variations, unlike the ones graphed in Figures 10.2 and 10.3, cannot be accounted for by changes in concentration or flux destiny; they reflect shifts in the type of magnetic mineral assemblage deposited. In all probability, complex interactions between weathering, and hydrological and sedimentological regimes are involved. As yet, case studies on more recent sediment suites have failed to provide a clear guide to interpretation.

From these case studies it may be concluded that mineral magnetic measurements, whether concentration dependent or not, can reflect palaeoclimatic conditions as a result of the effect that changing climate has on the environmental processes which control the concentrations and types of magnetic minerals deposited in lake sediments. In the context of the major theme of this chapter, these conclusions raise important issues concerning deep-sea sediments and their magnetic record:

(a) If the palaeoclimatic record in deep-sea sediments parallels not only the NRM but also the *mineral* magnetic variations, it should be possible, by analogy with the above, to look for a climatically modulated *sedimentologi-*

cal control of magnetic properties rather than for a geomagnetic–palaeoclimatic linkage.

(b) If there are palaeoclimatically related *normalised* mineral magnetic variations, these variations must reflect something more than the simple concentration and dilution effects of, for example, changing rates of calcium carbonate deposition.

10.4 Mineral magnetic measurements and palaeoclimatic indices in deep-sea sediments

Figure 10.5, from Kent (1982), plots NRM intensity as well as concentration-related mineral magnetic parameters (volume susceptibility and isothermal remanent magnetisation) against calcium carbonate content and oxygen isotope ratios for cores from the southern Indian Ocean (RC11–120) and the equatorial Pacific Ocean (RC11–209). In each core, NRM is very strongly correlated with the *mineral* magnetic parameter ($r = 0.85$ and 0.86 respectively), and the mineral magnetic parameter is strongly correlated with $CaCO_3$ ($r = -0.93$ in both cases), and with ^{18}O in the Indian Ocean core. Kent is able to show that in the Pacific case, the NRM and IRM minima correspond with maximum ice volume and minimum temperature during glacial intervals, whereas in the southern Indian Ocean, the converse is the case. The correspondence between NRM and the mineral magnetic parameters clearly indicates that the former is here largely controlled by changing magnetic mineral concentrations in the sediment, and not by past changes in the intensity of the Earth's magnetic field. At the same time, the opposite nature of the magnetic–palaeoclimatic relationship at the two sites reinforces the difficulty of sustaining any interference of a global geomagnetic–palaeoclimatic link from the evidence shown. This work thus constitutes a convincing refutation of the postulated long-term link between geomagnetic palaeointensity and palaeoclimate as recorded in the deep-sea sediment record. Directly comparable results have been reported recently by Robinson (1982) for late Pleistocene cores from North Atlantic sediments, and by Bloemendal (1980) for DSDP Core 514 from the South Atlantic. In both cases, NRM is closely related to mineral magnetic 'concentration' parameters (χ, SIRM and ARM), and in both cases the palaeoclimatic evidence parallels both the concentration-dependent and the concentration-independent mineral magnetic variations. Figure 10.6 shows plots of a section of Bloemendal's mineral magnetic results for DSDP core 514, and it can be seen that the 'harmonic' changes in NRM, χ, SIRM and ARM are paralleled by changes in $IRM_{-0.1T}/$ SIRM, a parameter (often termed the 'S' ratio) derived from the same type of measurements as those illustrated in Figure 10.4. Such large variations in this parameter confirm that the magnetic mineral assemblage is varying with regard to grain size and/or mineralogy; thus the NRM, χ, SIRM and ARM variations cannot be explained solely in terms of concentration and dilution effects. Indeed, the strength of variation in 'S' is indicative of a significant haematite component in the system.

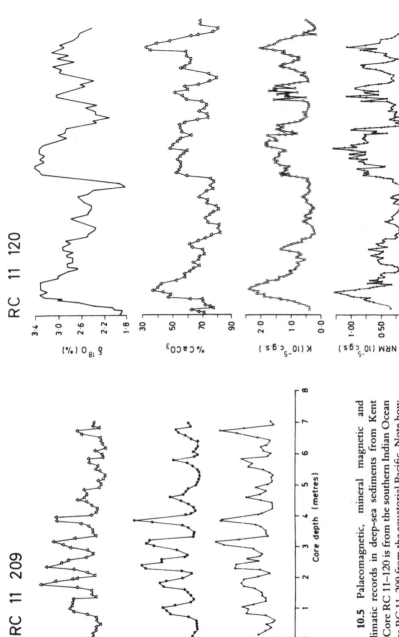

Figure 10.5 Palaeomagnetic, mineral magnetic and palaeoclimatic records in deep-sea sediments from Kent (1982). Core RC 11–120 is from the southern Indian Ocean and Core RC 11–209 from the equatorial Pacific. Note how both the *mineral magnetic* (susceptibility & SIRM) and the *palaeomagnetic* (intensity of NRM) measurements parallel the palaeoclimatic indices.

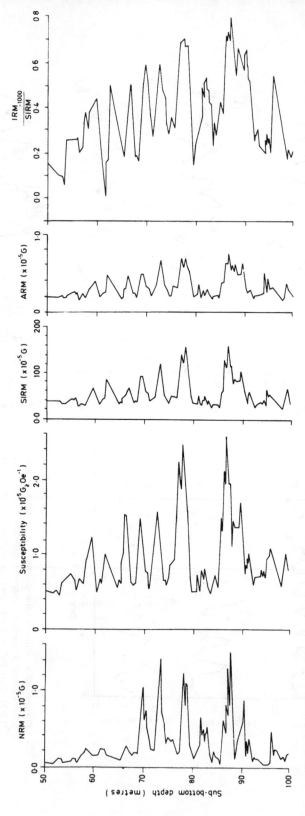

Figure 10.6 Part of the Pliocene palaeomagnetic and mineral magnetic record from DSDP core 514, southeast Argentine Basin (see Bloemendal 1980). Note how closely both the partially concentration-dependent (susceptibility, SIRM and ARM) as well as the normalised (IRM–1000/SIRM) parameters parallel the intensity of NRM record.

Preliminary results (Fig. 10.7) from a late Pleistocene North Atlantic deep-sea sediment core indicate that both concentration-related and concentration-independent mineral magnetic parameters exhibit a clear parallelism with the palaeoclimatic record of this core. In this case, the palaeoclimatic record is indicated by measurements of oxygen isotopes, calcium carbonate content, and temperature-sensitive foraminiferal species. This parallelism once more suggests that both the sedimentological processes controlling the influx of magnetic minerals to the depositional site, and the source of these magnetic minerals, are linked to climate. Concentration-related parameters (like χ, SIRM and ARM) which effectively monitor changes in the flux-density of terrigenous components to the sediment, are superimposed on a background rate of carbonate deposition, which also fluctuates in response to changes in climate (Volat et al). However, the concentration–independent parameter S-ratio ($IRM_{-0.1T}$/ SIRM), which indicates changes in the mineralogy and grain size of the assemblage of magnetic minerals in the sediment, also varies in direct response to glacial/interglacial fluctuations. Distinct peaks in the concentration-related parameters, indicating periods of intense magnetic mineral flux density, clearly correspond to abrupt changes in the composition of the magnetic fraction of the sediment, indicated by the S-ratio curve. This suggests that the source of the high glacial input of magnetic minerals differs from that of the interglacial magnetic mineral flux. Therefore, in this case, the mineral magnetic parameters used are effectively acting as indicators of both the rate of influx and the source of the terrigenous fraction of the sediment. To summarise, in this example, mineral magnetic measurements are potentially capable of discriminating:

(a) palaeoclimatically linked variations in the carbonate budget and rate of terrigenous sedimentation,

(b) changes in sediment source, between glacials and interglacials. The evidence for these changes serves to enhance and corroborate the palaeoclimatic significance of the covariance of concentration–related mineral magnetic parameters with the $CaCO_3$ content, foraminiferal and oxygen isotope record of this core.

From these references to the mineral magnetic record in deep-sea sediments, it may be concluded that:

(a) The apparent geomagnetic–palaeoclimatic linkage is an expression of changing magnetic mineral concentrations and types, and is essentially independent of changes in the Earth's magnetic field;

(b) the mineral magnetic–palaeoclimatic linkage is real, and reflects the control exercised by climate over the concentration, flux density and types of magnetic minerals deposited in deep-sea sediments;

(c) the nature of this climatic control is to be sought in the relationship between climate and the processes of weathering, erosion, transport and deposition which determine the flux of magnetic minerals to deep-sea sediments.

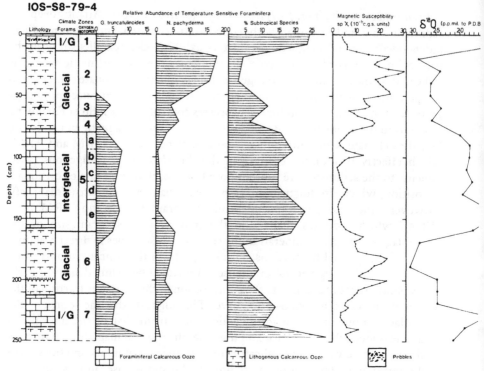

Figure 10.7   Oxygen isotope, foraminiferal, calcium carbonate and mineral magnetic indicators of climatic change for core IOS–S8–79–4 from the North Atlantic.

Beside these conclusions, may be set the apparently contradictory demonstrations of a relationship between palaeomagnetic *inclination* and palaeoclimate (e.g. Wollin *et al.* 1971, Fig. 10.1). Changes in inclination, since they are *directional* variations, cannot readily be explained in the same terms used to account for mineral magnetic and NRM intensity variations, though Stuiver (1972) outlines a possible mechanism. However, the apparent geomagnetic–palaeoclimatic linkage may appear very plausible where abrupt and major changes in not only NRM intensity, but also direction (for example, palaeo-magnetic 'excursions') correspond with evidence for climatic change. Even in long lake sediment sequences, this link has been proposed, for example in the case of Lake Biwa 'excursions' (Kawai *et al.* 1975). When the Biwa excursion sequence is set alongside the sedimentological evidence for the same core (Koyama 1975), the so-called 'excursions' correspond with changes in sediment type. The most convincing explanation for 'excursions' lies in the control exercised by sediment structure over the directional integrity and stability of the NRM signal. This aspect of palaeomagnetism has given rise to some lively debates in the literature over the past few years (Thompson & Berglund 1976), and the issue of how many palaeomagnetic excursions are recorded in late Pleistocene sediments from both Europe and North America remains

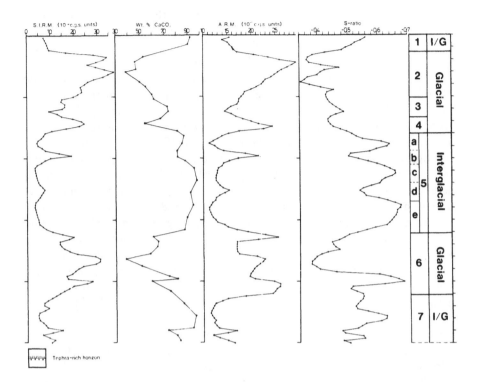

Tephra-rich horizon

unresolved. Much less open to doubt is the link between sediment structure and palaeomagnetic inclination, not only in many of the specific cases of alleged 'excursions', but also in the many cases where an 'inclination error' is consistently recorded over long stretches of undisturbed recent sediment (e.g. Granar 1957).

All this points to an explanation for inclination–palaeoclimate linkages consistent with the above account; namely, that the pattern of inclination variations is influenced by sediment structure in ways which modify the primary directional signal held in the sediments, and gives rise to changes reflecting climatically controlled sedimentation processes.

10.5   The origin of magnetic minerals in deep-sea sediments

In Figure 10.8, a partial model of the sources and fluxes of the magnetic minerals encountered in deep-sea sediments is presented. Little quantitative information is available about the relative importance of each source, either globally or locally. Nevertheless, in order to make further progress in exploring the link between mineral magnetic properties and climate, some preliminary appraisal of possibilities is necessary. It is important to recall that the mineral magnetic changes are not solely explicable in terms of concentration and dilution effects resulting from changes in the rate of accumulation of non-magnetic sediment. Changes in magnetic *mineralogy* are clearly implied.

At the present time, much of the Earth's atmosphere is quite rich in magnetic spherules. Until recently, it was believed that the majority of these were of extra-terrestrial origin, as part of a cosmic flux of matter entering the Earth's atmosphere. It is now known that the majority of these spherules are of industrial origin as a result of fossil fuel combustion (Doyle *et al.* 1976, Oldfield, Thompson & Barber 1978). Even in areas remote from contemporary industrial activity, magnetic measurements from ombrotrophic peat profiles show increases during the late 19th and 20th century resulting from industrial development and power generation (Oldfield *et al.* 1981). In view of this, it may be surmised that the extra-terrestrial input of magnetic minerals to marine sediments may be relatively important in only two types of context: very locally in the immediate neighbourhood of meteorite impact sites, and, on a larger scale, where all other types of input are at a minimum. Even then, the low flux density of cosmic spherules is unlikely to lead to significant concentrations in marine sediments, save where accumulation rates are exceptionally slow. Some mid–Pacific regions of pelagic sedimentation may fall into this category.

Most authorities accept that the major source of the magnetic mineral assemblages in deep–sea sediments is lithogenic and detrital in the form of volcanic dust, and the fine products of continental erosion. In marine sediments deposited close to zones of volcanic activity along continental margins, tephra may be frequently present, and these will often be rich in magnetic minerals (e.g. Radhakrishnamurty *et al.* 1968). Even in truly pelagic environments remote from continental margins, it is likely that there is a significant magnetic mineral input in the form of volcanic dust. Haggerty (1970), for example, regards this as the dominant source of remanence carrying detrital minerals in many deep–sea sediments.

Continental material also reaches deep–sea sediments as a result of wind, water and glacial erosion. Whereas quantitative estimates of the relative importance of each type of input in different areas are not yet possible, some qualitative inferences are possible. Clearly, glacial inputs have been restricted by the limits of ice rafting during the Pleistocene (Mullen *et al.* 1972, Ruddiman *et al.* 1972), whereas the relative contribution of fluvial inputs is at its greatest close to major river mouths (Currie & Bornhold 1983). Nevertheless, it is not possible to establish with complete confidence the relative importance of river- and current-borne as against wind-borne magnetic mineral inputs to pelagic and hemi-pelagic sediment, though for some areas, for example the mid-Atlantic between the Sahara and the Barbados, a major aeolian input can be confirmed for the present day as a result of the deflation of arid desert areas in North Africa (e.g. Chester 1972). It is rather more difficult to ascertain the significance of this relationship between deflation from arid desert areas and aeolian input to marine sediments during periods predating the recent desertification consequent upon

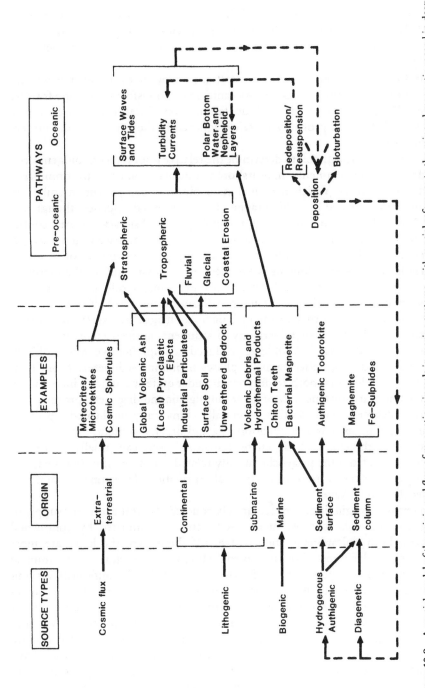

Figure 10.8 A partial model of the origin and flux of magnetic minerals in marine systems, with special reference to the mineral magnetic record in deep-sea sediments.

human overexploitation. It is worth recalling that the magnetic properties of eroded continental material will vary not only with lithology, but also with the depth in the regolith from which the material has been derived (cf. Oldfield *et al.* 1979).

Hydrothermal products are of greatest importance in the immediate vicinity of features such as seamounts, mid-oceanic ridges, transform faults and fracture zones. Their effect on marine sedimentation is very local (Henshaw & Merrill 1980), and they can probably be safely ignored in the context of the present discussion.

It is much more difficult to determine the relative importance of biogenic iron oxides in marine sediments. In nearshore sediments, as in lakes, the dominance of detrital sources (which may include a biogenic component in soils) can usually be confirmed by a wide range of circumstantial evidence, though it is sometimes difficult to eliminate all possibility of a subsidiary bacterial contribution from magnetotactic organisms living near the mud/water interface (cf. Blackmore 1981). The significance of such organisms as sources of magnetic minerals in deep-sea sediments is even less well known, and for the present it would seem unwise to regard them as playing a major role in the variations considered here. Despite recent interest (Lowenstam 1981), it is even more difficult to envisage other forms of biogenetic magnetite exercising a major control over the magnetic mineralogy of sediments in the deep oceans.

Henshaw and Merrill (1980) have produced a comprehensive review of both authigenic and diagenetic magnetic phases in marine sediments. From their account, it is clear that both are important under quite specific and increasingly well defined circumstances. Moreover, authigenic and diagenetic processes are to some degree related to other potentially climatically modulated aspects of sedimentation, such as accumulation rate and organic content. However, the effects of authigenic and diagenetic transformations can usually be readily distinguished from the palaeoclimatically related variations already noted, and neither process can be invoked as a general explanation of the mineral magnetic–palaeoclimatic linkage.

In terms of origin, it may be tentatively concluded that the primary supply of magnetic minerals to most marine sediments is dominated by continental sources, both volcanigenic and erosive. Of these two, the latter are most obviously under direct climatic control. Moreover, the nature of the mineral magnetic evidence, pointing as it does to a substantial and significantly varying haematite component, also suggests a continental erosive origin for an important fraction of the magnetic mineral assemblages recorded. Even more evident is the extent to which some of the major pathways identified are strongly dependent on climate. Changing patterns of general atmospheric circulation are inferred as essential components of climatic change, and these in turn will have strongly influenced the location, strength and direction of large-scale ocean currents. Using anisotropy of magnetic susceptibility measurements to establish the preferred orientation of magnetic fabrics in deep-sea sediments, Ellwood (1980) has shown that changes in bottom currents can lead to changes in magnetic mineral concentration in deep-sea sediments from the South Atlantic, and Bloemendal favours this mechanism as a possible explanation for the mineral magnetic changes in DSDP Core 514 plotted in Fig. 10.6. Conversely, it is tempting to invoke a more direct atmospheric control for those areas adjacent to major arid landmasses. It is sufficient for present purposes to note these possibilities without choosing from amongst them, qualifying them more closely, or even claiming that the list of possibilities is exhaustive. The system outlined above provides ample scope for generating climatically modulated changes in magnetic mineral assemblages sufficient to account for the apparent geomagnetic–palaeoclimatic linkages which were the starting point for this account. Future research may now concentrate on the mechanisms responsible for the linkages in specific environments, and on the scope offered for adding new dimensions to palaeoclimatic reconstruction through illuminating these mechanisms.

Acknowledgements

The authors are very grateful to the Institute of Oceanographic Sciences for providing samples from core S8–79–4, for magnetic analysis, the results of which are shown in Fig. 10.7. They are particularly indebted to Dr R.B. Kidd for permission to use I.O.S. oxygen isotope, carbonate-content and core-description data, and especially grateful to Dr P.P.E. Weaver for allowing the reproduction of the results of his micropalaeontological analysis of this core.

Magnetic measurements on core IOS–S8–79–4 were undertaken as part of a research project funded by the Natural Environment Research Council.

References

Barton, C. E. and M. W. McElhinny 1976. Detrital remanent magnetization in five slowly redeposited long cores of sediment. *Geophys. Res. Lett.* **6**, 229–32.

Blakemore, R. P. 1981. Magnetite formation by bacteria. *EOS* **62**, 849–50.

Bleomendal, J. 1980. *Palaeoenvironmental implications of the magnetic characteristics of sediments from Deep Sea Drilling Project Site 514, southeast Argentine Basin.* Washington: Initial Reports DSDP.

Chester, R. 1972. Geological, geochemical and environmental implications of the marine dust veil. In *Changing chemistry of the oceans: proceedings of the 20th Nobel Symposium 1971*, D. Dryssen and P. Jagner (eds), 291–305. New York: Wiley Interscience.

Currie, R. G. and B. D. Bornhold (1983). The magnetic susceptibility of continental shelf sediments, west coast Vancouver Island, Canada. *Marine Geol.* **51**, 115–27.

Doake, S. M. 1977. A possible effect of ice ages on the Earth's magnetic field. *Nature (Lond.)* **267**, 415–17.

Doyle, L. J., T. L. Hopkins and P. R. Betzer 1976. Black magnetic spherule fallout on the eastern Gulf of Mexico. *Science* **194**, 1157–9.

Ellwood, B. B. 1980. Induced and remanent magnetic properties of marine sediments as indicators of depositional processes. *Marine Geol.* **38**, 233–44.

Granar, L. 1957. Magnetic measurements on Swedish varved sediments. *Arkiv förr Geofysik* **3**, 1–40.

Haggerty, S. E. 1970. Magnetic minerals in pelagic sediments. *Carnegie Inst. Year Book* **68**, 332–6.

Hallam, J. S., B. J. W. Edwards, B. Barnes and A. J. Stuart 1973. A late-glacial elk associated with barbed points from High Furlong, Lancashire. *Proc. Prehist. Soc.* **39**, 100–28.

Harrison, C. G. A. and J. M. Prospero 1977. Reversals of the Earth's magnetic field and climatic changes. *Nature (Lond.)* **250**, 563–4.

Henshaw, P. C. and R. T. Merrill 1980. Magnetic and chemical changes in marine sediments. *Rev. Geophys. Space Phys.* **18**, 483–504.

Kawai, N., K. Yaskawa, T. Nakajima, M. Torri and N. Natsuhara 1975. Voice of geomagnetism from Lake Biwa. In *Palaeolimnology of Lake Biwa and the Japanese Pleistocene*, vol. 3, S. Horie (ed.), 143–60.

Kent, D. V. 1982. Apparent correlation of palaeomagnetic intensity and climatic records in deep-sea sediment. *Nature (Lond.)* **299**, 538–9.

Kershaw, A. P. 1978. Record of the last interglacial–glacial cycle from northeastern Queensland. *Nature (Lond.)* **272**, 159–61.

Koyama, T. 1975. Geochemical studies on a 200-meter core sample from Lake Biwa–the vertical distribution of carbon, nitrogen and phosphorus. In *Palaeolimnology of Lake Biwa and the Japanese Pleistocene*, vol. 3, S. Horie (ed.), 306–15.

Levi, S. and S. K. Banerjee 1976. On the possibility of obtaining relative palaeointensities from lake sediments. *Earth Planet. Sci. Lett.* **29**, 219–26.

Lowenstam, H. A. 1981. Magnetite biomineralization by organisms. *EOS* **62**, 849.

Mullen, R. E., D. A. Darby and D. L. Clark 1972. Significance of atmospheric dust and ice rafting for Arctic Ocean sediment. *Geol. Soc. Am. Bull.* **83**, 205–12.

Oldfield, F., C. Barnosky, E. B. Leopold and J. P. Smith 1983. Mineral magnetic studies of lake sediments – a brief review. *Proceedings of the Third International Symposium on Palaeolimnology: Developments in Hydrobiology*, 37–44. The Hague: Junk.

Oldfield, F., J. A. Dearing, R. Thompson and S. E. Garret-Jones 1978. Some magnetic properties of lake sediments and their possible links with erosion rates. *Polish Arch. Hydrobiol.* **25**, 321–31.

Oldfield, F., R. Thompson and K. E. Barber 1978. Changing atmospheric fallout of magnetic particles recorded in ombrotrophic peat sections. *Science* **199**, 679–80.

Oldfield, F., K. Tolonen and R. Thompson 1981. History of particulate atmospheric pollution from magnetic measurements in dated Finnish peat profiles. *Ambio* **10**, 185–8.

Oldfield, F., T. A. Rummery, R. Thompson and D. E. Walling 1979. Identification of suspended sediment sources by means of magnetic measurements: some preliminary results. *Water Resources Res.* **15**, 211–18.

Radhakrishnamurty, C., S. D. Likhite, B. S. Amin, and B. L. K. Somayajulu 1968. Magnetic susceptibility stratigraphy in ocean sediment cores. *Earth Planet. Sci. Lett.* **4**, 464–8.

Robinson, S. G. 1982. Two applications of mineral-magnetic techniques to deep-sea sediment studies. *Geophys. J. R. astr. Soc.* **69**, 294.

Ruddiman, W. F. 1971. Pleistocene sedimentation in the equatorial Atlantic – stratigraphy and faunal palaeoclimatology. *Geol Soc. Am. Bull.* **82**, 283–302.

Ruddiman, W. F. and L. K. Glover 1972. Vertical mixing of ice-rafted volcanic ash in North Atlantic sediments. *Geol Soc. Am. Bull.* **83**, 2817–36.

Stuiver, M. 1972. On climatic changes. *Quaternary Research* **2**, 409–11.

Thompson, R., R. W. Battarbee, P. E. O'Sullivan and F. Oldfield 1975. Magnetic susceptibility of lake sediments. *Limnol. & Oceanogr.* **20**, 687–98.

Thompson, R. and B. Berglund 1976. Late Weichselian geomagnetic 'reversal' as a possible example of reinforcement syndrome. *Nature (Lond.)* **263**, 490–91.

Volat, J-L., L. Pastouvet and C. Vergnaud-Grazzini 1980. Dissolution and carbonate fluctuations in Pleistocene deep-sea cores; a review *Marine Geol.* **34**, 1–28.

Wollin, G., D. B. Ericson and W. B. F. Ryan 1971. Magnetism of the Earth and climatic changes. *Earth Planet. Sci. Lett.* **29**, 219–26.

Wollin, G., W. B. F. Ryan, D. B. Ericson and J. H. Foster 1977. Palaeoclimate, palaeomagnetism and the eccentricity of the Earth's orbit. *Geophys. Res. Lett.* **4**, 267.

11

Climate, sea-level and coastal changes

MICHAEL J. TOOLEY

The processes affecting sea level, coastal changes and coastal sedimentation are described, and the Dutch model of coastal sedimentation during the Flandrian Age is tested against the record in south-west Lancashire, UK. The record of sea floods and blowing sand on the Lancashire coast is presented, drawing on archival and natural sources. The evidence for sea surges in the coastal stratigraphy is reviewed. Problems of correlating coastal sedimentary units are considerd and proposals are made to standardise data sets.

11.1 Introduction

It was entirely characteristic of Professor Gordon Manley to interrupt and to disturb a comfortable intellectual state by insinuating an idea or a novel proposal apparently unrelated to current research. In April 1968, my research on coastal and sea-level changes in north-west England was interrupted by his proposal that I should look at the weather record for 1690–1728 kept by Nicholas Blundell at Crosby Hall, near the coast north of Liverpool. He suggested that this record might provide evidence of the effects of the Little Ice Age on this part of England. In fact, Blundell's record, and the papers of others (Ashton 1909, Beck 1954), provided a record of the storms and the consequential marine inundations that had affected this part of the Lancashire coast. At the time this proposal represented a distraction which did not seem immediately relevant to an understanding of coastal sedimentation during the Flandrian Age. Yet, in a perceptive and idiosyncratic way, Manley had been able to draw to the attention of an obdurate investigator in a challenging context the possibility of a connection between unlikely and apparently irreconcilable data. He showed the value of historical data to an understanding of coastal change, and pointed to the importance of extreme weather events in considering coastal sedimentation. As the debate on sea-level changes grew in the 1970s and the early 1980s, the value of this distracting exercise concerning the Blundell papers was realised.

If storms were characteristic weather phenomena of the Little Ice Age, there was every possibility that they characterised all periods of climatic deterioration in middle latitudes during the Flandrian Age, and probably every preceding glacial and interglacial age of the Quaternary and beyond. The record of extreme events was something which had to be considered. What would be the effect of these extreme events in the stratigraphic record, and what was the relationship of these periodic storms to the general trends of sea level?

It was posited that the effects of such periodic storms at a particular altitude of the sea-level surface on the stability of natural coastal defences would be manifest in the coastal stratigraphy. Therefore this could provide a proxy record of weather and climatic changes, complementing the more refined record provided by tree rings, recurrence surfaces, annually laminated sediments, and pollen diagrams.

The rôle of climate in general, and of extreme weather conditions in particular, in coastal change and coastal sedimentation had not gone unnoticed. While the overall control of climate through the glacio-eustatic process was acknowledged, some authors emphasised the overriding importance of this process, whereas others attributed the detailed changes in coastal stratigraphy to climatic cyclicity and extreme weather events.

In an historical context, Lamb (1981) considered climatic fluctuations and coastal changes, but it is clearly desirable to extend this record back to a time when the amplitude of both climatic fluctuations and coastal changes would have been greater. It is important to evaluate the rôle that climate and extreme weather conditions play in establishing a record of sea level in coastal sedimentary environments.

The objectives of this contribution are:

(a) to describe some of the factors affecting sea-level and coastal changes;
(b) to describe the sedimentation in low-altitude coastal environments;
(c) to examine the coastal sedimentary record in terms of sea-level movements, climate and extreme weather events.

The substance of this paper was given on 20 March 1981 at the International Conference on Climatic Change in Late Prehistory, held at the University of Durham (Harding 1982).

11.2 Processes affecting sea-level and coastal changes

Fairbridge (1961) identified four processes affecting global sea level and hence coastal change. Although the concept of 'global eustasy' has been questioned and discredited (Mörner 1976), and the search for a global sea-level curve abandoned, it is worthwhile recalling these four processes. The first is a movement controlled by the deformation of the ocean basins, which may be expressed locally, regionally, or on the scale of the ocean basin itself. The second is a movement caused by the addition of sediments to the ocean basin resulting in a rise of sea level. The third is a movement caused by the accumulation and melting of high-latitude and high-altitude ice caps and ice sheets, resulting in a fall and rise of sea level. The fourth is a change in the volume of water in the oceans, caused by a rise or fall in the temperature of the whole water mass.

Of these four processes, the third and fourth are inextricably linked to climate, and will be affected by changes in the global climate. The third, glacio-eustatic, process has probably been the most important in the explanation of gross sea-level and coastal changes during the past 2 million years (my), as it has for earlier geological periods. The rise of sea level during the early Silurian was probably the consequence of the melting of the late Ordovician Saharan ice sheet, and the late Carboniferous cyclothems may reflect changes in the mass budget of Gondwanan ice (Hallam 1981).

The past 2 my have been characterised by rapid changes of climate at all latitudes. An alternation of glacial and interglacial conditions occurred in middle and high latitudes: in low latitudes an alternation of wet and dry conditions has been recorded (e.g. Flenley 1979, Street & Grove 1979). It is probable that bursts of cold air masses penetrated the Tropics and equatorial zone (Damuth & Fairbridge 1970) and effected changes in the extent and pattern of the tropical rainforests.

In East Anglia (UK), West (1980) identified seven cold stages, during five of which permafrost and periglacial sediments have been recorded. During only three stages glacigenic sediments have been recorded. However, in Germany, three periods of ice advance and four periods of cold have been identified during the past 1 my, and more than four cold periods before that (Menke & Behre 1973, Behre et al. 1979). Shackleton and Opdyke (1973) recognised eight major glaciations during the past 0.7 my, based on oxygen-isotope analyses from a deep-sea core from the Solomon Plateau in the Pacific Ocean.

The overriding characteristics of the Quaternary period were glaciation and periglacial conditions in middle and high latitudes, low-latitude aridity, and low sea level. Thus, the continental shelves of the globe would have been subjected to long periods of subaerial weathering, and land connections between continents and offshore islands would have been characteristic.

The addition of water to and abstraction of water from the world's oceans during a glacial/interglacial cycle have not produced equal effects in terms of the altitude at which the sea-level surface has intersected the continents. Clark and Lingle (1977), Clark et al. (1978) and Clark (1980) have shown that the rise in sea level due to the melting of high-latitude and high altitude ice is not uniform over the globe, because the ocean floor, the continental shelves and the land deform under the changing weight of ice and water loads. In addition, the shape of the ocean surface (the geoid) will change as mass is redistributed within the Earth. Clark (1980) developed a model in which six relative sea-level change zones since 16 000 BP were predicted: in four of these zones, sea levels higher than the present were predicted for the mid-Holocene, c. 5000 BP, and there are observational data to support these predictions. Thus, sea levels higher than the present no longer constitute a problem, but contribute to a solution of local and regional sea-level changes, explanations for which (Mörner 1980) embrace climatic change (glacio-eustasy), earth movements (tectonic-eustasy), and gravitational changes (geoidal eustasy).

At the scales of an individual site (an estuary, an enclosed bay, or an epicontinental sea), coastal and sea-level changes will also be affected by sediment supply, tidal patterns and variations, weather patterns, and vegetation patterns.

During the Flandrian Age the clearing of vegetation and the cultivation of the land carried out by man have initiated erosion and have led to an exponential increase in sediment influx to parts of the epicontinental seas in Western Europe. The record of microscopic and macroscopic charcoal in the sediments of Downholland Moss in south-west Lancashire bears witness to clearance and enhanced sediment influx from the Mesolithic onwards, and changes in sediment supply will have affected the stability of intertidal sand banks in this area.

Grant (1970) was probably the first investigator to draw attention to the rôle of tidal changes in explaining accelerated submergence and sea-level rise. He concluded that in the Bay of Fundy, about half the excessive rate of submergence can be explained by the differential rise of the high-tide datum through increasing tidal range.

Work on numerical tidal models at Durham (M. J. Davis, personal communication) for sea levels reduced by 20 m (c. 9500 BP), 10 m (c. 8500 BP) and 5 m (c. 7500 BP) showed significant changes in the amplitude of the M2 tide. In Liverpool Bay, for example, the amplitude of the M2 tide was reduced from c. 2.80 m to c. 1.80 m when mean sea level was lowered by 20 m. At the head of the Bristol Channel, the reduction was from c. 4.30 m to c. 1.50 m, but at the mouth of the Bristol Channel, there was a slight increase in amplitude from c. 2.20 m to c. 2.30 m. A reduction in the rate at which sea level fell yielded changes in amplitude that were also reduced, but the direction persisted. As the sea-level altitude approached that of the present day, so did the tidal amplitude. On the Dutch coast, Roep et al. (1975) attempted to demonstrate that tidal changes had not occurred for the past 4000 years. This conclusion is to be expected as past regional sea levels converge on those of the present, and the greatest expected changes would be before 6000 BP, when sea level was rising rapidly, water depths were increasing, and the geometry of epicontinental seas and bays was changing significantly.

Whereas the processes affecting global sea-level changes may be dominated by climate, local sea-level changes and the registration of sea-level index points may owe much to secondary effects which obscure a simple relationship between sea level and coastal changes and climate.

However, the starting point for resolving the relationship between the primary and secondary effects must be the objective recording of low-altitude coastal stratigraphy in the field for which empirical models exist to group the observed data.

11.3 Coastal sedimentation during the Flandrian Age

Dutch geologists (e.g. Hageman 1969, Jelgersma et al. 1979) have identified three genetic coastal landscapes in the Netherlands that together serve as a basis for grouping empirical data and as a useful model of the evolution of low-altitude coasts. The three landscapes are the coastal barriers and coastal dunes, the tidal flat and lagoonal areas, and the perimarine areas.

The coastal barriers and dunes consist of fine- to medium-grained sands, with shells forming the barriers and overlaid by blown sand. This sand is interrupted by palaeosols of mor soil and dune slack peat, permitting a chronostratigraphic subdivision into Older and Younger Dunes.

The tidal flat and lagoonal deposits are in part overlapped by the coastal barriers and dunes, and the lithostratigraphic units consist of low and high tidal flat deposits, marine, brackishwater and freshwater clays, intercalating salt-marsh, telmatic and terrestrial peats, and lake muds. The marine deposits comprise the Calais and Dunkerque Members of the Westland Formation (Doppert et al. 1975, quoted in Jelgersma et al. 1979), laid down during marine transgressive intervals; whereas the Holland Peat Member consists of organic layers laid down during marine regressions.

In the perimarine zone, the main lithostratigraphic units are sand, clay and peat, laid down under fluvial or limnic conditions at altitudes controlled by effective sea level to seaward in the tidal flat and lagoonal areas. The clastic layers are correlated chronostratigraphically with the marine transgression facies to seaward: the Gorkum Member of clay and sand with the Calais Member, and the Tiel Member of clay and sand with the Dunkerque Member.

Jelgersma (1961, 1966) has posited that increased river discharge during wet periods, together with an increase in storminess, could be responsible for marine transgressions (the Calais and Dunkerque Members) in the tidal flat and lagoonal zone, and clastic sedimentation in the perimarine zone (the Gorkum and Tiel Members). Furthermore, corroboration for this cyclicity of climate inferred from the pattern of transgressions and regressions was found from an examination of the lithostratigraphy of coastal dunes in the Netherlands (Jelgersma et al. 1970). Peaty horizons in the coastal dunes were thought to represent wet periods, and were therefore associated with marine transgressions; whereas the sand layers represented dry periods, and were associated with marine regressions.

Jelgersma (1961, 1966, 1979) argued that these patterns of coastal sedimentation were a result of a smoothly rising sea level, effected by climatic change and phases of extreme weather patterns, and denied the possibility of fluctuations of sea level to explain the stratigraphy of the Dutch coast. However, without re-opening the discussion on the form of the curve of sea-level rise during this interglacial age, it is worth considering the Dutch model of coastal landscapes and the detailed lithostratigraphy as a proxy record for climatic change and extreme weather patterns.

If the model is verified from other low-altitude coastlands in several ocean basins, the existence of an overriding control of coastal sedimentary processes will be established. If the model is verified at the scale of an enclosed epicontinental sea, and not at an oceanic or global scale, it is likely that a glacio-eustatic control has been modified by regional conditions, such as tectonic movements,

hydro-isostasy, tidal variations, and recurrent extreme weather patterns producing storm surge conditions. Verification at an oceanic or global scale implies
a global control, such as that provided by glacio-eustasy.

11.4   Coastal sedimentation in south-west Lancashire (UK)

In order to test the Dutch model of the evolution of low-altitude coasts, the
record of coastal sedimentation in south-west Lancashire can be elaborated.

T. Mellard Reade (1871) was probably the first to describe the coastal
landscapes and the sediments associated with them in south-west Lancashire,
although there had been a long history of investigations in this area (see Tooley
1978a). Reade recognised seven lithostratigraphic units: Boulder Clay, Washed
Sand, The Inferior Peat and Forest Bed, The Formby and Leasowe Marine
Beds, The Superior Peat and Forest Bed, Recent Silts, and Blown Sand. His
explanation for these units involved periods of depression and elevation of the
land, and in this explanation Reade followed Lyell (1875) and Picton (1849),
although he could equally well have followed Binney and Talbot (1843), who
explained the lithostratigraphy of part of the coastlands of south-west Lancashire in terms of the breaching of coastal barriers and the blocking of channels by
sand banks while sea level remained static (see Tooley 1978b).

South-west Lancashire has continued to yield information on coastal sedimentation (Tooley 1974, 1976, 1978a & b), and further investigations on the
sediments of Downholland Moss are reported here.

Downholland Moss (Fig. 11.1) is one of many sites in north-west England
and North Wales in the tidal flat and lagoonal zone, as defined by Hageman
(1969) and Jelgersma et al. (1979). It is flanked by Boulder Clay and Shirdley Hill
Sand (Reade's 'Washed Sand') to landward, by intertidal sand banks and blown
sand to seaward, and by identical palaeoenvironments to the north and south,
such as Halsall Moss and Altcar Moss. The stratigraphy of the Moss has been
described by Binney and Talbot (1843), Wray and Cope (1948), and Tooley
(1978a & b). Howell (1973) has shown the existence of a valley in the subdrift
surface at altitudes of 0 to − 30 m O. D. (Ordnance Datum). The morphology
of this valley is reflected in the Boulder Clay surface, but has been obliterated by
a fill of unconsolidated sediments of Flandrian Age, up to 18 m in thickness at
the seaward end of the axis of the buried valley.

Further eastward, at Downholland Moss–16 (DM–16), the surface of the
Boulder Clay lies at an altitude of − 9.8 m O. D., and is overlaid by 13 m of
sands, silts and clays with intercalated peat layers. Eleven beds are recognised
here (detailed in Table 11.1). Each bed (1–11) can be traced laterally in dyke
sections and temporary sections, and proved in borings (see Figs 6, 7 & 33 in
Tooley 1978a). The conditions under which some of the beds were laid down
can be further elaborated by reference to Figures 11.2 and 11.3.

Eight beds are shown in Figure 11.2: these are equivalent to the following
beds in Table 11.1: Bed 1 = Bed 8; Bed 2 = Bed 9; Beds 3 & 4 = Bed 10, and

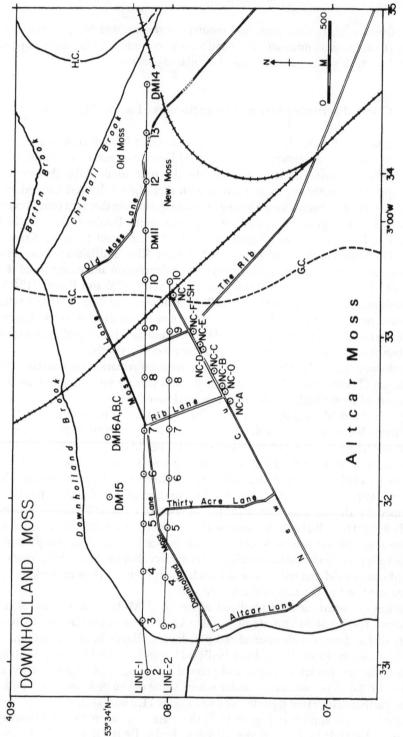

Figure 11.1 Map of Downholland Moss showing the location of sampling sites on the moss.

Table 11.1 The lithostratigraphy at Downholland Moss-16.

Bed	Depth from surface (cm)	Altitude (m O.D.)	Description
11	000–187	+3.20 to +1.33	peat
10	187–203	+1.33 to +1.23	clayey silt
9	203–218	+1.23 to +1.02	peat
8	218–293	+1.02 to +0.27	clayey silt
7	293–315	+0.27 to +0.05	peat
6	315–617	+0.05 to −2.97	fine sandy silt
5	617–645	−2.97 to −3.25	peat
4	645–833	−3.25 to −5.13	fine sandy silt
3	833–882	−5.13 to −5.62	peat
2	882–1212	−5.62 to −8.92	sand
1	1212–1300	−8.92 to −9.80	clay

Beds 5–8 = Bed 11. In Figure 11.3, Beds 2 and 5–8 have been further subdivided on the basis of a detailed examination of the plant macrofossil content of monolith samples taken from a free-face excavation. Standard volumetric samples of 50 cc were taken from each bed, and analysed quantitatively for their plant macrofossil content (by Dr P. A. GreatRex) in order to establish the environmental conditions under which the beds were laid down, and to confirm the marine origin of Beds 1, 4 and 5 (Figure 11.3). The results of this investigation are given in Figure 11.3.

The seed assemblage from Bed 1 was dominated by *Juncus maritimus*, with low frequencies of *Salicornia* spp. and other *Chenopodiaceae*. At 146 cm and 148 cm, pollen frequencies of 7% and 32% (Eland pollen) of Chenopodiaceae were recorded. There is no doubt that the silty clay of Bed 1 was deposited in the upper levels of a salt marsh in a community dominated by *Juncus maritimus*. Tansley (1939) commented that plants typical of the lower zones of salt marshes (such as *Salicornia* spp.) may reappear in *Juncus*-dominated communities, and this is the situation here.

Macrofossil remains were poorly preserved in Bed 2, and this, together with low seed numbers, made interpretation difficult. The deposit is a dark brown, granular *Phragmites*-rich peat, and charcoal is abundant at the base of the bed. It is probable that the water table had fallen to produce these effects.

In the overlying Bed 3, rising water levels are indicated by the occurrence of the seeds of brackish-tolerant plants such as *Lemna* cf. *trisulca*, *Ranunculus sceleratus* and *Typha* sp., succeeded by a community characterised by *Juncus subnodulosus*, *Cladium mariscus*, *Schoenoplectus tabernaemontanii*, *Scirpus maritimus* and *Samolus valerandi*, all of which are tolerant of brackish water conditions. A brackish water fen or marsh was probably the peat-forming community.

The organic sedimentation of Bed 3 is succeeded by predominantly inorganic sedimentation of Beds 4 and 5. Superficially, these beds of clayey silt with from 7% to 12% organic content appear to indicate a return to an upper salt marsh sedimentary regime, such as that recorded in Bed 1, but the seed assemblages

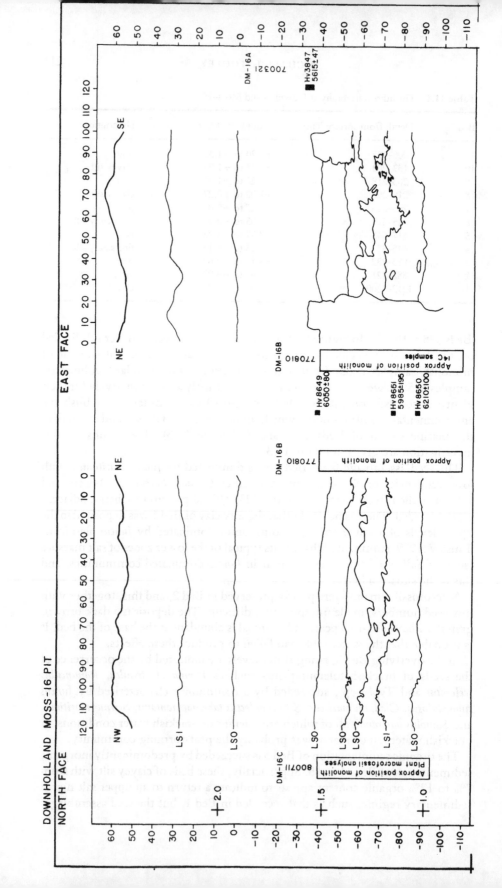

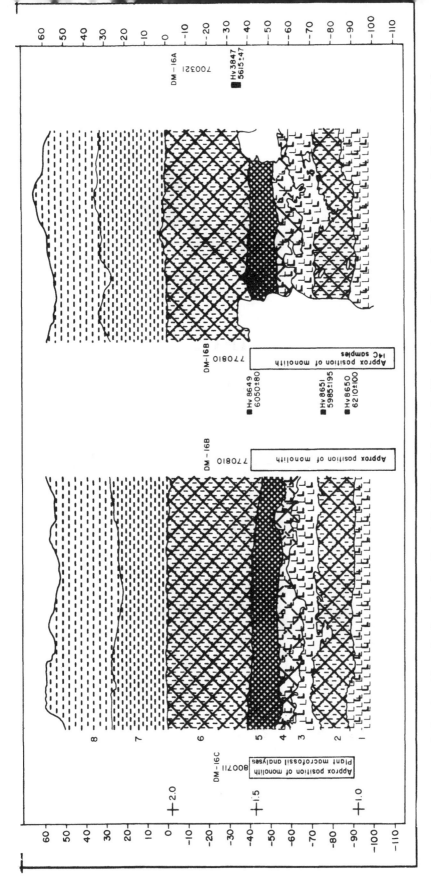

Figure 11.2 Stratigraphic section from the pit at Downholland Moss–16 showing the litho-stratigraphy. The symbols are those employed by Troels–Smith (1955).

DOWNHOLLAND MOSS – 6C : MACROFOSSILS

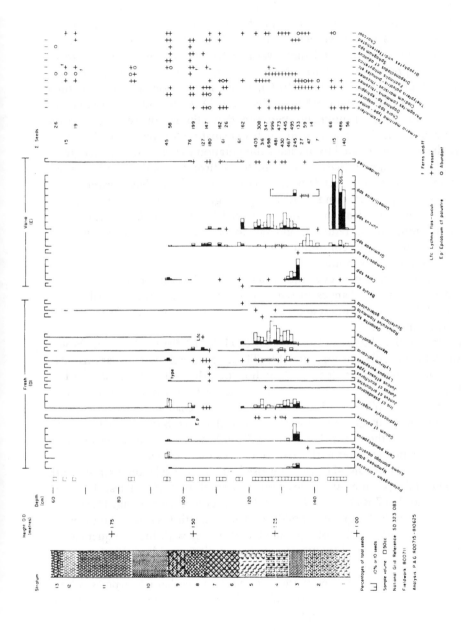

Figure 11.3 Macrofossil diagram from Downholland Moss–16C. Each taxon is shown as a percentage of total seeds counted. The taxa are grouped ecologically into Salt (A), Brackish (B), Fresh/Brackish (C), Fresh (D) and Varia (E) classes.

indicate a dominantly fresh-brackish to fresh character of the plant communities, in which *Juncus subnodulosus* is the dominant taxon. Charcoal is present sporadically throughout both beds. A rapid rate of sedimentation is testified by two ^{14}C dates that are scarcely distinguishable (see Fig. 11.2). A rapid influx of clastic sediments may have derived from a fluvial source, or have been the consequence of clearance and sheet wash. Alternatively, Beds 4 and 5 are the landward expression of a marine transgression, where it overlaps the organic beds. In Figure 33 of Tooley (1978a), this transgression is shown in three bore-hole records, pushing 900m landward of DM–16, and is overlapped by organic sediments. It is unlikely, therefore, that a fluvial source or soil erosion consequent upon clearance is an adequate explanation for these beds. It is more likely that they are related to marine sedimentation to seaward, and changes in the seed assemblages in Beds 6 and 7 tend to confirm this. In these beds there are absolute increases in the seeds of salt marsh taxa, such as *Juncus maritimus, J. gerardii*, Chenopodiaceae and *Salicornia* spp, together with brackish water taxa such as *Scirpus maritimus* and *Schoenoplectus tabernaemontanii*. Thereafter, the salt-water influence is reduced, and salt to brackish water communities are replaced by freshwater communities (of which *Nymphaea* is an element), and then fen and bog communities.

The changes in water level and water quality that can be inferred from a consideration of the changes in sedimentation and plant macrofossil assemblages at DM–16 are manifest in the stratigraphy of sampling sites further east in the perimarine zone.

At Downholland Moss–11 (Figs, 11.1 & 11.4), 12 distinct beds are shown, although further subdivision is possible. These beds are shown in relationship to three marine beds recorded at Downholland Moss–10, 300 m seaward of DM–11, of which only the lowest transgression is recorded directly at DM–11 and dated from 6980 ± 55 to 6760 ± 95. The changes in sedimentation succeeding this transgression, and closely correlated with two more transgressions recorded to the seaward on Downholland Moss, allow further elaboration of environmental changes in this part of the perimarine zone.

Beds 4 and 5 at DM–11 are comprised of monocotyledonous *turfa* with *Phragmites* (4), passing up in to woody detrital peat (5). These beds are equivalent to Bed 7 at DM–16 (Table 11.1). The pollen spectra (Fig. 11.4) show a rise, a fall, and a subsequent rise of the pollen of aquatic taxa, whereas the nadir of frequency of the pollen of aquatic taxa coincides with high frequencies of *Quercus* pollen, enhanced frequencies of *Pinus* pollen, and high frequencies of fern spores. At the same levels, the frequencies of the pollen of *Alnus* and *Betula* decline. In Bed 5, pollen preservation are poor and pollen frequencies is low. The number of traverses required to obtain 200 tree pollen increases from 11 at the top of Bed 4, to 44 in Bed 5, where the pollen frequency of *Quercus* attains a peak, returning to 11 in Bed 6 (see Fig.8 in Tooley 1978a).

Bed 6 is a *limus* with *Phragmites*. In this bed, lowered frequencies of *Quercus* are compensated by higher *Alnus* frequencies, and these levels coincide with a peak frequency of aquatic taxa.

At the top of Bed 7, the interruption in marine sedimentation shown by the break between Downholland II and III is reflected in the higher frequencies of Quercus pollen and the lower frequencies of Alnus pollen. Drier conditions are also reflected by poor pollen preservation and a reduction in pollen density: the number of traverses increased to 26 at the level at which Quercus pollen frequencies attained a maximum.

In Beds 8–10, the relationship between Quercus and Alnus does not show such nice effects, even though there is a persistent increase, followed by a decrease in obligate aquatic taxa such as Nymphaea, that synchronises with a marine transgression to seaward.

These patterns conform to what Shennan (1981) has described as the Godwin model. Shennan has drawn attention to the fact that Godwin (Godwin & Clifford 1938, Godwin 1978) used arboreal pollen frequencies as indicators of water-level movements. Dryness is testified by an increase in the frequencies of the pollen of Quercus and Pinus, poor pollen preservation, sparse pollen, and a ⁻rise in the frequency of fern spores. Wetness is testified by high Alnus pollen frequencies. Iversen (1960) also showed a clear negative correlation between the pollen curve for oak and the pollen curve for other trees such as Alnus and Betula participating in hydroseres. At the site on Lundergaard Mose in northern Jutland, where he worked, the Littorina Transgression deposits are overlaid by beds of turfa, limus and detritus similar to the succession of beds at Downholland Moss–11.

Further south on Downholland Moss, at the boundary with Altcar Moss, the number of beds is considerably reduced as the subcrop of Boulder Clay and Shirdley Hill Sand increases in altitude and approaches the surface. The number of marine beds is reduced from four to two.

Recent excavations along the New Cut have revealed considerable variations in the thickness and composition of the marine beds. In addition, tidal channels with levees have been recorded in section, and, flanking the New Cut, these levees can be traced as low sinuous banks in the fields. The channels are infilled with fine-grained, laminated sediments. They have been recorded for c. 1 km along the south side of the New Cut, and two sections have been drawn up (see Figs. 7 & 8 in Heptinstall 1983). They are recorded here for the first time as distinct morphological features in the tidal flat and lagoonal zone of south-west Lancashire, and appear to be equivalent to the raised banks of laminated silt in the Cambridgeshire and Lincolnshire Fenlands and Siltlands, first described by Skertchly (1877), defined as roddons by Fowler (1932, 1934), and re-examined by Godwin (1938).

These detailed stratigraphic, pollen and spore microfossil, and plant macro-fossil records, together with the geomorphological and sedimentary records of the roddons, allow a close interpretation of environmental changes at the boundary between the tidal flat and lagoonal zone and the perimarine zone of south-west Lancashire. They draw attention to the fact that gross changes in the stratigraphy and superficial similarities in lithology cannot be interpreted as the same sedimentary response to similar environmental changes without the

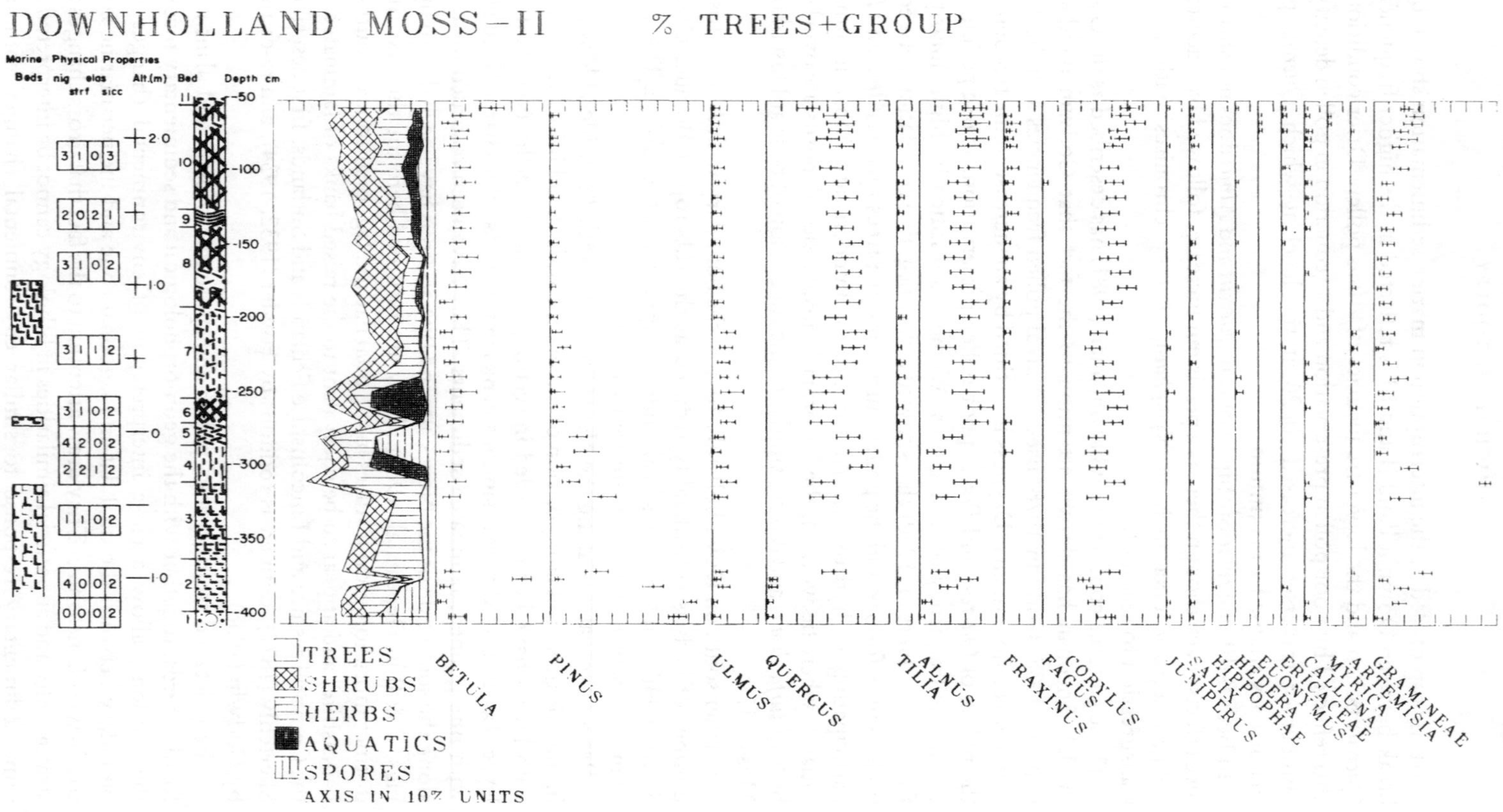

DOWNHOLLAND MOSS-II   % TREES+GROUP

GRAMINEAE
ARTEMISIA
MYRICA
CALLUNA
ERICACEAE
EUONYMUS
HEDERA
HIPPOPHAE
SALIX
JUNIPERUS

CORYLUS
FAGUS
FRAXINUS

ALNUS
TILIA

QUERCUS

ULMUS

PINUS

BETULA

TREES
SHRUBS
HERBS
AQUATICS
SPORES

AXIS IN 10% UNITS

Marine Physical Properties
Beds nig sica   Alt(m) Bed   Depth cm
        srrf sicc

Figure 11.4 Pollen diagram from Downholland Moss–11. Each taxon is expressed as a percentage of total tree pollen and taxonomic group.

support of plant microfossil and macrofossil analyses. Nevertheless, they provide evidence of substantial changes in water quality and water depth, that have also been recorded at more seaward sites (eg. Downholland Moss–15, Tooley 1978b) over a minimum period of 3000 years.

The lithostratigraphy described here and elsewhere in north–west England (Tooley 1978a) is similar to the coastal lithostratigraphy reported from elsewhere in Britain – for example, in the Fenland (Godwin 1978, Shennan 1981); in the Somerset Levels (Godwin 1941); in the Thames estuary (Devoy 1979); and in north–east Scotland (Haggart 1982). Furthermore, a similar lithostratigraphy has been described elsewhere in north–west Europe (see Section 11.7 and review in Tooley 1978a). This type of coastal lithostratigraphy is not confined to middle–latitude coasts, but has also been described from low–latitude coasts.

The results of research (by Stephen Ireland of the Department of Geography at the University of Durham) on the lagoons adjacent to the coast of Rio de Janeiro State Brazil have shown that clastic sediments are intercalated with organic sediments. Surrounding the Lagao de Itaipú, south–east of Niteroi, there is a large expanse of marshland, and borings into the marsh have proved the existence of four distinct peat beds: the surface peat bed is 50 cm thick and lies at about the mean sea level (MSL); the second bed dips landward and lies between -1.0 and -1.7 m MSL; the third bed is found between -3.0 and -4.0 m MSL (and within these limits each recorded bed is variable in altitude); and the fourth bed is found between -3.5 and -5.8 m MSL. Here, there is a close similarity to the genetic coastal landscapes proposed by the Dutch geologists: the barrier beaches, sand dunes, tidal flats, lagoons and salt marshes are characteristic contemporary features of this coast. The present coast around Salvador, Bahia State, appears to have been transgressed about 7000 years ago (Martin et al. 1979), and it is probable that the coast of Rio de Janeiro State was similarly transgressed at this time, initiating the coastal landscapes described.

There is, therefore, a body of empirical data that supports the Dutch model very strongly, and this implies that there is an overall control of coastal sedimentation.

However, in one respect the Dutch model is lacking. Godwin and Clifford (1938) and Godwin (1978) showed that, in the area that would now be defined as the tidal flat and lagoonal zone of the Fenland, the culminating plant communities in the vegetation succession following the end of the deposition of the Fen clays and silts were characteristically oligotrophic, typified by Sphagnum or Sphagnum–Eriophorum–Calluna peat. These are the constituents of raised bogs, nourished exclusively by rain water, and are unaffected directly or indirectly by sea level (see discussion in Tooley 1979). Although these plant communities and the peats they formed are not to be found in the Fenland, they do still occur in Wales (for example Borth Bog), and in north–west England (for example Ellerside-White Mosses, Wedholme Flow). Hence a fourth coastal landscape should be discriminated in low-lying coastal areas: the raised bog landscape.

The explanation of these lithostratigraphic changes lies at the heart of the controversy on sea-level movements and the relationship of marine transgressions to climatic change. Godwin and Clifford (1938) have noted that 'the interpretation of Fenland history is complicated by the fact that dryness and wetness of the surface is affected by the operation of at least three major factors – climate, relative movements of the land and sea, and the natural successional tendencies of vegetation'. This applies not only to the Fenland of the United Kingdom, but to every low lying humid coastal site.

The fact that Godwin and Clifford used the term climate rather than weather is significant, because it implies weather averaged over many years, and hence the summation of extreme events. To have an effect on vegetation succession, a repetitive pattern of extreme events (either river or sea floods, or frost or extreme dryness) is necessary: an isolated event will have little or no effect either on the vegetation structure and composition, or on the sedimentary regime, as has already been shown for a single sea flood (Tooley 1979). Whether or not the pattern is repetitive on the Lancashire coast can be assessed by reference to the historic record of sea floods and blowing sand considered in the following section.

11.5   Sea floods and blowing sand on the Lancashire coast

The record of sea floods and blowing sand on the Lancashire coast has not been systematically drawn together from archival and natural sources, but there is a hint of rich archive sources which Manley had identified (e.g. Ashton 1909, Beck 1954, Bagley, 1968, 1970, 1972) The sand dunes of the Lancashire coast contain a record of changes from stable to unstable conditions which correlates in broad outline with the record of the Dutch dunes (Tooley 1978a). In Holland, a close correlation has been demonstrated between marine transgressions and an increase in storminess and sea floods, and dune stability (Jelgersma 1966, Jelgersma et al. 1970), whereas blowing sand is associated with marine regressions.

Ashton (1909) summarised the main dates and periods of inundation of the coasts of Lancashire and North Wales as AD 331, 500–520, 563, 1279, 1311–24, 1395–1411, 1400–1500, 1553–4, 1655 (Tyrer 1970), and 1720. Further notable floods occurred in 1833, 1907 and 1927 (Barron 1938), and again in 1978.

Whereas Beck (1954) provides a scholarly account of the inundation of the Lancashire coast in 1720, and of the way in which money was raised to recoup the losses incurred, Nicholas Blundell (Bagley 1972) recorded it briefly in his Great Diurnal, and then at length in his Anecdote Book (Tyrer 1970):

18 December 1720. Never the like Thunder and Lightoning known at this time of the year as was this morning or rather after Midnight.

December wet and most extraordinary great winds with Thunder and Lightoning, never so much damage done by High Tides in these parts as now.

Till the 10th generally Raine, then to the 16 very Fair the four next days very wet and extremely Windy the like scarce ever known and never so high a

Tide known as was these four dayes especially the 18th and 19th chiefly at the Meales, Alker, Alt-Grange and towards Lancaster, there also was very great Thunder and Lightoning, towards the End of this Month generally faire but much Wind.

Notwithstanding the loss of life and property during the 1720 inundation, that covered an underestimated minimum acreage of 6500 acres (Beck 1954), Nicholas Blundell wrote in his *Anecdote Book*:

There was a most prodigious FLOOD in England, Ireland and Flanders, etc by the unusuall and Prodijuas overflowing of the Sea; in Lancashire it did very great damage in the Fild, in Meyles, at Alt-Grange etc. and Breefs were obtained to beg thorrow England for the great Losses sustained in Lancashire: in the Fild a Man got up into a Tree to save his life and a hare swam to the same tree for refuge whilst he was in it, in the Meyles a Swine of 40 shillings Valew got up to the top of a Turf-Stack where he lay till he was fetched down.

Although marks have been cut on buildings or milestones to record the maximum altitudes attained by these inundations in Lancashire, there is no evidence of their occurrence in the stratigraphy of the peat mosslands. On the contrary, there is a rich stratigraphic record of sand blowing in Lancashire, of which an example is given here from Lytham Common and Lytham Hall Park, and extended by the accounts of Lytham Priory.

The dunes on the Lancashire coast appear to have begun to accumulate, either on shingle banks or intertidal sand banks, some time before 4090 ± 170 BP (Tooley 1978a), and, hence, to coincide with the formation of the oldest dunes in Holland (Jelgersma *et al.* 1970). Dates on dune slack peat of 2335 ± 120, 805 ± 70 and 830 ± 50 imply two periods of dune stability, and three periods of dune instability. The last period of instability coincides with the foundation of a monastic cell at Lytham (an outpost of the Benedictine monastery of Durham).

Although there are gaps in the account record for Lytham, Alan Piper has absracted records from 1448 to 1515 which show losses in income, largely as the consequence of blowing sand from two areas of grazing land (Old Park and Newhay, and The Green) and the common kiln or oven. These are tabulated in Table 11.2.

The combination of meteorological and tidal conditions that leads to storm surges is particular to each epicontinental sea: the North Sea is affected by surges generated in the northern North Atlantic and within the sea itself; whereas the Irish Sea is affected by secondary depressions approaching from the south-west or west. It is not realistic, therefore, to expect a synchronous record in different epicontinental seas, and yet the record for the 15th century from the Irish Sea and the North Sea does hint at groups of years in which storms and floods were more characteristic and destructive than in other years.

In the Netherlands, Gottschalk (1975) drew together the evidence for storm surges and river floods from 1400 to 1600. The period from 1449 to 1452 appears

Table 11.2 A record of sand blowing from the accounts of Lytham Priory.

Year	Value of receipts	Location and remarks
1448–9	12s	Old Park
1449–50	Nichil	Old Park quia per flatum venti cum sabulo superflatur
1450–52	Nichil	Ditto
1452–3	Nichil	Old Park
1453–6	Nichil	Newhay et Old Park quia in manu propria
1457–8	12s	diversis tenentibus pro herbagio de Old Park
1461–2	1s	herbagio del Grene
1462–4	Nichil	herbagio del Grene
1464–6	Nichil	herbagio del Grene inter manerium et villam propter flatum zabuli
1467–72	4d	herbagio del Grene
1476–1510	Nichil	herbagio de le Grene inter manerium et villam quia totaliter vastatum per flatum zabuli
1490–1510	Nichil	communi thorali quia vastatur propter flatum sabuli
1514–15	2s	herbagio de le Grene inter manerium et villam

to have been characterised by storm surges, and this included a period during which the Starr Hills (the coastal dunes) at Lytham were unstable, and pastureland was lost because of blowing sand. The period from 1457 to 1462 was characterised by few storm surges, and this period yielded some income at Lytham because the dunes had stabilised. In the Netherlands: '1463 was unmistakably a storm surge year' (Gottschalk 1975), and apart from a respite in 1470, the second half of the 15th century was characterised by a very high frequency of storm surges, with a particularly ferocious surge on 27 September 1477, which Gottshalk regarded as the worst since 1404. It is perhaps not an unsingular coincidence that during the long period from 1476 to 1510 at Lytham, no income came from the pastureland known as The Green because it had been completely destroyed by blowing sand.

The 15th century was a climatically critical period. It was a century of climatic variability and great weather extremes. Lying on the boundary of the Medieval Warm Epoch (MWE) of AD 1000–1400 and the Little Ice Age epoch of AD 1400–1800, there was a 400-year period (AD 1200–1600) during which there was a significant cooling of those parts of the globe that had been warmer during the MWE (Ingram et al. 1981).

The critical year appears to be AD 1433, which coincided with a very high but brief period of sunspot activity, followed by a marked minimum, which initiated a period of great climatic instability, presaging the Little Ice Age epoch and the period of maximum cold in the 17th century (Fairbridge & Hillaire-Marcel 1977).

There is the kernel of a suggestion that groups of storm surges may be registered on different coasts at the same time, and that there is the real and exciting possibility of correlating natural and archival records. Clearly, this is a rich area for future research in assembling stratigraphic records from coastal dunes in Britain, particularly where sand overlaps peat, and, for the later medieval period onwards, assembling corroborative archival records.

11.6 The stratigraphic record of storms

Kumar and Sanders (1976) described the characteristics of sediments laid down in the near shore zone off Fire Island (in New York State, USA) under fair weather and storm conditions, and compared the storm-deposited sequences with older geologic records. Sediments laid down under fair weather conditions comprise ripple and laminated sands between the breaker zone and the position of the wave base. Seaward, at depths greater than the wave base, the sediments are bioturbated and encrusted with organic debris. Under storm conditions, a thin gravel lag pavement is overlaid by plane-laminated micaceous fine sands: the lag was formed during the peak of the storm and the sands during the waning stages of the storm. Kumar and Sanders concluded that the character-istic sediments of the near shore zone are storm deposits.

During the Flandrian Age, there were times when sea level was rising very rapidly, coastlines were open and being transgressed continuously, and deep water conditions obtained. Under these conditions, near shore zone sediments would be laid down landward of the present coast, and, within transgressive overlap sequences, the type of storm deposits described by Kumar and Sanders should have formed. The coasts of north-west Europe share with northeastern North America a similar type of storm wave environment (Davies 1964), and there is a greater likelihood of.such storm deposits occurring during the Flandrian. However, no sedimentary successions have been reported landward of the present low-altitude, mid-latitude coastlines, which are formed of unconsolidated materials and which approach, in structure and composition, the sediments laid down under storm conditions. Although sediment source and supply would result in regional variations, it should be expected that the record of storms would be manifest in sedimentary successions if storms were the overriding process.

In south-west Lancashire, the clastic beds contain structures and textures that indicate quiet-water sedimentation, whereas the diatom assemblages indicate changes in position of the sedimentary site in the intertidal zone (Tooley 1978b). Huddart (1978 unpublished) has shown from analyses of the foraminifera from the clastic sediments of Downholland Moss that sedimentary cycles from salt marsh through tidal flat to salt marsh are characteristic of the clastic layers.

There are stages when conditions were so open on the coast of south-west Lancashire that small foraminifera were transported from the open shelf – a pattern identical to the one reported by Murray and Hawkins (1976) from sites on the east side of the Severn estuary. These small foraminifera are incorporated into an assemblage that is characteristically the lower part of the low marsh, and covered during most tides. Nevertheless, there is as yet no evidence to suggest sedimentation under storm conditions.

It has been suggested that thin clastic layers, of limited extent and closely dated, may have been laid down under extreme tidal and meterological conditions. Such layers have been recorded in north-west England, and are usually formed of fine-grained sediments. For example, in Nancy's Bay on the north side of the Ribble estuary (Tooley 1978a), there are two thin clastic layers, 18 cm and 38 cm thick; they comprise organic blue clay and are separated by 21 cm of *Phragmites turfa* and *limus*. The oldest of the two transgressions has been dated from 6245 ± 115 to 6250 ± 55, and the youngest from 5950 ± 85 to 5775 ± 85. Clearly, both could be the landward expression of a single storm or a group of storms. But the pollen diagram (Fig. 18 in Tooley 1978a) shows persistent low frequencies of the pollen of salt marsh taxa (Chenopodiaceae), and a peak of the pollen of aquatic taxa within the intercalated organic bed. This succession is similar to that described from Downholland Moss–16, and the dates on both successions are similar. It is more likely that the thin clastic layers represent the culminating stages of a transgressive overlap, or the onset of regressive overlap conditions.

In Nancy's Bay, organic sediments are buried beneath a laminated silty/sandy deposit that passes up into a pebble bed. There is a sharp erosional boundary between the highest organic bed and the lowest laminated bed. It is more likely that these beds represent storm conditions that affected this coast at some time after 5000 BP.

The clearest evidence in the UK for a storm surge recorded in coastal stratigraphy of Flandrian Age, comes from eastern Scotland (Smith *et al.* 1980, 1982, Morrison *et al.* 1981, Haggart 1982). Smith *et al*, have described a grey, micaceous, silty fine sand intercalating peat in the Montrose area and Philorth valley of north-east Scotland. In the former area, it is dated between 7555 and 7180 BP, and in the latter more closely to 7140–7050 BP, whereas Haggart (1982) has dated it to 7270–7430 BP at the head of the Beauly Firth. Smith and Cullingford (1982) argued that the grey sand layer was probably deposited as a result of a North Sea surge, about 7000 BP. The discovery of similar clastic layers of similar age further north (Haggart 1982) reinforces the conclusion that it may be the product of a storm surge event, and encourages a closer examination of stratigraphic sequences in low-altitude coastal areas. The existence of these marker beds will facilitate correlation, and allow greater use to be made of coastal stratigraphy as a proxy record for climatic change and extreme weather events.

11.7   Correlation of low-altitude coastal sediments

In the past 20 years, following the seminal investigations in the Netherlands, a considerable volume of empirical data has been published on the lithostratigraphy of low-altitude coasts formed of unconsolidated sediments: e.g. in Belgium by Baeteman (1981), in France by Morzadec-Kerfourn (1974) and Ters (1973), in Germany by Barckhausen & Streif (1978) and Streif and Köster (1978), in the Netherlands by Jelgersma (1961), Roeleveld (1974), and van de Plassche (1981), in the UK by Devoy (1979), Godwin (1978), Kidson and Heyworth (1973, 1976), Shennan (1982a & b) and Tooley (1978a) and in the USA by Kraft *et al.* (1979). In addition, three books have been published on the Quaternary history of the Irish Sea (Kidson & Tooley 1977), the North Sea (Oele *et al.* 1979) and the Baltic Sea (Gudelis and Königsson 1979).

The superficial similarity of the lithostratigraphy encouraged the establishment of local and regional chronostratigraphic schemes (e.g. Brand *et al.* 1965, Tooley 1974, 1978a, Devoy 1979). Scrutiny of these schemes shows that they are no more than assemblies of published data, in which no attempt has been made to evaluate the data in terms of their comparability, or to match the boundaries statistically. Any further use of these schemes in terms of a proxy record of climatic change is not justified.

Whereas the stratigraphic records in low-altitude coastal zones appear to be similar, and to show alternating clastic and organic beds, the terminology of the beds has not aided correlation. In particular, the use of the terms 'marine transgression' and 'marine regression' has been ambiguous and inconsistent. These terms have been used to describe layers of different lithological character, as well as to describe a process such as sea level, tectonic, or climatic change. It is proposed to use the terms 'transgressive overlap' and 'regressive overlap' as descriptive lithostratigraphic terms in which no process is involved. In this way, if all low-lying coastal stratigraphic successions are described and classified consistently according to a common scheme, meaningful correlations become possible (see discussion in Shennan 1982 a, b and Tooley 1982), and these can serve as proxies for other natural and historical records.

The implications of this conclusion are that no further progress is possible until all existing and new sea-level data are reproduced to a common denominator, and at the moment the use of sea-level and coastal stratigraphic data as a proxy for climate is considerably reduced in value.

However, not only are there some encouraging signs that when sea-level data are reduced to a common denominator, meaningful correlations between tectonically unstable areas can be achieved (Shennan *et al.* 1983), but also, for some of the marked changes in sea level and coastlines, an overriding climatic cause can be identified.

In 1978, the climatic implications of sea level and coastal changes were considered at the behest of Professor Manley (Tooley 1978a), and it was possible to show, albeit qualitatively, linkages between ice budgets and sea-level changes. For example, the culminating stages of the decay of the Scandinavian

Ice Sheet and the attenuation of the Laurentide Ice Sheet are shown by the transgressive overlap tendencies between $c. -17$ m and $c. -10$ m O.D. in Liverpool and Morecambe Bays (Figs 4 & 5 in Tooley 1982). The regressive overlap tendency at 8575 ± 105 is supported by plant macrofossil evidence (Tooley & GreatRex, in preparation), and the coincidence in timing with the maximum of the Cockburn Re-advance is too close to attribute to a random effect. The subsequent transgressive overlap tendencies in both Liverpool and Morecambe Bays synchronise with the dates for the final disintegration of the Laurentide Ice Sheet.

11.8 Conclusions

There are four broad conclusions to arise from this consideration of climate, sea level and coastal changes.

The first conclusion is that, while in a general sense there is a close correlation between climate, sea level and coastal changes, it can only be demonstrated on a large scale, and invariably over a long time period. Where sea level and coastal changes synchronise with climatic changes at a different scale, they do so infrequently or equivocally. Geyh (1980) concluded from ^{14}C histogram analyses that whereas there was a negative correlation between coastal peat formation (marine regressions) and ^{14}C activity (global temperature), the global occurrence of short-term sea level changes could not be discriminated by this analytical method.

The second conclusion is that the reason for the lack of concurrence of sea level, coastal and climatic data arises from the failure to use and to apply a consistent, common scheme, such as the one proposed by Shennan (1981).

The third conclusion is to draw attention to the similarity of the lithostratigraphy in low-lying coastal areas at different latitudes, and at the margins of different ocean basins and epicontinental seas, which invites the suggestion of an overriding environmental control.

The fourth conclusion is to recommend the assembly of the natural record of coastal and sea level changes with the archival record of coastal and extreme weather events. The stratigraphic and documentary record of storm surges would provide a strong argument for a mechanism of coastal changes, but a recurrent theme in all these investigations is for an overriding control. The most satisfactory explanation of general coastal and sea-level changes is the glacio-eustatic control, but conditioned by local and regional events.

In 1961, and again in 1969, it appeared that investigations on sea level changes had been concluded, and that, at best, subsequent research would employ well-established methodologies in new areas to fit established models. Research since 1976 has taken novel directions, and has opened up exciting new areas of investigation, to the extent that the adage enunciated by Manley in 1950 has been proved correct:

> For those of us who like to seek fundamentals, the changing sea level offers a magnificent ground base for endless development.

Acknowledgements

I am grateful to Dr Ian Shennan for reading and commenting upon an earlier draft of this paper, and for making available and running the computer program for drawing and zoning the pollen diagram. Dr M. A. Geyh provided the ^{14}C dates, and it is a pleasure to acknowledge his help.

'I am also grateful to Dr P. A. GreatRex for allowing me to use the plant macrofossil diagram from Downholland Moss–16; to Dr D. Huddart for allowing me to refer to his foraminiferal results from Downholland Moss – 15; and to Mr A. Piper for abstracting information from the accounts of Lytham Priory. I am pleased to acknowledge the technical assistance of Mr David Hume and Mrs Elizabeth Pearson. The research, upon which this paper is based, was supported by a NERC grant.

References

Ashton, W. 1909. *The battle of land and sea on the Lancashire, Cheshire and North Wales coasts, with special reference to Lancashire Sandhills.* London: John Heywood; Southport: Ashton.

Baeteman, C. 1981. An alternative classification and profile type map applied to the Holocene deposits of the Belgian coastal plain. *Bull. Belg. Ver. voor Geologie* **90**, 257–80.

Bagley, J. J. (ed.) 1968. The Great diurnal of Nicholas Blundell of Little Crosby, Lancashire. Vol. 1: 1702–1711. *The Record Society of Lancashire and Cheshire* **110**, 1–350.

Bagley, J. J. (ed.) 1970. The Great diurnal of Nicholas Blundell of Little Crosby, Lancashire. Vol. 2: 1712–1719. *The Record Society of Lancashire and Cheshire* **112**, 1–328.

Bagley, J. J. (ed.) 1972. The Great diurnal of Nicholas Blundell of Little Crosby, Lancashire. Vol. 3: 1720–1728. *The Record Society of Lancashire and Cheshire* **114**, 1–289.

Barckhausen, J. and H. Streif 1978. *Erläuterungen zu Blatt Emden West Nr. 2608.* Hannover: Geologishe Kartt von Niedersachsen.

Barron, J. 1938. *A history of the Ribble navigation from Preston to the sea.* Preston: The Corporation of Preston.

Beck, J. 1954. The Church brief for the inundation of the Lancashire coast in 1720. *Trans Hist. Soc. Lancs & Ches.* **105**, 91–105.

Behre, K.-E., B. Menke and H. Streif 1979. The Quaternary geological development of the German part of the North Sea. In *The Quaternary history of the North Sea*, E. Oele, R. T. E. Shüttenhelm, A. J. Wiggers and L.-K. Königsson (eds.), 18–113. Uppsala: University of Uppsala.

Binney, E. W. and J. H. Talbot 1843. On the petroleum found in the Down Holland Moss. Manchester Geological Society Fifth Annual Meeting held on Thursday the 26th Day of October 1843, pp. 22–8, and *Trans Manchr. Geol. Soc.* **7**, 41–8. 1868.

Brand, von G., B.P. Hageman, S. Jelgersma and K. H. Sindowski 1965. Die lithostratigraphische Unterteilung des marinen Holozäns an der Nordseeküste. *Geol Jb.* **82**, 365–84.

Clark, J. A. 1980. A numerical model of worldwide sea level changes on a Viscoelastic Earth. In *Earth rheology, isostasy and eustasy*, N-A. Mörner (ed.), 525–34. Chichester: Wiley.

Clark, J. A. and C. S. Lingle 1977. Future sea level changes due to West Antarctic ice sheet fluctuations. *Nature* **269**, 206–9.

Clark, J. A., W. E. Farrell and W. R. Peltier 1978. Global changes in postglacial sea level: a numerical calculation. *Quaternary Research* **9**, 265–87.

Damuth, J. E. and R. W. Fairbridge 1970. Equatorial Atlantic deep-sea arkosic sands and Ice-Age aridity in tropical South America. *Geol Soc. Am. Bull.* **81**, 189–206.

Davies, J. L. 1964. A morphogenic approach to world shorelines. *Z. Geomorph.* **8**, 127–42.

Devoy, R. J. N. 1979. Flandrian sea level changes and vegetational history of the lower Thames estuary. *Phil Trans R. Soc. B* **285**, 355–407.

Fairbridge, R. W. 1961. Eustatic changes in sea level. In *Physics and chemistry of the Earth*, Vol. IV,: L. H. Ahrens, F. Press, K. Rankama and S. K. Runcorn (eds): 99–185. Oxford: Pergamon.

Fairbridge, R. W. and C. Hillaire-Marcel. 1977. An 8000 yr palaeoclimatic record of the 'Double-Hale' 45-yr solar cycle. *Nature* (Lond.) **268**, 413–6.

Flenley, J. 1979. *The equatorial rainforest: a geological history*. London: Butterworth.

Fowler, G. 1932. Old river-beds in the Fenlands. *Geog. J.* **79**, 210–12.

Fowler, G. 1934. The extinct waterways of the Fens. *Geog. J.* **83**, 30–6.

Geyh, M. 1980. Holocene sea level history: case study of the statistical evaluation of ^{14}C dates. *Radiocarbon* **22**, 695–704.

Godwin, H. 1938. The origin of roddons. *Geog. J.* **91**, 241–50.

Godwin, H. 1941. Studies of the post-glacial history of British vegetation. VI. Correlations in the Somerset levels. *New Phytol.* **40**, 108–32.

Godwin, H. 1978. *Fenland: its ancient past and uncertain future*. Cambridge: Cambridge University Press.

Godwin, H. and M. H. Clifford 1938. Studies of the post-glacial history of British vegetation. I. Origin and stratigraphy of Fenland deposits. II. Origin and stratigraphy of deposits in southern Fenland. *Phil Trans R. Soc. B.* **229**, 323–406.

Gottschalk, M. K. E. 1975. *Stormvloeden en rivieroversromingen in Nederland. Storm surges and river floods in the Netherlands. II de periode 1400–1600*. Assen: Van Gorcum.

Grant, D. R. 1970. Recent coastal submergence of the Maritime Provinces, Canada. *Can. J. Earth Sci.* **7**, 676–89.

Gudelis, V. and L.-K. Königsson 1979. *The Quaternary history of the Baltic*. Uppsala: Acta Universitatis Upsaliensis, Symposia Universitatis Upsaliensis, Annum Quingentesimum Celebrantis 1.

Hageman, B. P. 1969. Development of the western part of the Netherlands during the Holocene. *Geologie Mijn.* **48**, 373–88.

Haggart, B. A. 1982. *Flandrian sea level changes in the Moray Firth area*. Unpublished PhD thesis. Durham: University of Durham.

Hallam, A. 1981. *Facies interpretation and the stratigraphic record*. San Francisco: W. H. Freeman.

Harding, A. F. (ed.) 1982. *Climatic change in later prehistory*. Edinburgh: Edinburgh University Press.

Heptinstall, S. 1983. *Environmental change during the Flandrian on Altcar Moss, south-west Lancashire*. Unpublished BSc dissertation. Durham: University of Durham.

Howell, F. T. 1973. The sub-drift surface of the Mersey and Weaver catchment and adjacent areas. *Geol. J.* **8**, 285–96.

Ingram, M. J., G. Farmer and T. M. L. Wigley 1981. Past climates and their impact on Man: a review. In *Climate and history: studies in past climates and their impact on Man*, T. M. L. Wigley, M. J. Ingram and G. Farmer (eds), 3–50. Cambridge: Cambridge University Press.

Iversen, J. 1960. Problems of the early post-glacial forest development in Denmark. *Danm. geol. Unders* **4**, 1–32.

Jelgersma, S. 1961. Holocene sea level changes in the Netherlands. *Meded. Geol. Sticht.* Series C. VI. **7**, 1–100.

Jelgersma, S. 1966. Sea level changes during the last 10 000 years. In *Proceedings of the*

International Symposium on World Climate 8000 to 0 BC, J. S. Sawyer (ed.), 54–69. London: Royal Meterological Society.

Jelgersma, S. 1979. Sea-level changes in the North Sea basin. In *The Quaternary history of the North Sea*, E. Oele, R. T. E. Schuttenhelm and A. J. Wiggers (eds), 233–48. Uppsala: University of Uppsala.

Jelgersma, S., J. De Jong, W. H. Zagwijn and J. F. van Regteren Altena 1970. The coastal dunes of the Western Netherlands; geology, vegetational history and archeology. *Med. Rijks Geol. Dienst.* NS. **21**, 93–167.

Jelgersma, S., E. Oele and A. J. Wiggers, 1979. Depositional history and coastal development in the Netherlands and the adjacent North Sea since the Eemian. In *The Quaternary history of the North Sea*, E. Oele, R. T. E. Schüttenhelm and A. J. Wiggers (eds.), 115–42. Uppsala: Acta Universitatis Upsaliensis, Symposia Universitatis Upsaliensis Annum Quingentesimum Celebrantis: 2.

Kidson, C. and A. Heyworth 1973. The Flandrian sea level rise in the Bristol Channel. *Proc. Ussher Soc.* **2**, 565–84.

Kidson, C. and A. Heyworth 1976. The Quaternary deposits of the Somerset levels. *Q. J. Engng Geol.* **9**, 217–35.

Kidson, C. and M. J. Tooley (eds) 1977. *The Quaternary history of the Irish Sea.* Liverpool: Seel House Press.

Kraft, C., E. A. Allen, D. F. Belknap, C. J. John and E. M. Maurmeyer 1979. Processes and morphologic evolution of an estuarine and coastal barrier system. In *Barrier islands, from the Gulf of St Lawrence to the Gulf of Mexico*, S. P. Leatherman (ed.), 149–83. New York: Academic Press.

Kumar, H. and J. E. Sanders 1976. Characteristics of shoreface storm deposits: modern and ancient examples. *J. Sedim. Petrol.* **46**, 145–62.

Lamb, H. H. 1981. Climatic fluctuations in historical times and their connexion with transgressions of the sea, storm floods and other coastal changes. In *Transgressies en occupatiegeschiedenis in de Kustgebieden van Nederland en België*, A. Verhulst and M. K. E. Gottschalk (eds), 251–84. Ghent: Rijksuniversiteit.

Lyell, C. 1875. *Principles of geology, or the modern changes of the Earth and its inhabitants*, 12th edn. London: John Murray.

Manley, G. 1950. Degrees of freedom: An inaugural lecture given at Bedford College for Women, January 27th, 1949. London: Christopher Johnson.

Martin, L., J. M. Flexor, G. da Silva Vilas Boas, A. C. da Silva Pinto Bittencourt and M. M. Magalhães Guimarães 1979. Courbe de variations du niveau relatif de la mer a cours des 7000 dernières années sur un secteur homogène du littoral Brésilien (Nord de Salvador-Bahia. In *Proceedings of the 1978 International Symposium on Coastal Evolution in the Quaternary*, K. Suguio et al., 264–74. São Paolo: Instituto do Geociências, USP.

Menke, B. and K-E. Behre 1973. History of vegetation and biostratigraphy. In State of research on the Quaternary of the Federal Republic of Germany, K-E. Behre et al. (eds.) 251–67. *Eiszeitalter und Gegenwart* **23/24**, 219–370.

Mörner, N. -A. 1976. Eustasy and geoid changes. *J. Geol.* **84**, 123–51.

Mörner, N. -A. 1980. Eustasy and geoid changes as a function of core/mantle changes. In *Earth rheology, isostasy and eustasy*, N-A. Mörner (ed.), 535–53. Chichester: Wiley.

Morrison, J., D. E. Smith, R. A. Cullingford and R. L. Jones 1981. The culmination of the main postglacial transgression in the Firth of Tay area. *Proc. Geol. Assoc.* **92**, 197–209.

Morzadec-Kerfourn, M. -T. 1974. Variations de la ligne de rivage Armoricaine au Quaternaire. *Mém. Soc. Géol. Minér. Bretagne* **17**, 1–208.

Murray, J. W. and A. B. Hawkins 1976. Sediment transport in the Severn estuary during the past 8000–9000 years. *J. Geol. Soc. Lond.* **132**, 385–98.

Oele, E., R. T. E. Schüttenhelm, A. J. Wiggers and L. -K. Königsson (eds.) 1979. *The Quaternary of the North Sea*. Uppsala: University of Uppsala.

Picton, J. A. 1849. The changes of sea levels on the west coast of England during the Historic period. *Proc. Lit. Phil. Soc. Lpool*. **5**, 113–15.

van de Plassche, O. 1981. Sea Level, groundwater, and basal peat growth – a reassessment of data from the Netherlands. *Geologie Mijn*. **60**, 401–8.

Reade, T. M. 1871. The geology and physics of the post-glacial period, as shewn in the deposits and organic remains in Lancashire and Cheshire. *Proc. Lpool Geol Soc*. **2**, 36–88.

Roeleveld, W. 1974. The Holocene evolution of the Groningen marine clay district. *Berichten van de Rijksdienst voor het Oudheidkungid Bodemonderzoek* **24**, 1–132.

Roep, Th. B., D. J. Beets and G. H. J. Ruegg 1975. Wavebuilt structures in subrecent beach barriers of the Netherlands. In *Extraits des publications du IXme Congrès International de Sedimentologie, Nice*, 141–5.

Shackleton, N. J. and N. D. Opdyke 1973. Oxygen isotope and palaeomagnetic stratigraphy of Equatorial Pacific core V28–238: oxygen isotope temperatures and ice volumes on a 10^5 year and 10^6 year scale. *Quaternary Research* **3**, 39–55.

Shennan, I. 1981. *Flandrian sea level changes in the Fenland*. Unpublished PhD thesis. Durham: University of Durham.

Shennan, I. 1982a. Interpretation of Flandrian sea level data from the Fenland, England. *Proc. Geol Assoc*. **83**, 53–63.

Shennan, I. 1982b. Problems of correlating Flandrian sea level changes and climate. In *Climatic change in later Prehistory*, A. F. Harding (ed.) 52–67. Edinburgh: Edinburgh University Press.

Shennan, I., J. Tooley, M. J. Davis and B. A. Haggart 1983. Analysis and interpretation of Holocene sea level data. *Nature* (Lond.) **302**, 404–6.

Skertchly, B. J. 1877. *The geology of the Fenland*. Mem. geol. Surv. England and Wales. London: HMSO.

Smith, D. E. and R. A. Cullingford 1982. The culmination of the main postglacial transgression in eastern Scotland. In *Abstracts*, vol. II, p.302. XI INQUA Congress, Moscow.

Smith, D. E., R. A. Cullingford and W. P. Seymour 1982. Flandrian relative sea level changes in the Philorth valley, north-east Scotland. *Trans Inst. Br. Geogr. N. S.* **7**, 321–36.

Smith, D. E., J. Morrison, R. L. Jones and R. A. Cullingford 1980. Dating the main postglacial shoreline in the Montrose area, Scotland. In *Timescales in geomorphology*, R. A. Cullingford, D. A. Davidson and J. Lewin (eds), 225–45. Chichester: Wiley.

Street, F. A. and A. T. Grove. 1979. Global maps of lake-level fluctuations since 30 000 yr BP. *Quaternary Research* **12**, 83–118.

Streif, H. and R. Köster 1978. The geology of the German North Sea coast. In *Die Küste: Archiv für forschung und technik an der Nord-und Ostsee* **32**, 30–49.

Tansley, A. G. 1939. *The British Isles and their vegetation*. Cambridge: Cambridge University Press.

Ters, M. 1973. Les variations du niveau marin depuis 10000 ans le long du littoral atlantique français. In *Le Quaternaire: géodynamique stratigraphie et environnement*, 114–35. CNRS. Comité National français de L'INQUA.

Tooley, M. J. 1974. Sea level changes during the last 9000 years in North-West England. *Geog. J.* **140**, 18–42.

Tooley, M. J. 1976. Flandrian sea level changes in West Lancashire and their implications for the 'Hillhouse Coastline'. *Geol J.* **11**, 137–52.

Tooley, M. J. 1978a *Sea level changes in North-West England during the Flandrian stage*. Oxford: Clarendon Press.

Tooley, M. J. 1978b. Interpretation of Holocene sea level changes *Geol. För. Stockh. Förh.* **100**, 203–12.

Tooley, M. J. 1979. Sea level changes during the Flandrian stage and implications for coastal development. In *Proceedings of the 1978 International Symposium on Coastal Evolution in the Quaternary*, K. Suguio, T. R. Fairchild, L. Martin and J. M. Flexor (eds), 502–33. São Paulo: Instituto de Geociências, USP.

Tooley, M. J. 1982. Sea level changes in northern England. *Proc. Geol Assoc.* **92**, 43–51.

Tyrer, F. 1970. *Extracts from the Great Diurnal of Nicholas Blundell of Crosby Lancashire, and other of his manuscripts and footnotes.* typescript, 5 pp.

West, R. G. 1980. Pleistocene forest history in East Anglia. *New Phytol.* **85**, 571–622.

Wray, D. A. and F. Wolverson Cope. 1948. *Geology of Southport and Formby.* Mem. Geol. Surv. of Great Britain. London: HMSO.

12
The effect of climate on plant distributions

RICHARD N. CARTER and STEPHEN D. PRINCE

Many plant species are confined to, or are commonest in, either northern or southern Britain. The extent to which it is possible to explain the geographical distribution limits of plants in terms of plant physiological responses to climatic factors is considered. Investigations of the distributions of *Cirsium acaulon*, *Hordeum murinum* and *Tilia cordata* are used as examples.

12.1 Introduction

The climatic contrast between the cool, wet upland regions in the north and west of Britain and the warm, dry lowland regions in the south and east is reflected in the British distribution patterns of many plants. This can easily be seen from the dot distribution maps provided in the *Atlas of the British flora* (Perring & Walters 1962), which show that many species are confined to, or commonest in, either northern or southern Britain; eastern and western distribution patterns, though they occur, are relatively infrequent. The object of this contribution is to consider, with reference to species distributed in north-western and southeastern Britain, the extent to which it is possible at present to explain the geographical distribution limits of plants in terms of plant physiological responses to climatic factors. In dealing with these particular groups of plants, it is necessary to consider altitudinal as well as latitudinal limits, but little attention will be given here to the ecology of high montane species, about which much has already been written (see reviews by Billings & Mooney 1968, Grace 1977, Tranquillini 1979).

The geographical elements of the British flora (Salisbury 1932, Matthews 1937, 1955) are, for good reasons, defined in terms of the continental and world distributions of the species composing them, so that for flowering plants there are few, if any, systematic analyses of distribution patterns within Britain alone. Nevertheless, it has often been noted (Salisbury 1932, Perring & Walters 1962, Pigott 1970) that many plants (about 140) reach their northern limits in Britain along a line running south-west to north-east, typically from the Bristol Channel to the Humber, but often rather to the north or south of this. Some of these, e.g. *Thymus pulegioides*, are calcicoles which might be confined to the south of this line by the scarcity of chalk and limestone outcrops to the north of

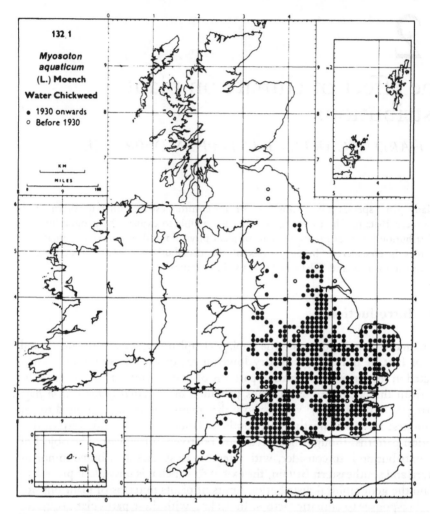

Figure 12.1 The distributions of *Myosoton aquaticum* and *Stellaria nemorum* in Britain. Note the south-west to north-east trends in the boundaries of both species. Each dot represents at least one record for a 10 × 10 km National Grid square. Reproduced from the *Atlas of the British flora* (Perring & Walters 1962) by permission.

it, but others, e.g. *Bryonia dioica* and *Myosoton aquaticum*, are non-calcicoles. In addition, many species, e.g. *Coronopus squamatus* and *Cardaria draba*, are common to the south of the line, but occur sparsely to the north and then mainly along the coasts. Complementary to these southern species are about 50 northern species (not including alpines), e.g. *Galium sterneri*, *Stellaria nemorum* and *Crepis paludosa*, that reach their southern limits along the same line (Fig. 12.1). Many climatic isolines run parallel to this line, and it is therefore generally supposed that climatic factors cause the south-west to north-east type of distribution limit. Salisbury (1932) and Pigott (1970) have listed circumstantial

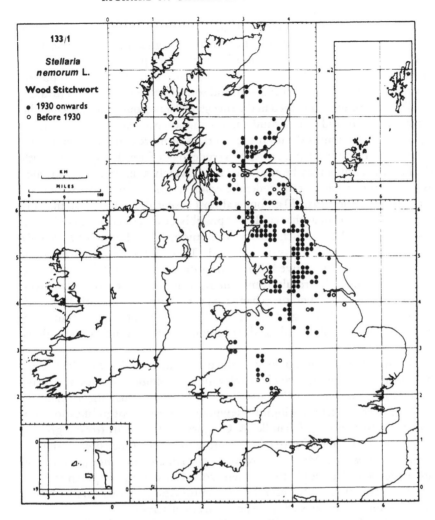

133/1

Stellaria nemorum L.

Wood Stitchwort

• 1930 onwards
○ Before 1930

reasons for supposing that even the limits of southeastern calcicoles are climatically controlled. But while there are many statements in the literature about the
climatic control of these and other limits, few are based on anything more than
circumstantial evidence; this, of course, may be compelling, but experimental
verification is usually lacking. In only a few species is the actual mechanism by
which the climate effects control of a distribution limit even superficially
understood.

It can generally be assumed that the altitudinal and latitudinal distribution
limits of northern European plants are determined in identical ways; this is
strongly suggested by the fact that the upper altitudinal limits of many southern
species become lower in the northern parts of their ranges (as do the lower limits
of many montane species). However, such reasoning must be applied with
caution to limits on high mountains (from 660 m upwards in Britain) where

there may be strong winds, high insolation, persistent snow cover and other conditions not necessarily paralleled at comparably high latitudes. But the main difficulty in explaining altitudinal limits lies in the very poor documentation of altitudinal distribution patterns. For British plants, there is a list of reported upper altitudinal limits (Wilson 1931–3), and some further information (usually anecdotal in character) is included in the *Biological flora* accounts published in the *Journal of Ecology*. In addition, altitudinal distribution patterns may be extracted from published distribution maps using overlays or more exhaustive methods, but the patterns thus revealed are often too crude to be of use. The precision with which altitudinal limits may be expressed is not widely appreciated. In Britain, for example, the south-eastern ruderal *Lactuca serriola* is common within its range at altitudes of less than 70 m, but is almost absent above 110 m, so that it is locally excluded from areas that could not possibly be described as upland. This is illustrated in Fig. 12.2, which shows the altitudinal distribution of *L. serriola* colonies on the M5 motorway in England. There are plenty of seemingly suitable sites for this plant above 110 m, and its absence is presumably due to the small climatic changes that result from an increase in altitude of about 50 m, over which range the mean temperature change expected from the adiabatic lapse rate is less than 0.5°C. The 10 km grid squares used for most plant mapping purposes are far too large to reveal altitudinal distribution patterns on such a fine scale as this.

The climate of Britain is extensively documented, and Manley (1952) provided a classic description, which is relevant to ecological studies at a geographical scale. Salisbury (1939) and Tansley (1949) discuss the climate of Britain from a specifically ecological viewpoint. More recent publications on the climate of Britain (e.g. Chandler & Gregory 1976) benefit from a greater accumulation of data and improved computing methods. One contribution to climatology for which Manley is renowned, namely, the interpolation of climatic conditions for sites between meteorological stations, is potentially of great use to ecologists; only recently have automatic methods (White 1979) enabled ecologists to make such interpolations themselves.

Two peculiar features of the British climate, which probably have broad effects on plant distributions, may be noted here. The first is the marked contrast between summer and winter temperature patterns: the warmer areas in summer do not coincide with the warmer areas in winter, which no doubt complicates the distribution patterns of many species. The second is the unusually high lapse rate, which probably sharpens the distribution boundaries of many characteristically upland or lowland species.

12.2   Strategies used in explaining distribution limits

The idea that a plant's distribution can be explained in terms of its physiological responses to geographically varying environmental factors (climatic or otherwise) took form at the end of the 19th century, when writers such as Haberlandt

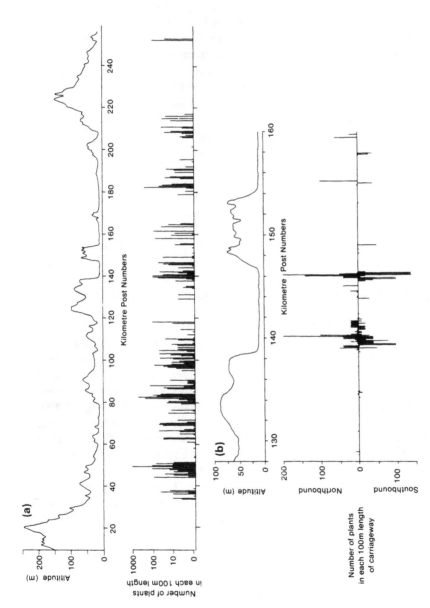

Figure 12.2 The distribution and abundance of prickly lettuce (*Lactuca serriola* L.) on the verges of the M5 motorway between Birmingham and Exeter. (a) Altitude of the carriageway and numbers of plants in 500 m stretches of both verges from Birmingham (0 km) to Exeter (250 km) in 1981. (b) A more detailed diagram showing altitude and numbers of plants in 1980 in 100 m stretches of the northbound and southbound carriageway verges separately, between kilometre posts 130 and 160 near Avonmouth (there are no verges between 141 km and 145 km).

(1884), Schimper (1903), Clements (1907), Warming (1909) and Drude (1913) placed ecophysiological factors alongside competition and historical factors in their attempts to explain plant distributions. Since then, several approaches to the task of explaining particular distributions in these terms have emerged. The simplest is that of identifying correlations between distribution limits and the isolines of synoptic climatic variables. A seminal example is the correlation between the eastern distribution limit of *Rubia peregrina* in Europe and the average January isotherm for 4°C described by Salisbury (1926). Other examples from the British flora are given by Salisbury (1932), Perring and Walters (1962), Conolly and Dahl (1970) and Pigott (1970). At a certain level, such correlations are in themselves explanations of the phenomena they describe – they convey a high level of predictability (Harper 1980) – and they should not be undervalued; some of those described for British plants are very striking, and in particular cases strongly suggest the type of control mechanisms that might be expected. But all too often such correlations are invested with causative significance, either explicitly or implicitly, when the climatic factor involved is called a 'limiting factor'. This is theoretically unsound because any correlation may after all be merely coincidental; and in addition there are many special difficulties attached to the interpretation of correlations between climatic iso-lines and distribution limits. Some (or all) of these have been discussed by Salisbury (1939). Boysen-Jensen (1949), Ratcliffe (1968), Conolly and Dahl (1970), Davison (1970) and Pigott (1970), and they will be summarised here. First, climatic variables are highly interdependent; for example, the distribution of cloudiness affects the duration of bright sunshine, which is correlated with temperature and negatively correlated with rainfall (Pigott 1970). If a limit is correlated with one isoline, it will therefore necessarily be correlated with others. Second, synoptic climatic variables may not relate in any simple way to conditions actually experienced by a plant. A plant does not, for example, experience a mean temperature; it experiences the continually changing temperatures from which the mean is computed. Third, the surface and tissue temperatures to which plants respond physiologically are scarcely ever the same as air temperatures; often they are vastly different (Salisbury 1939, Geiger 1965). Fourth, species frequently comprise physiologically distinct races, so that individuals in one district may respond to a climatic factor differently from those in another. And, fifth, plants may become acclimatised so that, according to the conditions they have experienced beforehand, their responses to particular levels of climatic factors may differ from time to time and from place to place. Finally, it should be noted that even in Britain, meteorological stations may be too sparse to allow an isoline to be defined well enough for a satisfactory correlation with a distribution limit to be established; this is a real problem in the remoter upland areas of Britain (Conolly & Dahl 1970). The type of detailed meteorological records that Manley (1936, 1980) collected for upland areas is much needed by plant geographers and ecologists.

Another approach to the task of explaining particular distributions is that of observing plant responses to environmental factors under experimental conditions. This approach may be an essential part of an integrated study but, like the observation of correlations between isolines and limits, it is subject to criticism if used alone. Even ecologists occasionally do this, and some plant physiologists (whose experiments are no doubt justified within their sphere) are unhappily prone to add unsupported comments about the causation of plant distributions to their otherwise legitimate conclusions. If a plant's response to an environmental factor is to be attributed with a rôle in controlling a limit, then it is at least necessary to show that the response can be observed in the field. But it is also necessary, though more difficult, to show that the response leads to sufficiently large effects on individuals or populations of the plant to explain why it should not grow beyond its limit. If this cannot be shown, then it remains possible that the response is purely coincidental in so far as the control of the limit is concerned – even though it be a response to some aspect of the climate that deteriorates towards the plant's limit. This is not a trivial matter. It is in the nature of the experimental method that one environmental factor has to be singled out from all others for investigation (the others being held constant), and under these circumstances the effects of that factor are almost certain to appear significant, so that they might account for a plant's limit; the explanation advanced for a limit may thus depend entirely upon which environmental factors the researcher chooses to investigate, a choice that usually depends upon his personal interests and expertise (Harper 1980).

One widely acclaimed, but relatively seldom used, method of studying distribution limits is to transplant plants beyond their limits so that their failure can be observed in climates more severe than those in which they normally grow. Transplant experiments which are special cases of 'perturbation' analysis have been carried out on certain British and Scandinavian plants by Dahl (1951), Pigott (1970), Woodward and Pigott (1975), Prince (1976) and Davison (1977). The main difficulty with transplant experiments is that if the climate of a site differs much from that obtaining at the limit being investigated, it is likely that the plant will show responses to aspects of the climate other than those that really control its limit; it follows that the easier the results are to obtain, the harder they are to interpret.

A frequently mentioned idea about distribution limits is that they may be controlled by exceptional weather conditions which occur at long intervals. For British plants, frost damage in severe winters is usually envisaged, but reproductive failure in wet summers may also be important; the idea was suggested by Salisbury (1926, 1939) for such plants as *Rubia peregrina* and *Erodium moschatum*. Actually there is remarkably little evidence that limits are ever really controlled in this way. Pigott (1970) recorded that *R. peregrina*, 'limited' by the January 4°C isotherm, was undamaged in the North of England in February 1969, when grass minimum temperatures of −16°C were recorded. Iversen (1944) and Savidge (1970) have documented frost damage and mortality during severe winters in several woody species of Atlantic distribution, but did not show that this altered the distribution pattern of any; in fact, mortality of this kind has the properties of a dosage response curve – some individuals almost always survive

(MacArthur 1972). Moreover, exceptional climatic conditions are erratic not only when but also where they occur, and the problem of how they might lead to the repeated distribution patterns found in the *Atlas of the British flora* has not been discussed. In fact, the idea is practically (though perhaps not theoretically) unfalsifiable (the limiting conditions are rare because you have not observed them, and you never observe them because they are rare!). It no doubt owes much of its popularity to the fact that it is so conveniently invoked in explanation of distribution limits in cases where no other plant responses can readily be identified. Even so, it is plausible for some species, and can never quite be dismissed.

It is no longer common to think of distribution limits controlled by single plant responses to single climatic factors. The need for multi-factor explanations was pointed out by Boysen-Jensen (1949), and has gained widespread acceptance with the development of the concept of the ecological niche as a multi-dimensional hypervolume. Pigott (1970), Davison (1977) and Pigott and Huntley (1978) suggest multi-factor explanations for the limits of some south-eastern plants in Britain.

Much of what has been said here may seem negative. This is largely because the difficulty of accounting for distribution limits in terms of physiological responses to climatic factors has been, in general, vastly underestimated. Much of the difficulty relates to the problem of defining what constitutes a satisfactory explanation for the natural distribution of a plant at any scale – a subject upon which the attention of plant ecologists has lately been focused by, for example, Harper (1977, 1980, 1982). Even so, it is, of course, possible to make headway towards understanding geographical limits by using judicious combinations of the approaches that have been described, and sometimes it is easy thus to form very probable hypotheses about how a particular limit is controlled.

12.3   Some physiological effects of the climate

Plants respond in diverse ways to the climate during all stages of their development from germination to the dispersal of propagules, and any response may be a cause, or a contributory cause, of a plant's limit. Reported physiological effects of the climate that might affect the distribution of southeastern and northeastern species in Britain will be reviewed, and the discussion of their actual importance will be left until later.

The temperature responses of plants have often been studied over ranges of temperature as wide, or wider, than those that plants experience normally. The resulting response curves generally show that, above some minimum temperature, plant activity starts, and increases linearly with temperature up to a broad optimum where it is relatively independent of temperature; finally, it declines rapidly at high temperatures (for examples see Dilks & Procter 1975, Larcher 1980). Temperature response curves demonstrate the acclimatory abilities of plants particularly well. Regehr and Bazzaz (1976), for example, obtained 10–15°C shifts between summer and winter in the temperature optima for photosynthesis in winter annuals belonging to the genera *Erigeron*, *Rorippa*, *Capsella* and *Lactuca*. Similar acclimatory shifts in the temperature minima for many physiological processes, following exposure to low temperatures for periods varying from a few hours to several days, have been demonstrated in many species. The acclimatisation process may involve changes in metabolic substrate concentrations, the replacement of enzymes by isozymes having the same functions but different temperature optima, and changes in the lipid content of cell membranes (Strain 1969, Mooney & Harrison 1970, Sawada & Miyachi 1974).

To some extent, temperature response curves must be treated with caution. The experimental conditions at the extreme temperatures used are often inevitably unlike any experienced by a plant in nature. Furness and Grime (1982), for example, showed that 40 species of bryophytes died at temperatures of 30–35°C, but concluded that bryophytes commonly survive in nature in a dry state at such temperatures, rather than remaining in the hydrated state enforced in their experiments. Another reason for caution is that different physiological races of a species can (and indeed very commonly do) have widely divergent temperature responses, so that response curves based on plants of limited provenance are most unlikely to be representative of the species as a whole.

It may fairly be supposed that individual metabolic processes in plants are affected by temperature in ways that follow the Arrhenius and Van't Hoff relationships, which describe the effects of temperature on the velocity of chemical reactions. A Q_{10} temperature coefficient may thus be used to describe any rate increase induced in such a process by a 10°C rise in temperature. But there is accumulating evidence that plant performance overall is linearly related to temperature over those ranges of temperature commonly experienced by growing plants (Hegarty 1973, Arnold & Monteith 1974, Gallagher & Biscoe 1978, Monteith 1981). On this basis, Arnold and Monteith (1974) showed that mean temperatures integrate, and therefore represent, the effects of fluctuating temperatures on plants if the rate of the physiological process in question is a linear function of temperature, the fluctuations do not exceed the range of the linear response, and the response time is short compared to that over which temperatures are averaged. They were thus able to predict leaf expansion rates successfully (from mean temperatures alone) for *Festuca ovina*, *Sesleria albicans* and *Kobresia simpliciuscula* in Upper Teesdale.

The direct effects of temperature on photosynthesis and respiration have been intensively studied, and respiration proves to be generally more affected than photosynthesis. It has been suggested that at high temperatures respiration may exceed photosynthesis, whereas at low temperatures a number of factors (of which a short growing season is the most important) may limit photosynthesis; in either case a decline in the carbon balance of plants at their climatic distribution limits may result (Boysen-Jensen 1949). Data from alpine timberlines are reviewed by Tranquillini (1979), and confirm that less growth is possible at the upper altitudinal limits of alpine trees.

Morphogenetic changes in leaf structure which affect growth rates have been demonstrated in a number of lowland species grown in upland sites, and the same effects have been reproduced in controlled environments at low temperatures. Woodward (1979) showed that the upland species *Phleum alpinum*, *Sesleria albicans*, *Saxifraga aizoides* and *Sedum rosea* are insensitive in this respect to the high temperatures typical of lowland sites; whereas at high temperatures the lowland species *Phleum bertolonii*, *Dactylis glomerata*, *Umbilicus rupestris* and *Sedum telephium*, show a marked increase in leaf area relative to plant weight (leaf area ratio). The component of leaf area ratio largely responsible for this is the ratio of leaf area to leaf weight (specific leaf area).

Short growing seasons in cold climates may lead to various secondary effects on plants which may be important at distribution boundaries. A striking example is the failure of maturation in shoots of woody plants above the alpine timberline, which leads to fatal dessication damage during the winter when the ground is frozen (Tranquillini 1979). However, perhaps more important for British plants are the delays in all aspects of reproduction which may follow upon delayed vegetative growth.

Reproduction failure is not uncommonly observed in plants at the edges of their ranges. *Juncus squarrosus*, for example, survives vegetatively but, for a variety of reasons, fails to reproduce at altitudes above 820 m in the Lake District (Pearsall 1950). Similarly, many arctic and alpine species show intermittent reproduction in severe environments (Callaghan 1974, Campbell 1981), and many upland species produce fewer flowers at high altitudes, e.g. the American species *Mimulus primuloides* (Douglas 1981). Various types of reproductive failure have been reported for southeastern plants at their northern limits in Britain: pollination failure may result from lack of pollinators, or slow pollen tube extension (Pigott & Huntley 1981); the fertilised ovule may fail to develop (Pigott 1970, 1981, Braak 1978, Pigott & Huntley 1981), or may do so only partially (Prince 1976); and dispersal of propagules may be impaired by wet conditions which, for example, alter the aerodynamic characteristics of cypselae in some members of the Compositae (Sheldon & Burrows 1973).

Annual cycles of frost and heat resistance are common in species which are exposed to extreme conditions during a part of each year. The degree of resistance is generally related to the severity of the conditions experienced (e.g. Kaku & Iwaya 1979, Pearce 1980, Bannister 1981). Freezing damage is sometimes reported for Atlantic species at their eastern limit in Britain, but only rarely for southern species at their northern limits.

Seed germination takes place over a limited range of temperatures in any one population of a species. It is delayed by low temperatures above the necessary minimum, but these often enable a higher proportion of seeds to germinate in the end (Prince 1980). In some species, the temperature range at which germination can occur changes during dormancy, and the timing of germination in seasonal climates has been attributed to after ripening changes of this sort (Baskin & Baskin 1972, 1976, 1981, Pemadasa & Lovell 1975, Willemsen 1975, Karssen 1980). The amplitude of temperature fluctuations appears to be detected by some species, and low fluctuations have been shown to inhibit germination in these (Thompson et al. 1977). Temperature can interact with other environmental controls on seed germination; for example the light requirement for germination in lettuce is overcome at temperatures in the lower part of its normal range for germination.

Direct effects of the climate on plants cannot easily be separated from secondary effects resulting from the primary effects of the climate on soils. Soils are strongly influenced by rainfall and temperature, and therefore similar parent materials may produce different soils in the south-east and north-west of Britain. In upland areas, low temperatures, poor aeration and acid conditions reduce the activity of soil organisms (Scott Russell 1940), and high rainfall reduces the nutrient content of soils which often have a poor exchange capacity to start with. The most obvious deficiency of many upland soils is their low nitrogen content (Munro et al. 1973).

12.4   Other effects of climate

It might be expected that information about geographical variation in agricultural crop yields would be pertinent to an understanding of the distribution limits of wild plants, since the reduction in crop yields in northern and upland regions compared with the lowlands of southern Britain (Munro & Davies 1973) does indicate a strong relationship between yields and climate. Even local variations in yields may be climatically determined: areas at low altitudes in the drier eastern parts of Scotland, for example, are no more productive for grass than those at intermediate altitudes, owing to the summer moisture deficits which develop (Hunter & Grant 1971).

Crop yields are often deemed to be correlated with the length of the 'growing season', i.e. the number of days when mean temperatures rise above a 'base temperature' below which growth is negligible. Improved correlations can be achieved using 'thermal time' (Monteith 1981), which is calculated by subtracting the base temperature from each day's mean temperature and accumulating the positive values as 'day degrees'. In Britain, this method is much less successful than it is in continental climates, because of the low amplitude of annual temperature fluctuations and the greater length of time for which temperatures are near the base temperature. Furthermore, the adoption of a particular base temperature by convention rather than by experiment with each individual crop leads to significant errors (Monteith 1981), since base temperatures vary widely between species and between populations of the same species (e.g. Ollerenshaw & Baker 1981).

Studies in which components of yield are measured for individual plants are also informative. For example, Prince (1976) showed that an excess of photosynthetic capacity was available for barley grown at an upland site; the reduction in grain yield found there was due to climatic effects on the physiology of grain development.

It is evident from studies of infra-specific variation in wide-ranging species that the climate exerts a dominant effect on the evolution of locally adapted races (Heslop-Harrison 1964). For example, Clausen *et al.* (1948) showed that in species growing along a transect across central California, many physiological characteristics vary clinally in a way that is related to the climate. Analysis of the selective effects of climate is beset with difficulties, which arise from the inter-correlation of climatic factors and from differences in the potential for variation among different characters in different species. There is consequently a need for caution in attributing adaptive significance to the patterns of variation that have so often been demonstrated (Harper 1982).

Variation has been described for all stages of the life-cycle, e.g. germination (Clebsch & Billings 1976), vegetative growth (Cooper 1964), photosynthesis (Mooney & Billings 1961), and reproduction (McNaughton 1966). Comprehensive studies have been made of locally adapted varieties in forage grasses; the seasonal cycles of flowering and vegetative growth in several species (e.g. *Lolium perenne*, and *Dactylis glomerata*) have been studied in collections of plants obtained from the Mediterranean to North and Central Europe (Cooper 1963). Mediterranean plants have a winter growing season, and the timing of flowering varies according to the timing of the onset of summer drought. These plants have little winter-hardiness (as do those from oceanic areas of Western Europe). All non-Mediterranean races have a summer growing season, and northern and continental races are winter dormant. Therefore, a cline in the variation of local races can be traced from the Mediterranean, along the coast of Western Europe, to northern and central Europe. Studies on the physiology of these locally adapted races revealed differences in photosynthesis, leaf expansion and the timing of germination and flowering.

12.5 Discussion

Much of the known variation between geographical races of plants at the infra-specific level is connected with developmental processes (such as seed germination and flowerbud formation), which proceed only if a plant is exposed to certain sets of environmental conditions. Different races tend to require exposure to different sets of conditions, so that each develops normally in response to the particular conditions it meets in the area it inhabits. The failure of some plants to survive beyond their limits may be due to the fact that an environmental stimulus essential for normal development is not encountered there, but this has seldom (if ever) been explicitly suggested as a mechanism of distribution limit control. The very ubiquitousness of adaptability among plant requirements for environmental stimuli makes it seem likely that plants would evolve to develop normally in regions beyond their limits if other factors did not supervene to prevent them from growing there. The observation that a plant is adapted to develop normally within its range should not in itself lead us to presume that it cannot adapt to develop normally in other places (though there are presumably limits to such adaptability).

Relationships between temperature and plant growth have been studied extensively. From what is known, it does not seem likely that low temperatures in northwestern Britain would ever render southeastern plants unable to survive there in consequence of poor vegetative growth alone (except, of course, at high altitudes). However, carbon deficits, resulting from an excess of respiration over photosynthesis, and curtailed growth due to short growing seasons, are implicated in the control of alpine timberlines (Tranquillini 1979), and might therefore conceivably affect the upper altitudinal limits of other plants in upland Britain – especially those that reach relatively high altitudes. This remains speculative because the fairly extensive work that has been done on carbon balances in British upland species (e.g. Bannister 1981) has not been directly linked to the question of plant distributions.

The vegetative growth of southeastern species in northwestern Britain may nevertheless be impaired a little, and this may lead to secondary effects more clearly related to the control of distribution limits. These effects fall into two principal categories: first, effects on a plant's competitive ability, and second, effects on its reproduction. The climate may, of course, affect reproduction directly in several ways, but in most plants reproduction is the culmination of the life-cycle, so that small effects on growth during preceding stages in the life-cycle may accrue to produce a large effect at this stage.

It is widely believed that on a world scale, temperate species are controlled at their northern limits by the direct effects of cold climates on their growth and reproduction, whereas at their southern limits they are out-competed by species better able to take advantage of the warm climates that prevail towards the Tropics (MacArthur 1972). However, Salisbury (1932) noted that many European plants tend to behave increasingly as ruderals towards their northern limits in Britain, and he suggested that this may be due to an increasing inability to compete with other plants in unfavourable climates; reduced competitive vigour may therefore be involved in the control of northern limits as well. The rôle of competition in eliminating plants at their limits is extremely hard to demonstrate in the field, but much may be deduced from the results of an investigation (Woodward & Pigott 1975, Woodward 1975) into how climatic factors affect competition in the species *Sedum telephium* and *S. rosea*. In Britain, *S. telephium* is a lowland species, and *S. rosea* is a northwestern species confined to high mountains in the southern part of its range. When the two were grown in competition at a range of altitudes, *S. telephium* proved to be the larger plant at low altitudes, whereas *S. rosea* was the larger at high altitudes. The growth of *S. rosea* varied little with altitude, whereas that of *S. telephium* declined markedly with increasing altitude in a way that suggested control by the deteriorating climate. Controlled environment experiments confirmed that the growth of *S. telephium* is stimulated more by increased temperature than is that of *S. rosea*, and showed that a decrease in temperature commensurate with an increase in altitude of 250 m is sufficient to reverse the outcome of competition between the two species. Though these two species probably never compete with each other in nature, the outcome of competition between them probably does indicate the relative chances of their competing successfully with other species. It is therefore a plausible hypothesis that *S. rosea* grows in the north because its vigour in cold climates is reduced less than that of its competitors; while *S. telephium* might be eliminated at its limit for any or all of several reasons, including the direct effects of reduced growth, as well as the reductions in competitive ability and reproductive output that result from reduced growth (it produces fewer flowers at high altitudes).

It is a matter of observation that many British plants of southeastern distribution fail to set seed in the northern parts of their ranges. Matthews (1937), for example, lists species that grow well in Scotland, but scarcely ever fruit there. This type of failure has been investigated experimentally, and has been identified more often than any other as a potential cause of distribution limits. Investigations on *Cirsium acaulon* (Pigott 1970), *Hordeum murinum* (Davison 1970, 1977), and *Tilia cordata*. (Pigott 1981, Pigott & Huntley 1978, 1981) will now be discussed; these are all south-eastern species in Britain.

Cirsium acaulon, a perennial herb of calcareous grasslands, reaches its northern limit in Derbyshire, where it is confined to slopes of southern aspect. Various summer isotherms, isohyets of summer rainfall, and isolines for the duration of bright sunshine are correlated with its limit. Pigott (1970) found that the proportion of the plant's mature seeds that contain fully developed embryos declines towards its limit from 24 per inflorescence in Surrey to 1 or 2 per inflorescence at the limit. This decline appears to be related to temperature: any treatment which reduces the temperature of the inflorescence (e.g. shading or spraying) reduces the number of viable seeds produced. Other effects evident towards the plant's limit include a delay in flowering, from June in Surrey to August at the limit; a decline in seedling establishment; and a greatly increased susceptibility of the inflorescence to fungal infection by *Botrytis*. Since rain-wetting both cools the inflorescences and renders them more liable to fungal infection, and since sunshine both warms and dries them, it can be seen that all the climatic factors that correlate with the plant's limit may affect the development of its fruits.

The annual grass *Hordeum murinum* is a ruderal having a distribution pattern very similar to that of *Cirsium acaulon*. Davison (1970) analysed correlations between environmental variables and the distribution of *H. murinum* in considerable detail, and showed that on this basis temperature appears to be of overriding importance in controlling its limit. In transplant experiments (Davison 1977), he compared its growth at a site near to its limit at Newcastle-upon-Tyne with that at two upland sites which were on average about 3°C cooler than the Newcastle site, and much wetter. Over a period of about two years, the plant failed at the upland sites owing to a complex of causes, which included slow growth, late development of flowers and fruits, poor fruit ripening, and poor fruit dispersal; to some extent delays and failures at each stage in the life-cycle tended to result in further delays and failures at later stages.

Tilia cordata is an infrequent tree that reaches its northern limit in Britain in the Lake District. At its limit it rarely produces good seed, but in Worcestershire it regularly does so. Its northern limit throughout Europe is correlated with the mean July isotherm for 16°C. Pigott & Huntley (1978, 1981) found that ovules from trees in the Lake District are seldom fertilised, and they showed that pollen-tube growth (which is essential to fertilisation) can only take place at temperatures above 18–20°C. Temperatures in the Lake District during early August (when the trees are in flower) are frequently lower than this. Moreover, if fertilisation does happen to occur in the Lake District, the resulting seeds may fail to complete their development before the onset of winter; whereas they usually ripen before this in the South of England (where flowering takes place in July). Pigott (1981) also found that fertilisation does normally occur in the more continental climate that prevails at the plant's northern limit in Finland (no doubt because August temperatures are higher than in the Lake District), but the seeds almost always fail to complete their development because temperatures fall more rapidly than in the Lake District during September and October. This illustrates the important, and all too often neglected, point that the principal causes of a plant's limit may vary from place to place.

The results of these investigations clearly implicate diverse reproductive failures caused by adverse climatic conditions as major factors controlling the limits of these three plants. The explanations of these limits advanced by the investigators concerned are among the most satisfying yet advanced for any, because they relate climatic conditions at the limit to potentially limit-controlling plant responses observed in the field. However, it is remarkable that so few distribution limits should have been explained in this way, and this fact seems to indicate some prevailing difficulty in producing explanations of this type. Are there, then, types of distribution limit that cannot be so easily explained?

There are plants that seem to show no decline in performance towards their limits. Melzack (1982), for example, measured several characters in *Taxus baccata* towards the limit of its natural range in Britain, but could only find a decline (not necessarily critical) in seed size. More important, however, there are many plants that grow well if they are planted beyond their limits: in particular, many annual ruderals are able to produce copious seed and persist locally. South-eastern British examples are *Mercurialis annua*, *Melampyrum arvense*, *Kickxia elatine*, *K. spuria*, *Lathyrus nissolia* and *Lactuca serriola*. This problem is clearly presaged by the fact that many plants can be grown far beyond their natural ranges. It is probably true that most ecologists only feel satisfied by explanations of distribution limits that invoke observable declines in the performance of the individual plant at or beyond its limit; and yet many commonplace evidences suggest that such declines are not to be found in a large number of plants. Information about the performance of such plants beyond their limits is scarce, perhaps because it is normally deemed to be inconclusive. The limits of such plants can always be dismissed, in a hypothetical way, as being due to the inscrutable effects of competition, or they may be attributed to unobserved events that take place during exceptional weather, or at some stage in a plant's life history that is hard to investigate (e.g. seedling establishment). No doubt such hypotheses sometimes come near the truth, especially as they are often supported by anecdotal evidence. But many of the difficulties encountered in explaining these and other limits may result from theoretical inadequacies in understanding these limits.

It is, for example, far from clear how partial reproductive failures (such as those observed in Cirsium acaulon, Hordeum murinum and Tilia cordata) might operate at the population level to cause a plant's limit; this is briefly discussed by Pigott (1970). Hitherto, studies seeking to explain distribution limits in this way have not been forthcoming, no doubt because they pose immense, and probably as yet insuperable, practical problems. But until they are forthcoming, it will remain difficult to quantify the relative contributions made to the control of a plant's limit by each of the several physiological responses that may appear to control it.

In essence, the problem is one of integrating information pertaining to different hierarchical levels of biological organisation. To achieve a comprehensive explanation of a plant's distribution limit, it is necessary to know something about the plant's physiological responses to the climate; to know something about its population dynamics; and to know something about metapopulations (sensu Levins 1970) of the plant, i.e. spatially separated populations that constitute a plant's presence in a geographical area. The dynamics of metapopulations can, of course, be studied independently of those of the populations that compose them. A distribution limit is a phenomenon defined only at the metapopulation level; to one side of a limit there is a metapopulation, and to the other there is not. A plant therefore reaches its limit not at the point where the climate renders the individual unable to survive (i.e. the point where the plant reaches its physiological 'tolerance limits'), but rather at some point (usually well before this) where the metapopulation is unable to survive because the rate at which new colonies are founded has fallen below that at which existing ones become extinct. These rates are, of course, determined at several removes by the plant's responses to the climate, which therefore remain the ultimate controlling factors. Carter and Prince (1981) applied a simple epidemic model to this process, and predicted that a very small change in the performance of a plant might be sufficient to determine whether a metapopulation could survive or not, so that limits might be caused by declines in performance that intuitively seem too small to explain them. This in itself probably accounts for some of the difficulties so commonly encountered in explaining distribution limits in the terms set out at the beginning of this chapter. The inability to predict the effects of a biological response at one hierarchical level (such as the physiological response level) on processes taking place at the next level up remains the most serious obstacle to achieving such explanations.

In conclusion it may be said that, in a qualitative way, it is often possible to identify plant responses to climatic factors that control distribution limits. Failure to set seed in cool, wet climates has most often been cited as the principal cause of limits for southeastern plants in Britain; and inability to compete with southeastern plants in warm dry climates has most often been cited as the principal cause of limits for northwestern plants. However, the effects of these responses are hard to quantify, and this means, first, that the relative importance to any species of complementary controls on its limit cannot be assessed, and, second, that predictive explanations of why limits occur where they do remain beyond our grasp.

References

Arnold, S. M. and J. L. Monteith 1974. Plant development and mean temperature in a Teesdale habitat. *J. Ecol.* **62**, 711–20.

Bannister, P. 1981. Carbohydrate concentration of heath plants of different geographical origin. *J. Ecol.* **69**, 769–80.

Baskin, J. M. and C. C. Baskin 1972. Ecological life cycle and physiological ecology of seed germination of *Arabidopsis thaliana*. *Can J. Bot.* **50**, 353–60.

Baskin, J. M. and C. C. Baskin 1976. High temperature requirement for after ripening in seeds of winter annuals. *New Phytol.* **77**, 619–24.

Baskin, J. M. and C. C. Baskin 1981. Seasonal changes in germination responses of buried seeds of *Verbascum thapsus* and *V. blattaria* and ecological implications. *Can. J. Bot.* **59**, 1769–80.

Billings, W. D. and H. A. Mooney 1968. The ecology of arctic and alpine plants. *Biol Rev.* **43**, 481–529.

Boysen-Jensen, P. 1949. Causal plant geography. *Biol Meddr.* **21**, 1–19.

Braak, J. P. 1978. The effect of flowering date and temperature on embryo development in sweet cherry (*Prunus avium* L.). *Neth. J. agric. Sci.* **26**, 13–30.

Callaghan, T. V. 1974. Intraspecific variation in *Phleum alpinum* L. with special reference to polar populations. *Arct. Alp. Res.* **6**, 361–401.

Campbell, A. D. 1981. Flowering records for *Chionocloa*, *Aciphylla* and *Celmisia* species in the Cragieburn Range, South Island, New Zealand. *N. Z. J. Bot.* **19**, 97–103.

Carter, R. N. and S. D. Prince 1981. Epidemic models used to explain biogeographical distribution limits. *Nature* (Lond.) **293**, 644–5.

Chandler, T. J. and S. Gregory (eds.) 1976. *The climate of the British Isles.* London: Longman.

Clausen, J., D. D. Keck and W. M. Hiesey 1948. Experimental studies on the nature of species. III. Environmental responses of climatic races of *Achillea*. *Publ. Carneg. Instn* no. 581.

Clebsch, E. E. C. and W. D. Billings 1976. Seed germination and vivipary from a latitudinal series of populations of the arctic–alpine grass *Trisetum spicatum*. *Arct. Alp. Res.* **8**, 255–62.

Clements, F. E. 1907. *Plant physiology and ecology.* New York: Holt.

Conolly, A. P. and E. Dahl 1970. Maximum summer temperatures in relation to the modern and Quaternary distribution of certain arctic–montane species in the British Isles. In *Studies in the vegetational history of the British Isles*, D. Walker and R. G. West (eds), 159–223. Cambridge: Cambridge University Press.

Cooper, J. P. 1963. Species and population differences in climatic response. In *Environmental control of plant growth*, L. T. Evans (ed.), 381–403. New York: Academic Press.

Cooper, J. P. 1964. Climatic variation in forage grasses. I. Leaf development in climatic races of *Lolium* and *Dactylis*. *J. Appl. Ecol.* **1**, 45–61.

Dahl, E. 1951. On the relation between summer temperature and the distribution of alpine vascular plants in the lowlands of Fennoscandia. *Oikos* **3**, 22–52.

Davison, A. W. 1970. The ecology of *Hordeum murinum* L.I. Analysis of the distribution in Britain. *J. Ecol.* **58**, 453–66.

Davison, A. W. 1977. The ecology of *Hordeum murinum* L. III. Some effects of adverse climate. *J. Ecol.* **65**, 523–30.

Dilks, T. J. K. and M. C. F. Procter 1975. Comparative experiments on temperature responses of bryophytes; assimilation, respiration and freezing damage. *J. Bryol.* **8**, 317–36.

Douglas, D. A. 1981. The balance between vegetative and sexual reproduction in *Mimulus primuloides* (Scrophulariaceae) at different altitudes in California. *J. Ecol.* **69**, 295–310.

Drude, O. 1913. *Die Ökologie der Pflangen*. Braunschweig: F. Vieweg.

Furness, S. B. and J. P. Grime 1982. Growth rate and temperature responses in bryophytes. II. A comparative study of species of contrasted ecology. *J. Ecol.* **70**, 525–36.

Gallagher, J. N. and P. V. Biscoe 1978. Radiation absorption, growth and yield in cereals. *J. Agric. Sci. Camb.* **91**, 103–16.

Geiger, R. 1965. *The climate near the ground*, (transl. from the 4th German edn). New York: Harvard University Press.

Grace, J. 1977. *Plant response to wind*. London: Academic Press.

Haberlandt, G. 1884. *Physiologische Pflanzenanatomie*. Leipzig: Engelmann.

Harper, J. L. 1977. *Population biology of plants*. London: Academic Press.

Harper, J. L. 1980. Plant demography and ecological theory. *Oikos* **35**, 244–53.

Harper, J. L. 1982. After description. In *The plant community as a working mechanism*, E. J. Newman (ed.), 11–25. Special publications series of the British Ecological Society No. 1. Oxford: Blackwell.

Hegarty, T. W. 1973. Temperature coefficient (Q_{10}), seed germination and other biological processes. *Nature* (Lond.) **243**, 305–6.

Heslop-Harrison, J. 1964. Forty years of genecology. *Adv. Ecol. Res.* **2**, 159–247.

Hunter, R. F. and S. A. Grant 1971. The effect of altitude on grass growth in east Scotland, *J. Appl. Ecol.* **8**, 1–19.

Iversen, J. 1944. *Viscum, Hedera* and *Ilex* as climate indicators. *Geol. För. Stockh. Förh.* **66**, 463–83.

Kaku, S. and M. Iwaya 1979. Deep supercooling in xylems and ecological distribution in the genera *Ilex, Viburnum* and *Quercus* in Japan. *Oikos* **33**, 402–11.

Karssen, C. M. 1980. Patterns of change in dormancy during burial of seeds in soil. *Israel J. Bot*, **29**, 65–73.

Larcher, W. 1980. *Physiological plant ecology*, 2nd edn. Berlin: Springer-Verlag.

Levins, R. 1970. Extinction. In *Some mathematical questions in biology*, M. Gerstenhaber (ed.), 77–107. *Lect. Math. Life Sci.* **2**.

MacArthur, R. H. 1972. *Geographical ecology*. New York: Harper & Row.

McNaughton, S. J. 1966. Ecotype function in the *Typha* community-type. *Ecol. Monogr.* **36**, 297–325.

Manley, G. 1936. The climate of the northern Pennines: the coldest part of England. *Q. J. R. Met. Soc.* **62**, 103–113.

Manley, G. 1952. *Climate and the British scene*. London: Collins.

Manley, G. 1980. The northern Pennines revisited: Moor House, 1932–1978. *Met. Mag. Lond.* **109**, 281–92.

Matthews, J. R. 1937. Geographical relationships of the British flora. *J. Ecol.* **25**, 1–90.

Matthews, J. R. 1955. *Origin and distribution of the British flora*. London: Hutchinson.

Melzack, R. N. 1982. Variations in seed weight, germination, and seedling vigour in the yew (*Taxus baccata* L.) in England. *J. Biogeog.* **9**, 55–63.

Mooney, H. A. and W. D. Billings 1961. Comparative physiological ecology of arctic and alpine populations of *Oxyria digyna*. *Ecol. Monogr.* **31**, 1–29.

Mooney, H. A. and A. T. Harrison 1970. The influence of conditioning temperature on subsequent temperature-related photosynthetic capacity in higher plants. In *Prediction and measurement of photosynthetic productivity*, C. T. de Wit (ed.), 411–17. Wageningen: Centre for Agricultural Publishing and Documentation.

Monteith, J. L. 1981. Climatic variation and the growth of crops. *Q. J. R. Met. Soc.* **107**, 749–74.

Munro, J. M. M. and D. A. Davies 1973. Potential pasture production in the uplands of Wales. 2. Climatic limitations on production. *J. Br. Grassld Soc.* **28**, 161–9.

Munro, J. M. M., D. A. Davies and T. A. Thomas 1973. Potential pasture production in the uplands of Wales. 3. Soil nutrient resources and limitations. *J. Br. Grassld Soc.* **28**, 247–55.

Ollerenshaw, J. H. and R. H. Baker 1981. Low temperature growth in a controlled environment of *Trifolium repens* plants from northern latitudes. *J. Appl. Ecol.*, **18**, 229–39.

Pearce, R. S. 1980. Relative hardiness to freezing of laminae, roots and tillers of tall fescue. *New Phytol.* **84**, 449–63.

Pearsall, W. H. 1950. *Mountains and moorlands*. London: Collins.

Pemadasa, M. A. and P. H. Lovell 1975. Factors controlling germination of some dune annuals. *J. Ecol.* **63**, 41–59.

Perring, F. H. and S. M. Walters 1962. *Atlas of the British flora*. London: Nelson.

Pigott, C. D. 1970. The response of plants to climate and climatic change. In *The flora of a changing Britain*, F. H. Perring (ed.), 32–44. BSBI Conference Reports no. 11. Faringdon: E. W. Classey.

Pigott, C. D. 1981. Nature of seed sterility and natural regeneration of *Tilia cordata* near its northern limit in Finland. *Ann. Bot. Fennici* **18**, 255–63.

Pigott, C. D. and J. P. Huntley 1978. Factors controlling the distribution of *Tilia cordata* at the northern limits of its geographical range. I. Distribution in north-west England. *New Phytol.* **81**, 429–41.

Pigott, C. D. and J. P. Huntley 1981. Factors controlling the distribution of *Tilia cordata* at the northern limits of its geographical range. III. Nature and causes of seed sterility. *New Phytol.* **87**, 817–39.

Prince, S. D. 1976. The effect of climate on grain development in barley at an upland site. *New Phytol.* **76**, 377–89.

Prince, S. D. 1980. Ecophysiology of wild lettuce. In *Proceedings Eucarpia meeting on leafy vegetables*, J. W. Maxon-Smith and F. A. Langton (eds), 63–73. Littlehampton: Glasshouse Crops Research Institute.

Ratcliffe, D. A. 1968. An ecological account of atlantic bryophytes in the British Isles. *New Phytol.*, **67**, 365–439.

Regehr, D. L. and F. A. Bazzaz 1976. Low temperature photosynthesis in successional winter annuals. *Ecol.* **57**, 1297–303.

Salisbury, E. J. 1926. The geographical distribution of plants in relation to climatic factors. *Geog. J.* **57**, 312–35.

Salisbury, E. J. 1932. The East-Anglian flora. *Trans Norfolk Norwich Nat. Soc.* **13**, 191–263.

Salisbury, E. J. 1939. Ecological aspects of meteorology. *Q. J. R. Met. Soc.* **65**, 337–58.

Savidge, J. P. 1970. Changes in plant distribution following changes in local climate. In *The flora of a changing Britain*, F. H. Perring (ed.), 25–31. BSBI Conference Reports no. 11. Faringdon: E. W. Classey.

Sawada, S. and S. Miyachi 1974. Effects of growth temperature on photosynthetic carbon metabolism in green plants. I. Photosynthetic activities of various plants acclimatized to varied temperatures. *Plant Cell Physiol.* **15**, 111–20.

Schimper, A. F. W. 1903. *Plant geography upon a physiological basis* (trans. W. R. Fisher from the German edn of 1898). Oxford: Clarendon Press.

Scott Russell, R. 1940. Physiological and ecological studies on an arctic vegetation. II. The development of vegetation in relation to nitrogen supply and soil micro-organisms on Jan Mayen Island. *J. Ecol.* **28**, 269–88.

Sheldon, J. C. and F. M. Burrows 1973. The dispersal effectiveness of the achene-pappus units of selected Compositae in steady winds with convection. *New Phytol.* **72**, 665–75.

Strain, B. R. 1969. Seasonal adaptations in photosynthesis and respiration in four desert shrubs growing *in situ*. *Ecol.* **50**, 511–13.

Tansley, A. G. 1949. *The British Islands and their vegetation*. Cambridge: Cambridge University Press.

Thompson, K., J. P. Grime and G. Mason 1977. Seed germination in response to diurnal fluctuations of temperature. *Nature* (Lond.) **67**, 147–9.

Tranquillini, W. 1979. *Physiological ecology of the alpine timberline*. Berlin: Springer-Verlag.

Warming, E. 1909. *Oecology of plants* (trans. P. Groom and I. B. Balfour from the Danish edn of 1895). Oxford: Clarendon Press.

White, E. J. 1979. The prediction and selection of climatological data for ecological purposes in Great Britain. *J. Appl. Ecol.* **16**, 141–60.

Willemsen, R. W. 1975. Effect of stratification temperature and germination temperature on germination and the induction of secondary dormancy in common ragweed seeds. *Am. J. Bot.* **62**, 1–5.

Wilson, A. 1931–3. The altitudinal range of British plants. *North West Nat.* **5 & 6**: suppl.

Woodward, F. I. 1975. The climatic control of the altitudinal distribution of *Sedum rosea* (L.) Scop. and *S. telephium* L. II. The analysis of plant growth in controlled environments. *New Phytol.* **74**, 335–48.

Woodward, F. I. 1979. The differential temperature responses of the growth of certain plant species from different altitudes. II. Analyses of the control and morphology of leaf extension and specific leaf area of *Phleum bertolonii* D. C. and *P. alpinum* L. *New Phytol.* **82**, 397–405.

Woodward, F. I. and C. D. Pigott 1975. The climatic control of the altitudinal distribution of *Sedum rosea* (L.) Scop. and *S. telephium* L. I. Field observations. *New Phytol.* **74**, 323–34.

13

Climate and the diseases and pests of agriculture

AUSTIN BOURKE

In recent years meteorology and climatology have been playing an important and growing rôle in the struggle against the spread and virulence of agricultural pests and diseases. The problems encountered in evaluating the links between the weather conditions and agricultural symptoms are reviewed, and case studies are presented of potato blight and liver fluke. The importance of the rôle of wind for the transport of inoculum is noted. Climatic changes have influenced the occurrence of diseases and pests, and many of the new cultivars may not survive even small climatic fluctuations.

13.1 Introduction

Climate and weather are the predominant factors which determine agricultural production. The validity of this claim is repeatedly and painfully brought home in those seasons when frost, drought or floods lead to widespread food shortages, and to consequent famine in vulnerable areas, although it tends to be quickly forgotten again in the lulls between climatic extremes.

Apart from direct effects on yield and quality, an important aspect of climatic influence on agricultural production is its effect on the uneasy balance between crops and farm animals on the one hand and, on the other, a malevolent army of diseases and pests whose distribution and life-cycles are largely controlled by conditions in the atmosphere or in the soil. It has come to be recognised increasingly that a reduction in the heavy losses caused by these parasites during growth and in storage represents the most promising and readily available potential for increasing food supplies, particularly in developing countries.

In this struggle against the spread and virulence of pest and disease, meteorology and climatology have been playing an important and growing part in recent years. Perhaps the best known example concerns the international campaign for control of the desert locust (Rainey 1963, World Meteorological Organization 1965). In the early stages of this campaign, a detailed study was undertaken (for the critical year 1954–5) of weather and locust infestation in the entire area in Africa and the Middle East which is liable to be affected by the pest (Aspliden & Rainey 1961). The primary objective of this study was to pinpoint particular aspects of synoptic meteorology of significance to the multiplication, distribution and movement of locusts. An important finding was that low-level wind fields were the determinant factor governing the large-scale movements and distribution of the desert locusts, and that such movements generally result in displacement towards zones of convergent low-level airflow, and hence of probable rainfall. The meteorological association, which links the assembling together of the locusts with the rains which are essential for their successful breeding, is characteristic of the survival mechanism often found to be associated with other organisms for example, sporulation in fungi is triggered off by high levels of air humidity. If penetration and infection are to follow, the subsequent presence of free water on the plant surface is required, i.e. a sequence which is commonly found in the atmosphere. The results of the research into the links between meteorology and the life-cycle of the locusts provided the basis whereby the locust control organisation was supplied with guidance by the meteorologists, particularly on the hour-to-hour and day-to-day movement of flying swarms, so that spraying operations from aircraft might be carried out more effectively and economically.

In the longer term, the continued studies threw new light on climatic influence on the periodic upsurges and recessions which are such a marked feature of the desert locust plague (Pedgley 1979). The occurrence of several spells of widespread and heavy rains in normally arid areas is a major factor in the breeding of the vast number of locusts involved in a plague upsurge. The wind patterns, which in some formations can serve to keep the scattered groups of locusts apart, can, in other cases, crowd them together into massive swarms. As these processes become better understood, and advantage is taken of improved monitoring of the relevant rain storms with the help of satellites, not only will warning bells be rung in advance of the resurgence of the pest, but also it will become easier to pinpoint the locust breeding sites. Thus a quicker, more effective and more economical counter attack by the control authorities can be organised.

13.2 Climate and the spread of parasites to new areas

The case of the desert locust illustrates that the tactical war against the seasonal impact of pests or diseases (as affected by weather) calls for the assistance of the synoptic meteorologist. The help of the climatologist, as such, is more often sought in broad strategic problems, such as defining the areas which have so far escaped a parasite but which are climatically liable to invasion. An alert of this kind, when heeded, can lead to a strengthening of the phytosanitary defences and the preparation in good time of plans to repel an attack; ignored, it leaves the climatologist only the scant satisfaction of 'I told you so'. Thus, experience in North America of tobacco blue mould (*Peronospora tabacina*) indicated clearly, taking account of the European climate, that it would ravage the tobacco crop if it ever reached the Old World. A warning to this effect was in fact formulated fully five years before the disease appeared in Europe (Miller 1964). Notwithstanding, an industrial firm introduced the fungus into England in 1958 for use in fungicide experiments, without the consent of the responsible authorities. In the following years, the disease spread like wildfire through the tobacco fields of Europe (Populer 1964). The introduction of tobacco blue mould into Europe 'will remain, in the eyes of posterity, a serious error of judgement' (Zadoks 1967).

A happier illustration of meteorological and phytopathological co-operation concerns the possible transportation of an insect pest from North America to Europe (Bourke 1961). During the First World War, the Japanese beetle, *Popillia japonica*, found its way (presumably in cargo ships) from its homeland in Japan to the hinterland of the great ports on the north-east coast of the USA. The pest established itself in its new home in the following half century, and caused considerable damage to grassland, plants and trees. The failure of the beetle to set up a similar bridgehead in Europe was attributed to the absence of suitable climatic conditions near the transatlantic shipping terminals. However, the development of fast, direct aircraft flights across the Atlantic introduced a new risk of the inadvertent carriage of the pest. Before long, a number of live beetles was discovered on aircraft on their arrival in the heart of Europe. The European and Mediterranean Plant Protection Organization (EPPO) sought the help of the World Meteorological Organization in defining the 'danger areas' in Europe where climatic conditions were such as to permit the build up of permanent colonies of the beetle. This information was to help EPPO in planning the distribution of insect traps to locate primary bridgeheads of the enemy, and in general to concentrate its countermeasures with maximum intensity and efficiency.

The influence of climatic factors on successive growth stages in the life-cycle of the Japanese beetle was reviewed. The greater part of the insect's life is spent underground. Laboratory studies indicated the critical importance of soil temperature and soil moisture at certain stages of development of the grubs in the ground. A model was drawn up of the climatic conditions suitable for permanent settlement by the beetle:

(a) the total mean rainfall in the three summer months must exceed 250 mm;
(b) the mean soil temperature at a depth of 5–10 cm in a moist loam under sod must lie in the range 20–28°C in the warmest month;
(c) the corresponding mean soil temperature of the coldest month must be above − 2°C.

The general validity of this rough model was verified against climatic conditions in the geographical areas in Asia and North America in which the beetle had already established itself.

The requirement for a combination of moist soil with high soil temperature in the summer months rules out large parts of Europe as a permanent home for the beetle. The Mediterranean area is eliminated by reason of its dry summers. The north-west coastal fringe, the continent north of about 53°N, and the Russian

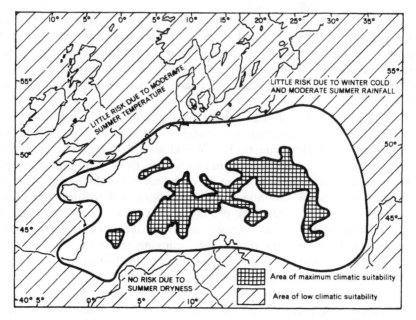

Figure 13.1 Map of climatic suitability of areas in Europe for the establishment of the Japanese beetle (*Popillia japonica* N).

interior all fail to meet one or other of the required temperature conditions. Inside the heavy boundary line in Figure 13.1, the unhatched area indicates the regions where the dryness of the soil in midsummer would normally be a restricting factor; here, the risk is not considered high, except in moist marshy areas or as temporary infestation in wet years, readily eradicable by chemical means. The region where the conditions of the model are fully met in years of normal climatic conditions are the valleys of the mountainous regions of Central Europe; here, there is a high risk that beetles introduced inadvertently through international airports such as Zurich or Geneva could, if unopposed, establish permanent colonies.

A more recent study considered the climatic aspects of the possible establishment and multiplication of the Japanese beetle in India (Dubey & Venkataraman 1980).

The climatic requirements of another plant pest, the Colorado potato beetle, are rather more modest, and its spread has been correspondingly more spectacular (Hurst 1975). The beetle, *Leptinotarsa decemlineata*, expanding from obscurity in Colorado in the mid-1820s, gained notoriety some 30 years later by the vigour of its attacks on potato foliage as that crop began to be cultivated in central USA. Later, it ate its way steadily eastwards across the USA, gained entry into Europe during the relaxed vigilance of wartime, and spread eastwards through most of that continent. Today it is continuing its advance into western Asia, still feeding mainly on the leaves of the potato plant. A recent climatic analysis was designed to define the areas most at risk from the beetle in India and to plan countermeasures against its ultimate arrival (Dubey & Godvindaswamy 1978).

During its period of hibernation in the earth, the Colorado beetle favours a moderately dry soil with temperatures above $-8°C$. It takes to the air in spring when the soil temperature is about 11–12°C, thus timing its emergence with cunning synchronisation to match the progress of the potato crop. The beetle is active when the air temperatures are between about 16°C and 27°C, with an optimum of 22°C. It thrives well in general in areas with an annual rainfall of between 600 and 1500 mm. Outside these temperatures and rainfall ranges, the activity of the beetle is restricted, and it is subject to an increased mortality rate. In an area such as France, where the pest is well established, a cool moist summer like that of 1946 leads to a sharp, if temporary, decline in numbers. Severe winters with limited snow cover leading to very low soil temperatures, such as were encountered, for example, over much of Europe in 1962–3, cause high mortality not only in the case of the Colorado beetle, but among all insects which overwinter as eggs or larvae in the ground. On the other hand, in borderline areas where the normal climate permits the Colorado beetle no more than a precarious foothold in an average year, the pest has proved its ability to multiply explosively in seasons of unusually favourable meteorological and biological conditions. On such occasions, particularly careful monitoring is required to detect windborne migration into fresh areas.

The wax and wane of populations of many other insects are also controlled by meteorological factors, so that monitoring of the relevant factors can give adequate advance warning of the upsurge of the pest, and so permit timely planning of control measures. During outbreak years of the cotton leaf worm (*Spodoptera littoralis*) in Egypt, for example, mean air and soil temperatures during a critical period are about 2°C lower than in the quiescent years (Omar 1980)

13.3   The bioclimatic approach to pathogens and pests, and some of its limitations

Studies of the spread of the Japanese and Colorado beetles shared a common approach to the fundamental problem of building up a simple model of the climatic conditions which are favourable to a pest. The type of empirical analysis which characterised these studies has also been commonly used in deriving models or indices to describe the relationship between meteorological factors and certain diseases of agricultural crops and animals.

The empirical approach to meteopathological problems may be schematically illustrated (as in Fig. 13.2, in which the area enclosed within the curve P stands for the range of environmental conditions in which a parasite can flourish, and the areas enclosed by the curves C_1, C_2 . . . represent, for each of a number of different areas, the totality of its regional climate). Several possibilities may be

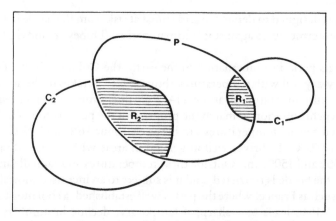

Figure 13.2 Schematic representation of the relationships between the environmental requirements of a parasite (P) and the climates (C_1 and C_2) of different regions (R_1 and R_2).

distinguished. If there is no overlap whatever between P and a particular C curve, then the range of environmental conditions which favour the pathogen are entirely outside those encountered in the climatic area concerned, and consequently the disease or pest represents no threat there. Conversely, if a C curve is entirely enclosed within P, it implies that weather conditions within the area concerned are never a restraining factor for the pathogen, and that the climatologist can contribute little or nothing to the struggle against disease or pest. In perhaps the most common case (as illustrated in Fig. 13.2), a certain measure of overlap occurs between the P and C curves, the overlapping area R representing the climatic conditions in that particular region which permit the pathogen to flourish.

In practice, the determination of the R regions, and hence the derivation of disease–climate rules, has usually involved the comparison of records of meteorological data (considered as input) with corresponding records of disease (considered as output) over a considerable number of years, and determination of the more significant relationships by inspection, statistical analysis or other means. The objective is to segregate from a complex intermix of relationships the vital climatic factors which possibly control a critical stage or bottleneck in the development of the disease, and to express their effect in simple numerical form.

The comparison of historical records of disease with corresponding climatic data presents great difficulties in practice. Accurate, representative and objective reports of disease impact, expressed in quantitative terms and not in broad subjective categories such as 'slight', 'average' or 'severe', are usually difficult to find. Normally climatic information is more readily available, although particular categories, such as soil temperatures and soil moisture, may be scarce. In the case of many plant diseases, the presence of free moisture on the foliage of the host crop over a period of time is essential to infection; information on the 'duration of leaf wetness' usually involves indirect deductions from observations of rainfall and air humidity. Even when reasonably complete files of information are available on the disease impact over a large number of seasons, and on a wide range of primary and secondary climatic parameters in the corresponding years, interrelating the first set of data with various possible groupings of the second in the search for meaningful links is a formidable task. Some pre-knowledge of the environmental requirements of the pathogen, as determined in the laboratory or elsewhere, may be essential in penetrating the labyrinth. The progressive evolution of early models for the development of potato blight (which is discussed later) is a good example of how the purely empirical approach may be illuminated and refined with the help of laboratory findings. Nevertheless, it must be conceded that simple empiricism has had its triumphs too, usually in cases where the climatic factor is so dominant and clear cut as to be almost obvious to the alert observer. Experience had taught growers of sweet corn in Connecticut that bacterial wilt caused by *Bacterium stewarti* developed as a serious problem following a warm winter, but was scarce or absent after an average or cold winter. So dominant was this factor, that it proved that a simple arithmetical index based on nothing more than the mean air temperature of the months December, January and February was a reliable guide to disease levels in the following season (Miller 1953). Why this should be so was a complete mystery at the time the index was developed; it was only later that it was discovered that the critical element was the winter survival of the insect vector, *Chaetocnema pulicularia*, whose rôle in spreading the disease was previously not known. The success rate of the winter temperature index was increased when it was recognised that more than a single favourable winter was required if both the vector and the bacterium were to build up to levels which would result in maximum disease incidence.

There are obvious dangers in the oversimplification of using a single numerical value to characterise the climate of a whole season, for example where the average temperature is taken to represent the severity of the winter season. In the case of many insects which overwinter in the soil, the rate at which cooling sets in may be as critical as the ultimate temperature level which is reached: a gradual cooling will permit many organisms to take refuge at safer, warmer depths, whereas a single, sharp, rapid cold snap (not perhaps reflected in the monthly or seasonal temperature averages) can lead to high mortality. Again, two seasons of the same mean temperature may incorporate very different frequencies of extreme temperatures. No simple index should be blindly or mechanically applied.

At the end of the winter, another recurrent phenomenon which may escape the broad mesh of seasonal or even monthly mean temperatures is the occurrence of an early mild spell which serves to awaken both plants and parasites from dormancy into premature and vulnerable growth, only for them to be struck down by the return of cold weather. Even when the cold itself is not fatal, the development of the parasites may be severely arrested, leaving them at the mercy of predators and disease, with consequent high mortality (MacHattie 1979).

An important reservation applies to the exportability of models which have been empirically derived: the fact that they operate well in the region where they were developed is no guarantee that they will prove to be valid in other areas. One reason for this (Fig. 13.2) is that the overlap, R1, between the entire environmental range, P, of the disease and the climate of one region, C1, which is the basis of the model developed there, may be quite different from that, R2, to be found in another climatic area, C2. Another factor which tends to restrict the general validity of a particular model is the possibility that it may be expressed, not directly in terms of the climatic element essential to disease, but in terms of another parameter which is closely related in the area of derivation but not necessarily in other areas. Thus in some areas of recurrent frontal activity, air humidity and rainfall may be closely correlated, and hence a model for a disease which is favoured by periods of high air humidity may successfully be expressed in terms of rainfall amounts. However, if applied in a humid region of low rainfall, such as parts of the Chilean and Californian coasts where wetting sea fogs are common, or in a continental climate where the rainfall is predominantly of shower type, such a model fails completely.

In a discussion of fireblight (*Erwynia amylovora*) Sutton and Jones (1975), noting that disease prediction based on monitoring inoculum potential or weather data is probably most effective in the geographical area in which it was developed, stressed the need to seek a comprehensive system applicable to a wide variety of climates. The question of whether a group of diverse models (e.g. based on R1, R2 etc. in Fig. 13.2) might not be integrated into a universal formula corresponding to P was taken up by Billing (1980), who compared four warning systems for fireblight which had been developed in New York, Illinois and California as well as south-east England.

A more direct path to a generally applicable model is to use the results of laboratory experiments (where such are available) on the reaction of the pathogen to different environmental conditions, i.e. to attempt to base the model on the complete curve P in Figure 13.2, rather than on segments of it. This approach also has its difficulties: it requires that the often complex results of laboratory experiments be reduced to a simple set of numerical criteria which are used as the definition of the weather which favours the progress of the pathogen. The data are rarely so simple as to point to a single unique model. Furthermore, the model, based as it is in this case exclusively on the requirements of the pathogen, will mirror the progress of the disease only if the weather–pathogen interrelationship is the dominant factor. No account is taken of the weather influence on the susceptibility of the host, nor of possible competition between the pathogen and other organisms. In some cases, a pathogen will make most progress, not in the environmental conditions which favour it most, but in those which most restrict its enemies.

Clearly, no matter which approach is used to derive a model of the climatic–disease relationship, it requires checking over a number of years before it can safely be put to practical use. A model derived from laboratory tests of the pathogen, when tested and found satisfactory, is in principle of universal application; it might, for instance, be used to assess the degree of risk of a new disease being successfully introduced into a particular climatic area. This application involves two further assumptions. One is that the agent of disease retains the same environmental requirements in space and time. The question of the acclimatisation of parasites or pests, or their adaption to changed climatic conditions, is controversial (Parry 1979). It is, however, an important consideration; if, for instance, the Japanese beetle adapted itself to conditions radically different from those previously found necessary, it would negate the deductions illustrated in Figure 13.1. The second assumption is that no effective climatic change or sustained fluctuation radically alters the position *vis-à-vis* the pathogen (see p. 273).

13.4 Climate and the diseases of agricultural plants: in particular, potato blight

Practical attempts to use climatology and meteorology in the fight against a number of plant diseases date from the end of the First World War. In the following six decades, considerable progress has been made, which is fully described in a number of review articles (Miller & O'Brien 1952, 1957, Bourke 1955, 1970, World Meteorological Organization 1963, 1969, Krause & Massie 1975, Miller 1967).

Potato blight (*Phytophthora infestans*) has been chosen as a representative example to outline the progress achieved. It is a major plant disease, in the impact of which climate plays a dominant part; indeed, in the 1840s when blight first appeared in Europe, it was almost universally diagnosed as being an effect of the weather alone (Graham 1847). It was not until some 20 years later that the fact that blight was due to a fungus won general acceptance.

The first attempt to define the climatic conditions which favour the blight fungus was carried out in a pioneer investigation in the Netherlands in the years 1919–23 (Bourke 1955). Miss Löhnis tried without success to correlate outbreaks of the disease in those five years with individual weather parameters. Professor van Everdingen (then Director of the Royal Netherlands Meteorological Institute) reworked the same data with various groupings of the weather factors, and concluded that the following four conditions were always met within the two weeks which preceded an outbreak of potato blight:

(a) at least four hours of dew in the course of a night;
(b) minimum temperature not below 10°C during the same night;
(c) mean cloudiness in the following day at least 8/10ths;
(d) a rainfall of not less than 0.1 mm in the 24 hours following the dew night.

These rules, operated with the help of a network of special weather-observing stations located within or very close to the potato fields of the main growing areas, were the foundation on which the pioneer Dutch potato blight warning system operated with some success for 20 years.

Nowadays, the primary defect of van Everdingen's rules would be seen to be a certain cumbrousness arising from their complexity, and the recurrent problem of near misses, in which three of the rules were met while the fourth was narrowly missed. In general, approximate models serve best when they are kept as simple as possible. Again, the use of special weather observations within the growing crop, while often necessary at the research stage, imposes a heavy burden in costs and communications if they are to be continued during the operational stage. Indeed, investigations into the microclimate of potato crops have shown that while in sunny weather there can be wide divergence between conditions in the meteorological instrument screen and those within the growing crop, they approximate closely in the cloudy and moist weather critical for potato blight (Broadbent 1949). Today, almost all practical forecast systems for potato blight and other plant diseases make use of standard meteorological observations.

The first reaction of Beaumont (1940, 1947), who tested van Everdingen's rules in Devonshire for the ten years 1929–38, was to add yet one further requirement in order to ensure that the foliage wetness caused by dew lasted well into the following day:

(e)   relative humidity at 1500 h on the afternoon after the dew night must be at least 75%.

Further work convinced him that relative humidity was the crucial factor, and, abandoning the first, third and fourth of the Dutch rules, he evolved the critical period which bears his name and which, with minor changes, was used until quite recently as the basis for the British potato blight warning service. In modified form, a Beaumont period was defined as commencing with a spell of 48 hours during which the air temperature had been 10°C or more and the relative humidity 75% or above on at least 46 of the 48 hourly observations. The period continued until two consecutive hourly observations failed to meet one or both of the Beaumont criteria.

Initially, both the Dutch and British rules were designed as automatic warnings, following which an outbreak of potato blight was to be expected within two weeks. Indeed, an automatic recording instrument for the use of farmers was developed in Scotland which measured dry and wet bulb temperatures within or near the potato crop, and enabled the grower to see at a glance when a Beaumont 'critical period' had occurred, and to take countermeasures against the expected onset of disease (Grainger 1953).

The general trend in countries with longer experience in applying weather data to plant disease forecasting, such as Holland, UK and Ireland, has been to move away from a rigid or mechanistic application of a set of rules, and towards a more flexible approach which regards the rules as merely one stage in meteopathological analysis. The correct interpretation of the data calls for considerable experience, skill and judgement. Constant reference to the course of disease in response to weather conditions in past seasons, as well as local knowledge and expertise are required (Billing 1980). Best results follow from close collaboration between meteorologists and plant pathologists. A review of the success of Beaumont's rule when it was the basis of the British blight forecasting system made no extreme claims: 'although the occurrence of weather that satisfies the rule, whether recorded in screens or in crops, is not invariably followed by blight outbreaks, when the rule is combined with experience and knowledge of the state of potato crops and the behaviour of the fungus it provides a workable basis for forecasting' (Hirst & Stedman 1956).

Tests of the Beaumont rule in the more maritime climate of Ireland proved disappointing: in particular, on the west coast, where the average relative humidity is high, it is possible for the afternoon values in a long series of successive days not to fall below 75% without any appreciable change in the blight situation. Nor was it possible to develop an alternative empirical model in the absence of an adequate series of annual disease reports. It was decided to attempt to base a model on available laboratory determination of the meteorological requirements of the pathogen (as represented by curve P in Fig. 13.2). A comprehensive American report on the relationship between weather conditions and the development of blight on potato foliage had been published by Crosier (1934). The blight–weather model which was developed in Ireland in 1949 on the basis of Crosier's work makes use of hourly weather reports from standard meteorological observing stations, and requires as an absolute minimum:

(a) a humid period of at least 12 hours with the temperature not falling below 10°C nor the relative humidity below 90% (conditions favourable for the formation of sporangia to begin);
(b) free water on the leaves for a subsequent period of at least 4 hours (conditions favourable for the germination of sporangia and for reinfection).

The second requirement usually follows automatically because of persistent rain or drizzle when condition (a) has been met. Should there be no precipitation ('dry cases', normally of fog), the alternative requirement is a further 4 hours beyond the initial 12 in which the relative humidity remains at 90% or above.

Each period in which the rules are met is considered to correspond to a single generation of the fungus. Between three and five generations are normally required before the disease is visible in the potato fields (Bourke 1953).

The rules were tested and found satisfactory in the 1950 and 1951 Irish potato growing seasons, and were introduced as the basis of a public blight warning system in 1952. They have continued in successful operation without modification since that time. Their method of derivation suggests that they might be valid in diverse climatic regions, and in fact they have been successfully used in Chile and Tasmania. A model, which despite some differences of emphasis is basically similar, was independently derived by Wallin (1962) from Crosier's data.

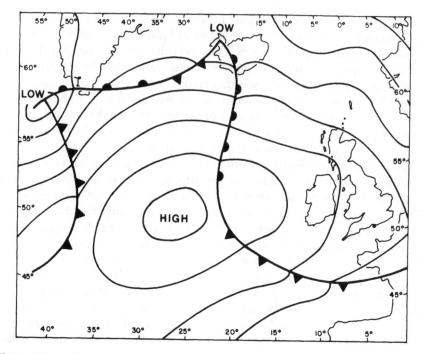

Figure 13.3 Surface synoptic weather chart for 06.00 GMT 4 July 1964, on the basis of which a potato late blight warning for Ireland was issued.

Two aspects of the Irish system of blight forecasting which were novel when first introduced have proved of value in other localities, and in respect of other diseases and pests. One is the emphasis on fairly lengthy periods of 90% or higher relative humidities in the screen as a measure of the duration of air saturation within the crop. The criterion of 90% relative humidity has been widely adopted; it is, for example, used in the blight forecasting technique which has replaced the Beaumont rule in the UK.

A more important innovation was the use of synoptic weather map and airmass analysis as an aid to phytopathological forecasting (Bourke 1957). The fact that conditions favourable to potato blight in the coastal fringe of north-west Europe are associated mainly with incursions of broad currents of moist warm air of tropical or subtropical origin (maritime tropical airmasses), provides a basis for converting disease warnings into true forecasts, and thereby considerably improving their value. On 4 July 1964, for example, it was clear that the synoptic situation (Fig. 13.3) which had earlier been unfavourable to potato blight was about to change to a set-up which would locate tropical air over Ireland at intervals during the next two weeks. On this basis, a spraying recommendation was issued which gave a full four days' notice of the onset of recurrent blight weather in Ireland (Fig. 13.4). It is a typical example of how a confident, broad long-term weather forecast, in terms adequate for plant disease

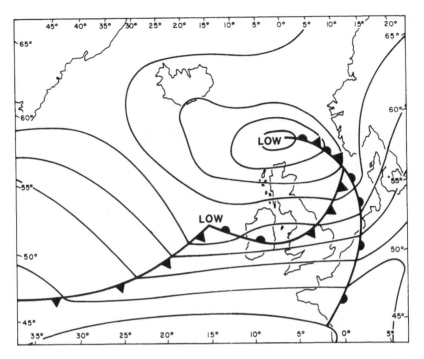

Figure 13.4 One of the successive weather situations highly favourable to potato late blight, which developed in the 2 weeks following 4 July 1964.

epidemiology, is often possible in situations where the precise timing necessary for a day-to-day forecast would be extremely difficult.

The use of weather map analysis in forecasting potato blight (as well as the occurrence of other plant diseases and pests) has proved fruitful elsewhere, particularly in the USA. The direction of upper air winds, as deduced from upper level charts, can indicate the possibility of an influx of maritime tropical air, favouring potato blight and similar diseases (Wallin & Riley 1960). Long-term prognostic weather charts have also been found useful in phytopathological forecasting.

An unexpected feedback from the use of weather map analysis as an aid to plant disease epidemiology has been the development of improved numerical models of the disease–weather relationship. Scarpa and Raniere (1964) found that outbreaks of downy mildew on lima beans in New Jersey depend on the frequency and persistence of incursions of maritime tropical air over the region, and, relating this to the conservative property of dewpoint values in this airmass, evolved a rule which calls for 40 or more hours of dewpoint values greater than 20°C.

The frequent recurrence of outbreaks of maritime tropical air seems to favour not only many fungal diseases (notably the downy mildews), but also certain plant pests. Huff (1963) found that a sustained flow of tropical air from the Gulf of Mexico into the American Midwest led to an influx into the area of numbers of potato leafhoppers (*Empoasca fabae*).

In the last few years, the wide availability of microcomputers has led almost inevitably to a potentially less happy development. There has been a tendency to return to the use of simple disease–weather models as automatic mechanised disease predictors from which programmes of the measures which should be taken against the disease can be derived. In the early 1970s, 'Blitecast' was developed as a computer program which incorporated two previous forecasting systems for the onset and development of potato blight, and which also drew up a recommended programme of fungicide applications (Krause *et al.* 1975). A similar electronic instrument has been designed to monitor temperature, relative humidity, and leaf wetness in apple orchards, and to detect primary infection periods of apple scab, *Venturia inaequalis* (Jones *et al.* 1980). Goltz *et al.* (1978) have developed a similar computerised device to measure temperature, relative humidity, and rainfall as a basis for predicting *Phytophthora* blight of tomatoes, and for scheduling spray applications against the disease. Adaptations of this basic system are outlined as a basis for drawing up spray programmes for other plant diseases.

The trend towards the complete computerisation of the disease–weather relationship and the control programme, without seasonal access to the skill and experience of plant pathologists and meteorologists, is not without its dangers.

13.5 Climate and the diseases of livestock; in particular, liver fluke

Although it has long been recognised that climate has a considerable influence on many diseases of farm animals, it was not until quite recently that attention has focused on making practical use of the relationship to help veterinarians in their work. Attempts to draw up usable climatic models of animal diseases have taken place mainly in the last 30 years. There are a number of excellent summaries of progress to date, much of which has occurred in the UK (Smith 1970, Taylor & Muller 1974, Gibson 1978, Gettinby & Gardiner 1980).

The major pioneering work in this connection was undertaken by Ollerenshaw (of the UK Central Veterinary Laboratory), who had noted that liver fluke disease caused by *Fasciola hepatica* is most severe after a wet summer. He set out to quantify the connection in order to forecast the extent of the countermeasures which would be necessary (Ollerenshaw & Rowlands 1959). The stage in the life-cycle of the parasite during which weather is critical is clearly that when it no longer shelters safely within the body of the farm animal, having been passed out of the host as eggs in its dung. The eggs hatch in a few weeks, given favourable conditions of moisture and temperature. At this vulnerable stage, the larvae must find a suitable mud snail in which to take refuge while they develop further. They emerge from the snail in their final form, attach themselves to grass or other convenient vegetation, and are taken once again into the bodies of grazing animals to complete their damaging life-cycle.

The same environmental conditions favour the hatching of the eggs which have been deposited in dung and the breeding of a large population of the snails which give the larvae temporary refuge, i.e. temperature of at least 10°C and a suitably high level of soil moisture. The temperature requirement is consistent with the fact that the disease does not cause problems in the cool sheep-raising areas of Iceland and northern Scandinavia. The moisture requirement for big thriving snails capable of supporting a large parasite population is that the upper layer of soil under pasture should have a consistently high water content without being permanently waterlogged.

Initial trials were made of the index $R - P$ as a rough parameter for measuring each month's disease potential, where $R =$ rainfall amount as measured, and $P =$ potential evapotranspiration as calculated by Penman's formula. The method showed distinct promise but it was found necessary to take into account the distribution, as well as the total amount, of rain during the month. This was achieved by introducing the number of days of appreciable rain into the analysis. An arbitrary constant was added to $R - P$ to avoid the inconvenience of negative values. The final Ollerenshaw monthly index for liver fluke potential took the form $(R - P + 5)n$, where $n =$ number of days during the month on which at least 0.01 inches of rain fell.

Summed over the months during which average air temperature is 10°C or above, this index, when applied to climatic data for England and Wales, helps to explain the broad pattern of liver fluke occurrence in these countries. The disease is not a serious problem in southern and eastern England, where summer rainfall rarely exceeds potential evapotranspiration. Over high country where rainfall is heavy, waterlogging tends to occur; this, combined with the acidity of peat-lands which is inimical to mud snails, restricts the impact of the disease. The disease is endemic in the rainy lowlands of Wales and western England, the actual level varying considerably with weather conditions from year to year. It is in the latter areas that the index, summed over the appropriate months, shows a close relationship with later disease levels, and has proved valuable in advising farmers on the extent and nature of desirable control measures. The Oller-enshaw system has now been in operation in England and Wales for over 20 years; it has also been adopted in Ireland, and a similar approach has been employed with success in France and the USSR. A simpler system, based entirely on the number of raindays, was later developed by Ross (1970) in Northern Ireland and was found to be valid also in Scotland.

The establishment of a numerical interconnection between meteorological data and the level of liver fluke disease has encouraged similar developments in the case of a number of other parasitic diseases of livestock.

13.6 Towards a climatology of disease potential

A quantitative weather–disease index which has proved its value in assessing the seasonal course of a disease or pest may, when averaged over a number of years, serve as a basis for drawing up a climatology of the weather-induced risk of damage by that disease or pest in a given area. At first sight, it might appear that information on the actual occurrence of attacks of the pathogen would serve the purpose better, but in fact such occurrence depends on a complex of factors apart from climate – density of crops, cultivars in use and their susceptibility, dates of sowing and emergence, management practices, countermeasures taken etc. – so that reports of actual damage by disease or pest frequently obscure the basic suitability of a given area to facilitate or discourage epidemic development. On the other hand, the demarcation of climate zones on the basis of a valid weather–pathogen model is related exclusively to the *potential* for disease or pest impact which is inherent in the climate of an area. Such climatic zoning of disease risks can be of value in larger scale planning of agricultural land use and in developing control strategies.

Figure 13.5 shows the distribution of climatic suitability in Ireland for the development of (a) liver fluke, measured in terms of the mean Ollerenshaw index for the months May–October, 1957–81, and (b) potato blight, measured in terms of the mean duration in hours of effective periods of blight weather, as determined by the Irish rules during the months May–September, 1957–81. For weighting humid spells of different lengths which meet the Irish rules, the 'effective' length of the minimum period of 12 hours is, by convention, taken as 1; accordingly, the 'effective period' in longer cases is determined by deducting 11 hours from the total duration when rain or drizzle occurs, or 15 hours in dry cases.

13.7 Meteorology and the transport of inoculum

The rôle of the wind is a climatic factor which is important in the case of certain pests and diseases (Hirst & Hurst 1967, Johnson 1969, Gregory 1973). The flight activity of many insects is greatly reduced in windy conditions, a fact which can restrict the damage they cause either directly or as vectors of disease. In the case of haematophagous insects, wind can even affect population levels, since the suppression of flight restricts blood-feeding, and thus reduces the insects' reproductive capacity (Service 1980). As in the case of the desert locust and the potato leafhopper, suitable wind patterns can assist in dispersal and in the colonisation of new areas.

That many plant diseases are spread locally on the wind has long been known; more recently, the important rôle of the wind in the short-distance extension of animal diseases such as fowl pest and foot-and-mouth has been established in England.

While the local movement of airborne pathogens is significant when seeking out and destroying secondary infections in the course of a disease outbreak, greater overall importance attaches to the long-distance transport of viable disease spores or insect pests which might thus be carried from one continent to another.

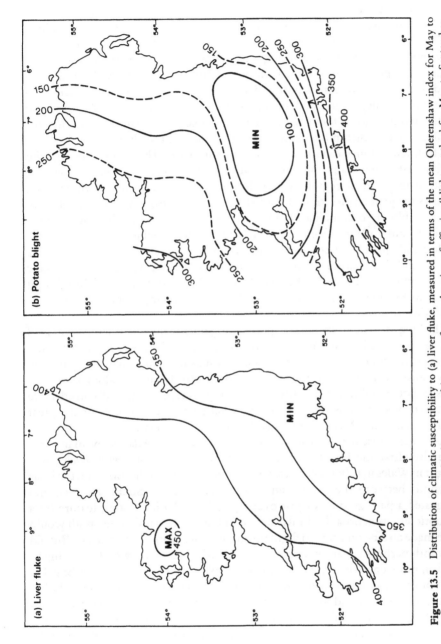

Figure 13.5 Distribution of climatic susceptibility to (a) liver fluke, measured in terms of the mean Ollerenshaw index for May to October 1957–81, and (b) potato blight, measured in terms of mean duration of effective 'blight weather' for May to September, 1957–81.

The best established regular long-distance commuter among disease spores is the black or stem rust of wheat, *Puccinia graminis* (Zadoks 1965, 1967). Hogg (1970) described the wind trajectories which have been shown to bring black rust spores from continental Europe to England, a country not subject to the disease except when inoculum is imported from elsewhere. Each instance of uredospore introduction was found to involve a precise synchronisation of liberation and deposition with the movement of weather systems and the speed of the winds within them. A similar atmospheric mechanism underlies the familiar south–north movement of rust inoculum in North America.

The broader European situation has been summarised by Zadoks (1965), who differentiates two main summer migration tracks of black rust in Europe. The western track, in a region where the prevailing winds are southwesterly, follows the general direction of the Atlantic coastline from Morocco to Britain, and possibly on to Scandinavia. The eastern track has its probable source in the lower Danube plains, from which a lesser branch reaches westwards to the middle and upper Danube plains, while the main track fans out over the Ukraine and Poland and continues northwards into Scandinavia. Clearly, any climatic change which substantially changed the frequency and strength of the relevant winds in Europe could lead to major changes in the impact of this important grain disease.

The long-distance transport of spores along tracks which are mainly over land can be confirmed by surveys made *en route*, and, if the host crop is generally grown in the countries traversed, the journey may be facilitated by being broken down into shorter stages as the crops are consecutively infected along the route. This is not possible if the journey is mainly over the sea, and considerable controversy surrounds the question of whether viable crossings of the great oceans such as the Atlantic are possible for spores or pests.

A plausible case for an airborne crossing of the North Atlantic by an insect is made by Hirst and Hurst (1967). An American moth, *Phytometra biloba*, was caught in Wales in July 1954. Back tracking the journey of the moth, on the basis of winds between the surface and 600 m, led back to South Virginia, and suggested a passage time of just over three days. An interesting feature (in a synoptic situation broadly similar to that of Fig. 13.4) was that the moth would have made an air-conditioned flight throughout in maritime tropical air. But did the moth perhaps cheat by making at least part of its journey on board ship?

Even more controversy has been aroused by the discovery in part of Brazil in 1970 of the devastating coffee rust disease, *Himeleia vastatrix* (Schieber 1972). This fungal disease had long been spreading and ravaging coffee crops in Asia and Africa, and when the impact of the disease in coastal West Africa became marked, Rayner (1960) wrote that coffee rust disease 'presented a threat to the coffee-growing regions of the New World as uredospores might be carried there from Africa by the north-east trade winds'. Bowden *et al.* (1971) calculated that the trade winds could plausibly have transported spores across the South Atlantic in 5–7 days, from Angola, where the disease was active in 1966, to Bahia in Brazil where it made its first appearance in South America.

The subsequent progress of the disease in Latin America from its initial foothold in Bahia has been generally consistent with atmospheric dissemination by wind and storms.

13.8   Climatic change and the impact of disease

In recent years, there has been a growing awareness that the climate of a region is far from being a fixed entity. There are obvious implications for plant protection in even temporary fluctuations of climate; an unusual series of mild winters, for example, may lead to a greater survival rate for insect pests, and a progressive build-up over several years to abnormal levels of population. Some 20 years ago, Whyte drew attention to the risk that a change of climate may radically alter the frequency and virulence of crop disease attacks, and characterised the problem as 'an aspect of great economic significance to land use and agricultural production which merits a full review by a competent authority' (Whyte 1963).

No such comprehensive review has yet been undertaken, although accounts of some scattered evidence of change have appeared. It is normally difficult to segregate with certainty the rôle played by the atmosphere as distinct from other factors of importance for disease, e.g. changes in acreage, cultural practices, cultivars, crop protection measures etc. (Coakley 1978, 1979). The fact that common rust of maize has become a bigger problem in Britain in recent years is due in large measure to intensive maize cultivation (Mahindapala 1978). Increased use of weedkillers has removed one reason for earthing up potato crops; failure to carry out this operation has increased the exposure of tubers to potato blight.

A reasonably convincing case of weather-induced disease is related to the succession of hot summers in north-west Europe which began in 1933, and to which is confidently attributed the sudden spread and increased destruction of rust disease of snapdragon (Moore 1940). Slow climatic change is held by Woods (1953) to have been largely responsible, firstly, for permitting the extension of the geographical distribution of the American chestnut and, later, in the period 1825–75, for the virtual elimination of the tree from the southern states of the USA by reason of *Phytophthora* root rot. The damaging upsurge of coffee berry disease in East Africa in the early 1960s has been attributed to 'a number of changing factors, probably the most important being a climatic change towards wetter, cooler conditions which favour infection' (Bent 1969). In the coffee-growing areas of tropical Africa 'the disease hazards will vary with the occurrence of wet and dry weather cycles' (Nutman & Roberts 1969).

The most spectacular cases of climatic control of disease or pest are clearly to be sought in areas which are marginal for the parasites and where any drift towards or away from critical threshold values of the relevant meteorological elements may be expected to have significant consequences. Iceland, where, despite the low summer temperatures and comparatively short growing season, potatoes are an important food crop, provides a remarkable example of the climatic control of potato blight. When Cox and Large (1960) carried out their comprehensive review of potato blight epidemics throughout the world, they commented (p. 89) that 'the occurrence of potato blight in Iceland is of special interest as the temperature conditions are here at the lower limit for any epidemic development of the disease'. The review indicated that, under favourable humidity conditions, the disease attacked the crop when the mean daily temperatures in July and August reached about 12–13°C, but appeared to be inhibited when these temperatures ranged about 10–11°C. Indeed, an examination of the disease data included in the report shows that even the crude indicators of mean monthly temperature could serve to differentiate the years of serious damage. In the period 1936–56, there were six years of severe blight; in fact, this was the period when the disease became a serious and recurrent problem in Iceland. In the six years of severe blight, the mean air temperatures were consistently higher in July (average 12.3°C) and in August (average 11.7°C) than they were in the years of little or no disease (11.2°C and 10.7°C respectively).

The shape of later developments in the potato blight situation in Iceland is clearly foreshadowed in Table 13.1, which gives mean monthly air temperatures during the potato-growing season in Reykjavik for each decade in the present century.

The gradual rise of temperature early in the century to the peak of warmth in the 1930s stands out clearly, as does the subsequent cooling, gradual at first, but sharp since 1960. Indeed, the mean July–August temperatures in each of the last two decades were markedly lower than those of the preceding years of the century. Simultaneously, the impact of potato blight in Iceland has dropped spectacularly, so that in the last few years the disease has been difficult to detect at all. Does this constitute a remarkable case of the success of a climatic fluctuation in wiping out a troublesome plant disease? Experience elsewhere suggests that the disease almost certainly lives on in Iceland at a subclinical level, ready to build up again if and when suitable climatic conditions return.

A point of some importance in this connection is that a successful pathogen

Table 13.1 Mean monthly air temperatures (°C) during the summer half-year in Reykjavik in successive decades, 1901–80.

	May	June	July	August	September	October
1901–10	6.2	9.8	11.4	10.3	8.4	4.2
1911–20	6.6	9.8	11.6	11.0	7.8	4.9
1921–30	6.1	9.2	11.1	10.4	7.2	3.7
1931–40	7.6	9.9	12.0	11.3	9.2	4.9
1941–50	6.8	9.8	11.5	11.1	8.3	5.1
1951–60	6.8	9.6	11.2	10.7	8.7	5.1
1961–70	6.7	9.6	10.6	10.5	7.8	4.8
1971–80	6.2	8.7	10.6	10.2	7.4	4.7

tends to have a life pattern well adapted to the growth cycle of its host (Billing 1980); consequently, conditions which are trying for the pathogen are unlikely to favour the crop upon which it feeds. The temperature drop which has blunted the edge of blight in Iceland has simultaneously reduced crop yields because of a shortened growing period. A similar phenomenon occurs in potato-growing areas, where soil moisture is apt to be the element in short supply: here, the wet years increase the potential yield, but also lead to more severe blight attacks (Cox & Large 1960). It is therefore not in areas where the crop is grown, and the parasite survives, in climatic conditions close to certain borderline values that the net consequences of climatic change may be greatest. It is, in fact, in regions where the climatic situation looks superficially more secure that the greater danger may lie. The advent of still-moister weather to a potato-growing area which is already moist enough for maximum potential yield will increase losses due to blight without benefiting the potential harvest, and may indeed put it at further risk by reason of waterlogging.

The marked rise of mean temperature which occurred in many northern areas in the first half of the present century has had consequences for the distribution of pests as well as plant diseases. In the case of insects, the controlling factor is often the intensity of winter cold and its effect on the survival rate of the pests. An insect pest of forest trees, the European pine shoot (*Rhyacionia buoliana*), was observed to spread northwards in Ontario in the period 1925–50; afterwards, it apparently ceased to expand. The northern limit of continuous distribution approximates to the $-26°C$ isotherm for mean annual minimum temperature (MacHattie 1979).

A recent increase in the frequency and severity of epidemics of yellow rust, *Puccinia striiformis*, on winter wheat crops in the Pacific Northwest of the USA has been attributed to climatic change (Coakley 1978, 1979). Mean air temperatures in January–February averaged 1.2°C higher in 1961–74 compared with the period 1935–60, whereas mean April temperatures averaged 1.3°C lower in the later period. The milder winters helped the survival of the fungus and shortened its latent period. The effect of colder springs is on the host crop. The cultivars which are grown need springtime warmth to trigger their resistance mechanism; accordingly, the consequence of the lower April temperatures was to render the crop more vulnerable to disease.

13.9 The future

It seems certain that the struggle to counter the major diseases and pests of agriculture will intensify sharply in the coming years. Avoidable losses in production will become increasingly intolerable as the demand for food rises. Further, there is the paradox that the risk of serious loss caused by plant parasites has been increased by many of the practices which are being used today to promote food production: the introduction of new cultivars for widespread use, more intensive cropping, the growing of more than one crop per year, the wider use of irrigation, and increased fertiliser application. There is further danger in the fact that many of the new cultivars, finely tuned to a narrow belt of temperature and water needs, are bred to give much enhanced yields in the climatic conditions for which they are programmed, but are likely, in more extreme weather, to fare worse than the older 'fail-safe' varieties which they replaced. A climatic fluctuation which simultaneously depressed yields over a wide area while favouring losses due to disease or pest could be disastrous. It is a risk for which students of climatic change should be alert.

References

Aspliden, C. I. H. and R. C. Rainey 1961. Desert Locust control. *WMO Bulletin* **10**, 155–61.

Beaumont, A. 1940. Potato blight and the weather. *Abstr. Trans Br. Mycol Soc.* **24**, 266.

Beaumont, A. 1947. The dependence on the weather of the dates of outbreak of potato blight epidemics. *Trans Br. Mycol Soc.* **31**, 45–53.

Bent, K. J. 1969. Fungicides in perspective. *Endeavour* **28**, 129–34.

Billing, E. 1980. Fireblight (*Erwinia Amylora*) and weather: a comparison of warning systems. *Ann. Appl. Biol.* **95**, 365–77.

Bourke, P. M. A. 1953. *Potato blight and the weather: a fresh approach*. Irish Met. Serv. Tech. Note 12.

Bourke, P. M. A. 1955. *The forecasting from weather data of potato blight and other plant diseases and pests*. WMO Tech. Note 10.

Bourke, P. M. A. 1957. *The use of synoptic weather maps in potato blight epidemiology*. Irish Met. Serv. Tech. Note 23.

Bourke, P. M. A. 1961. *Climatic aspects of the possible establishment of the Japanese beetle in Europe*. WMO Tech. Note 41.

Bourke, P. M. A. 1970. Use of weather information in the prediction of plant disease epiphytotics. *Ann. Rev. Phytopathol.* **12**, 345–70.

Bowden, J., P. H. Gregory and C. G. Johnson 1971. Possible wind transport of coffee leaf rust across the Atlantic Ocean. *Nature* (Lond.) **229**, 500–01.

Broadbent, L. 1949. Potatoes and the weather. *Q. J. R. Met. Soc.* **75**, 302–09.

Coakley, S. M. 1978. The effect of climate variability on stripe rust of wheat in the Pacific Northwest. *Phytopathology* **68**, 207–12.

Coakley, S. M. 1979. Climatic variability in the Pacific Northwest and its effect on stripe rust disease of winter wheat. *Climatic Change* **2**, 33–51.

Cox, A. E. and E. C. Large 1960. *Potato blight epidemics throughout the world*. Agriculture Handbook No 174. Washington, DC: US Department of Agriculture.

Crosier, W. 1934. *Studies in the biology of* Phytophthora infestans *(Mont.) de Bary*. Cornell University Agric. Exp. Sta. Memoir 155.

Dubey, R. C. and T. S. Godvindaswamy 1978. Regions in India favourable for the incidence and establishment of the Colorado beetle. *Ind. J. Met. Hydrol. Geophys.* **29**, 564–5.

Dubey, R. C. and S. Venkataraman 1980. A note on the likely areas of establishment of the Japanese beetle in India. *Mausam* **31**, 175–7.

Gettinby, G. and W. P. Gardiner 1980. Disease and incidence forecasts by means of climatic data. *Biomet.* **7**, 87–103.

Gibson, T. E. (ed.) 1978. *Weather and parasite animal disease*. WMO Tech. Note 159.

Goltz, S. M., F. Manzer and H. G. Schimmelpfennig 1978. Microcomputers – an application to potato late blight forecasting. *Am. Met. Soc. Bull.* **59**, 1511.

Graham, F. J. 1847. The potato disease. *J. R. Agric. Soc. Eng.* **7**, 357–91.

Grainger, J. 1953. Potato blight forecasting and its mechanism. *Nature* (Lond.) **171**, 1012–14.

Gregory, P. H. 1973. *Microbiology of the atmosphere.* London: Leonard Hill.

Hirst, J. M. and G. W. Hurst 1967. Long-distance spore transport. *Symp. Soc. Gen. Microbiol.* **17**, 307–44.

Hirst, J. M. and O. J. Stedman 1956. The effect of height of observation in forecasting potato blight by Beaumont's method. *Plant Pathol.* **5**, 35–40.

Hogg, W. H. 1970. Weather, climate and plant disease. *Met Mag.* **99**, 317–26.

Huff, F. A. 1963. Relation between leafhopper influxes and synoptic weather conditions. *J. Appl. Met.* **2**, 39–43.

Hurst, G. W. 1975. *Meteorology and the Colorado potato beetle.* WMO Tech. Note 137.

Johnson, C. G. 1969. *Migration and dispersal of insects by flight.* London: Methuen.

Jones, A. L., S. L. Lillevik, P. D. Fisher and T. C. Stebtaus 1980. A micro-computer based instrument to predict primary apple scab infection periods. *Plant Dis.* **64**, 69–71.

Krause, R. A. and L. B. Massie 1975. Predictive systems: modern approaches to disease control. *Ann. Rev. Phytopathol* **13**, 31–47.

Krause, R. A., L. B. Massie and R. A. Hyre 1975. Blitecast: a computerized forecast of potato late blight. *Plant Dis. Reptr.* **59**, 95–98.

MacHattie, L. B. 1979. Meteorology and forest insect control in Canada. *WMO Symp. on Forest Met.* **527**, 139–52.

Mahindapala, R. 1978. Epidemiology of maize rust. *Ann. Appl. Biol.* **90**, 155–61.

Miller, P. R. 1953. The effect of weather on diseases. *US Department of Agriculture Yearbook 1953*, 83–93.

Miller, P. R. 1964. Disease warning system organised for blue mold control. *Agricultural Chemicals* **19**(4), 22–3 and 131.

Miller, P. R. 1967. Plant disease epidemics: their analysis and forecasting. In *Papers presented at FAO Symposium Crop losses, Rome 1967*, 9–37.

Miller, P. R. and M. O'Brien 1952. Plant disease forecasting. *Bot. Rev.* **18**, 547–601.

Miller, P. R. and M. O'Brien 1957. Prediction of plant disease epidemics. *Ann. Rev. Microbiol.* **11**, 77–110.

Moore, W. C. 1940. Weather in relation to plant disease survey records. *Abstr. Trans Br. Mycol Soc.* **24**, 264.

Nutman, F. J. and F. M. Roberts 1969. Climatic conditions in relation to the spread of coffee berry disease since 1962 in the East Rift districts of Kenya. *E. Afr. Agric. & Forest J.* **37**, 118–27.

Ollerenshaw, C. B. and W. T. Rowlands 1959. A method of forecasting the incidence of fascioliasis in Anglesey. *Vet. Rec.* **71**, 591.

Omar, M. H. 1980. *Meteorological factors affecting the epidemiology of the cotton leaf worm and the pink bollworm.* WMO Tech. Note **167**.

Parry, W. H. 1979. Acclimatisation in the green spruce aphid. *Ann. Appl. Biol.* **92**, 299–306.

Pedgley, D. E. 1979. Weather during desert locust plague upsurges. *Phil Trans R. Soc. Lond.* **287**, 387–91.

Populer, Ch. 1964. Le comportement des épidémies de mildiou du tabac, *Peronospera tabacina* Adam. Part 1 la situation en Europe. *Bull. Inst. Agr. et Stat. Rech. Gembloux* **32**, 339–78.

Rainey, R. C. 1963. *Meteorology and the migration of desert locusts.* WMO Tech. Note 54.

Rayner, R. W. 1960. Rust disease of coffee. III. Spread of the disease. *World Crops* **12**, 222–4.

Ross, J. G. 1970. The Stormont 'wet day' forecasting system for fascioliasis *Br. Vet. J.* **126**, 401–08.

Scarpa, M. J. and L. C. Raniere 1964. The use of consecutive hourly dewpoints in forecasting downy mildew of lima beans. *Plant Dis. Reptr.* **48**, 77–81.

Schieber, E. 1972. Economic impact of coffee rust in Latin America. *Ann. Rev. Phytopathol.* **10**, 491–512.

Service, M. W. 1980. Effects of wind on the behaviour and distribution of mosquitoes and blackflies. *Int. J. Biomet.* **24**, 347–53.

Smith, L. P. 1970. *Weather and animal diseases.* WMO Tech. Note 113.

Sutton, T. B. and A. L. Jones 1975. Monitoring *Erwinia amylorara* populations on apple in relation to disease incidence. *Phytopathol.* **65**, 1009–12.

Taylor, A. E. R. and R. Muller (eds) 1974. The effect of meteorological factors upon parasites. *Symp. Brit. Soc. Parasit.* **12**.

Wallin, J. R. 1962. Summary of recent progress in predicting late blight epidemics in United States and Canada. *Am. Potato J.* **39**, 306–12.

Wallin, J. R. and J. A. Riley, 1960. Weather map analysis – an aid in forecasting potato late blight. *Plant Dis. Reptr.* **44**, 227–34.

Whyte, R. O. 1963. The significance of climatic change for natural vegetation and agriculture. In *Changes of climate*, UNESCO Arid Zone Res. Series. **20**, 381–6.

Woods, F. W. 1953. Disease as a factor in the evolution of forest composition. *J. Forest.* **51**, 871–3.

World Meteorological Organization 1963. *The influence of weather conditions on the occurrence of apple scab.* WMO Tech. Note 55.

World Meteorological Organization 1965. *Meteorology and the desert locust.* WMO Tech. Note 69.

World Meteorological Organization 1969. *Meteorological factors affecting the epidemiology of wheat rusts.* WMO Tech. Note 99.

Zadoks, J. C. 1965. Epidemiology of wheat rusts in Europe. *FAO Plant Prot. Bull.* **13**, 97–108.

Zadoks, J. C. 1967. International dispersal of fungi. *Neth. J. Plant Path.* **73** Suppl., 61–81.

Bibliography of papers by Professor Gordon Manley

GILLIAN M. SHEAIL, JOAN M. KENWORTHY and MICHAEL J. TOOLEY

This bibliography is based on a search in three libraries: Cambridge University Library, Durham University Library, and the National Meteorological Library, Bracknell. It extends the bibliography attached to the obituary in the *Transactions of the Institute of British Geographers* (1980, New Series Vol.5, No.4, 513–17). The absence of references for particular years led to a redoubled effort that was invariably rewarded, but there are probably gaps. The catalogued deposit of Professor Manley's papers in Cambridge University Library has yielded papers published in hitherto unsuspected places, as well as much unpublished material. Attention is drawn to this rich resource, and to the unpublished material in the National Meteorological Library (Manley 1969a, b, c, 1973). It had been hoped also to list all Professor Manley's contributions to *The Manchester Guardian* and *The Guardian* but the index to *The Guardian* ends in 1962, and the list at the end of the bibliography must be regarded as provisional.

1927 Appendix I. Survey. In Wordie, J. M. The Cambridge expedition to East Greenland in 1926. *Geog. J.* **70**, 241–4.

1927 Appendix IV. Weather. In Wordie, J. M. The Cambridge expedition to East Greenland in 1926. *Geog. J.* **70**, 249–52.

1927 Appendix VII. The pendulum observations. In Wordie, J. M. The Cambridge expedition to East Greenland in 1926. *Geog. J.* **70**, 260–62.

1930 Appendix III. Meteorology. In Wordie, J. M. Cambridge East Greenland expedition 1929: ascent of Petermann Peak. *Geog. J.* **75**, 498–502.

1931 Some notes on the maps of the county of Durham in the University Library. *Durham Univ. J.* **27** (2), 127–33.

1932 Pendulum observations at Sabine Island. *Meddr Grønland* **92**(2), 1–16.

1932 The weather of the High Pennines. *Durham Univ. J.* **28**(1), 31–2.

1932 Meteorological records from the northern Pennines. *Met. Mag., Lond.* **67**, 206–8.

1933 Notes on the weather in Upper Teesdale, 1933. *Durham Univ. J.* **28**(4), 304–7.

1933 Further records from the northern Pennines. *Met. Mag., Lond.* **68**, 180–81.

1934 On a means of transport. *Durham Univ. J.* **28**(6), 481–4.

1934 Saxton's survey of northern England. *Geog. J.* **83**, 308–16.

1934 The Plancius map of England, Wales and Ireland, 1592. *Geog. J.* **83**, 252–3.

1935 Some notes on the climate of north-east England. *Q. J. R. Met. Soc.* **61**, 405–10.

1936 The climate of the northern Pennines: the coldest part of England. *Q. J. R. Met. Soc.* **62**, 103–13; Discussion. *ibid.* 114–15.

1936 On atlases. *Durham Univ. J.* **29**(6), 439–43.

1936 On the instrumental climatology of the Durham district. *Durham Univ. J.* **30**(1), 43–8.

280   BIBLIOGRAPHY

1936   The earliest extant map of the County of Durham. *Trans Archaeol. Antiq. Soc. Durham* **7**, 278–87.

1937   Appendix III. Meteorology. In Wager, L. R. The Kangerdlugssuak region of East Greenland. *Geog. J.* **90**, 418–20.

1937   Snowfall in Scotland: problems and suggestions. *Scott. Geog. Mag.* **53**, 394–7.

1937   The weather of 1936 in the northern Pennines. *Met. Mag., Lond.* **72**, 8–10.

1937   Systematic records of British snowfall. *Met. Mag., Lond.* **72**, 231–2.

1938   Scottish snow. *Scott. Mountaineering Club J.* **21**, 366–7.

1938   Meteorological observations of the British East Greenland Expedition, 1935–1936, at Kangerdlugssuak, 68°10′N, 31°44′W. *Q. J. R. Met. Soc.* **64**, 253–76.

1938   The weather of 1937 in the northern Pennines. *Met. Mag., Lond.* **73**, 69–72.

1938   Appendix: snowfall and its relation to transport problems, with special reference to northern England. In RGS evidence to the Royal Commission. Memorandum on the geographical factors relevant to the location of industry. *Geog. J.* **92**, 522–6.

1938   Observations on the early cartography of the English hills. In *Comptes Rendus du Congrès International de Geographie*, Tome 11, Section IV, E. J. Brill (ed.), 36–42. Leiden: E. J. Brill.

1938   The City of Durham. *Geography* **23**, 147–55.

1939   On the occurrence of snow-cover in Great Britain. *Q. J. R. Met. Soc.* **65**, 2–24; Discussion. *ibid.* 25–7.

1939   The frequency of snow in Great Britain. *Wat. & Wat. Engng* **41**, 22–3.

1939   The helm wind of Crossfell. *Nature, Lond.* **143**, 377.

1939   High altitude records from the northern Pennines, 1938–39. *Met. Mag., Lond,* **74**, 114–17.

1939   Report on the snow survey of the British Hills, July 1938–July 1939. *Association for the study of Snow and Ice, Papers and Discussions* **1**, 18–20.

1940   The helm wind of the northern Pennines. *Met. Office Misc.* **387** (unpublished).

1940   Snowfall in the British Isles. *Met. Mag., Lond.* **75**, 41–48.

1941   Observations of snow-cover on the British mountains. *Q. J. R. Met. Soc.* **67**, 1–4.

1941   The Durham meteorological record 1847–1940. *Q. J. R. Met. Soc.* **67**, 363–80.

1941   Climate and agriculture in County Durham. In *The land of Britain. The report of the Land Utilisation Survey of Britain. Part 47. County Durham*, Ada Temple (ed.), 193–200. London: Geographical Publications.

1942   Meteorological observations on Dun Fell. *Q. J. R. Met. Soc.* **68**, 151–62; Discussion. *ibid.* 162–5.

1943   Further climatological averages for the northern Pennines, with a note on topographical effects. *Q. J. R. Met. Soc.* **69**, 251–61.

1943   Snow cover in the British Isles. *Met. Mag., Lond.* Limited typescript edition, 1–12. Repr. 1947 *Met. Mag., Lond.* **76**, 28–36.

1943   The climate of Cumberland and Westmorland. In *The land of Britain. The Report of the Land Utilisation Survey of Britain. Parts 49 and 50. Cumberland and Westmorland*, L. D. Stamp, G. Manley and E. Davies (eds), 274–80. London: Geographical Publications.

1944   Topographical features and the climate of Britain: a review of some outstanding effects. *Geog. J.* **103**, 241–58; Discussion. *ibid.* 258–63.

1944   John Dalton: 1766–1844. *Q. J. R. Met. Soc.* **70**, 235–9.

1944   Some recent contributions to the study of climatic change. *Q. J. R. Met. Soc.* **70**, 197–219.

1945 The effective rate of altitudinal change in temperate Atlantic climates. *Geog. Rev.* **35**, 408–17.

1945 Has the climate of N. W. Scotland deteriorated since 1700? *Q. J. R. Met. Soc.* **71**, 154–6.

1945 The helm wind of Crossfell 1937–1939. *Q. J. R. Met Soc.* **71**, 197–215; Discussion. *ibid.* 215–19.

1945 Problems of Scottish climatology. *Scott Geogr. Mag.* **61**, 73–6.

1946 Temperature trend in Lancashire, 1735–1945. *Q. J. R. Met. Soc.* **72**, 1–13.

1946 Variations in the length of the frost-free season. *Q. J. R. Met. Soc.* **72**, 180–84.

1946 Recent antarctic discoveries and some speculations thereupon. *Q. J. R. Met. Soc.* **72**, 307–17.

1946 Foreword by the President of the Royal Meteorological Society. *Weather, Lond.* **1**, 1.

1946 The centenary of rainfall observations at Seathwaite. *Weather, Lond.* **1**, 163–8.

1946 The climate of Northumberland, with notes on agriculture. In *The land of Britain. The report of the Land Utilisation Survey of Britain. Part 52. Northumberland*, L. D. Stamp (ed.), 429–39. London: Geographical Publications.

1946 The climate of Lancashire – past and present. *Trans Rochdale Lit. Sci. Soc.* **22**, 84–9.

1946 The climate of Lancashire. *Mem. Proc. Manchr. Lit. Phil Soc.* **87**, 73–95.

1947 A remarkable winter day on the High Pennines. *Weather, Lond.* **2**, 6–8.

1947 Snowfall and snow cover in the British Isles. *Weather, Lond.* **2**, 149.

1947 Tornadoes in England. *Weather, Lond.* **2**, 238–40.

1947 Looking back at last winter (a) February 1947: its place in meteorological history. *Weather, Lond.* **2**, 267–72.

1947 Dorothy Wordsworth's weather. *Weather, Lond.* **2**, 361–4.

1947 The geographer's contribution to meteorology. *Q. J. R. Met. Soc.* **73**, 1–10.

1948 Records of snow cover on the Scottish mountains. *Met. Mag., Lond.* **77**, 270–72.

1948 On the trend of temperature in N. W. Europe, 1720–1750. *Q. J. R. Met. Soc.* **74**, 119–22.

1948 Glacier variations and climatic fluctuations in Britain, U.G.G.I. *Assoc. Hydrol. Sci., Assoc. Gen., Oslo* **2**, 304–05.

1948 George Smith the geographer and his ascent of Crossfell. *Trans Cumb. Westm. Antiq. Archaeol Soc.* **48**, 135–44.

1949 Microclimatology – local variations of climate likely to affect the design and siting of buildings. *J. R. Inst. Br. Archit.* **56**, 317–23.

1949 The snowline in Britain. *Geog. Annlr.* **31**, 179–93.

1941 The extent of the fluctuations shown during the 'instrumental' period in relation to post-glacial events in N. W. Europe. *Q. J. R. Met Soc.* **75**, 165–71.

1949 Fanaråken: the mountain station in Norway. *Weather, Lond.* **4**, 352–4.

1949 Part I. The atmosphere. In *Physical geography*, 2nd edn, P. Lake, (ed.), revised and enlarged by J. A. Steers, W. V. Lewis and G. Manley, 3–137. Cambridge: Cambridge University Press.

1949 Glaciological research on the North Atlantic coasts: a review. *Geog. Rev.* **39**, 136–8.

1950 *Degrees of freedom. An inaugural lecture given at Bedford College for Women, January 27th 1949.* London: Christopher Johnson.

1950 Some consequences of the relation between glacier variations and climatic fluctuations in Britain. *J. Glaciol.* **1**, 352–6.

1950   On British climatic fluctuations since Queen Elizabeth's day. *Weather, Lond,* **5**, 312–18.

1950   Discussion: climatic change. In *Centenary Proceedings of the Royal Meteorological Society,* 225–8. London.

1951   The range of variation of the British climate. Part I: climatic fluctuations in modern times. Part II: the dimensions of the changes in the British climate before the 'climatic optimum'. *Geog. J.* **117**, 43–65; Discussion. *ibid.* 65–8.

1951   Climatic fluctuations: a review. *Geog. Rev.* **41**, 656–60.

1951   Royal Society's Conversazione, 24 May 1951. Conversazione included item 26 'Past climatic change as revealed by deep sea cores'. *Weather, Lond.* **6**, 208–09.

1952   John Dalton's snowdrift. *Weather, Lond.* **7**, 210–12.

1952   *Climate and the British scene.* London: Collins.

1952   Thomas Barker's meteorological journals, 1748–63 and 1777–89. *Q. J. R. Met. Soc.* **78**, 255–9.

1952   Britain's sunshine. *Geog. Mag.* **25**, 59–63.

1952   Variations in the mean temperature of Britain since glacial times. *Geol Rdsch.* **40**, 125–7.

1952   The weather and diseases: some 18th century contributions to observational meteorology. *Notes and Records. Royal Society of London* **9**, 300–07.

1952   A warm decade. *Weather, Lond.* **7**, 242–3.

1953   The mean temperature of central England, 1698–1952. *Q. J. R. Met. Soc.* **79**, 242–261; Discussion. *ibid.* 558–67.

1953   A prospect of winter. *Weather, Lond.* **8**, 357–60.

1953   Reviews of modern meteorology – 9. Climatic variation. *Q. J. R. Met Soc.* **79**, 185–209.

1953   (with Whiten, A. J.) World-wide distribution of pressure. *Q. J. R. Met Soc.* **79**, 419.

1954   Climatic change: the problem of transatlantic correlation. In *Proceedings of the Toronto Meteorological Conference 1953,* 199–206. London: Royal Meteorological Society, and Boston, Mass: American Meteorological Society.

1954   Changes in world glaciation. A geophysical discussion of the Royal Astronomical Society in conjunction with the British Glaciological Society held on January 29, 1954. *Nature, Lond.* **173**, 1206–08.

1954   The winter of 1798–9. *Weather, Lond.* **9**, 254.

1955   A climatological survey of the retreat of the Laurentide Ice Sheet. *Am. J. Sci.* **253**, 256–73.

1955   On the occurrence of ice domes and permanently snow-covered summits. *J. Glaciol.* **2**, 453–6.

1955   Constantia Orlebar's weather book 1786–1808. *Q. J. R. Met. Soc.* **81**, 622–5.

1955   The remarkable weather of 1954. *Manchr Guardian Weekly* **72**(1), 14.

1955   Arctic air over England. *Manchr Guardian Weekly* **72**(11), 7.

1955   On disappointing summers. *Scott. Mountaineering Club J.* **25**, 334–41.

1956   The climate at Malham Tarn. In *Annual Report of the Field Studies Council, 1955–1956,* 43–53.

1956   Future lines of progress in glaciology. *J. Glaciol.* **2**, 697–8.

1956   British summers of the past. *Weather, Lond.* **11**, 335.

1956   (with Jackson, P. S.) Severe winters in 1813 and 1860. *Weather, Lond.* **11**, 134–5.

1957   Climatic fluctuations and fuel requirements. *Scott. Geog. Mag.* **73**, 19–28.

1957   Climatic fluctuations and fuel requirements. *Advmt Sci., Lond.* **13**, 324–6.

1957   Glaciers and the changing climate. *New Scientist* **1** (17), 33–5.

1957   The Meteorological Office Report for 1955–56. *Nature, Lond.* **179**, 765–6.

1958   On the frequency of snowfall in metropolitan England. *Q. J. R. Met. Soc.* **84**, 70–72.

1958   Early meteorological journals; some further notes. *Q. J. R. Met. Soc.* **84**, 75–7.

1958   (with Bonacina, L. C. W.) Snowfall frequency in metropolitan England. *Q. J. R. Met. Soc.* **84**, 283–4.

1958   Bad summers. *New Scientist* **4**, 1054–6.

1958   Nuclear tests and the weather, reply by Professor Gordon Manley. *New Scientist* **4** 1339.

1958   The great winter of 1740. *Weather, Lond.* **13**, 11–17.

1958   The revival of climatic determinism. *Geog. Rev.* **48**, 98–105.

1958   Studies on the frequency of snowfall in England, 1668–1956, UGGI. *Assoc. Int. Hydrol. Sci. C. R. Rapp. Assoc. Gen. Toronto, 1957* **4**, 40–45.

1959   Temperature trends in England, 1698–1957. *Arch. Met. Geophys. Bioklim. Ser. B* **9**, 413–33.

1959   The Late-Glacial climate of north-west England. *Lpool Manchr Geol J.* **2**, 188–215.

1960   Climate and the International Geophysical Year. *Science Survey* **1**, 267–78.

1961   Late and Postglacial climatic fluctuations and their relationship to those shown by the instrumental record of the past 300 years. *Ann. NY Acad. Sci.* **95**, 162–72.

1961   A preliminary note on early meteorological observations in the London region, 1680–1717, with estimates of the monthly mean temperatures, 1680–1706. *Met. Mag., Lond.* **90**, 303–10.

1961   The Earth's climate. In *Discovery, fifteen talks given on television in the Discovery Series for the Granada Television Network*, 87–97. London: Methuen.

1961   Meteorological factors in the great glacier advance (1690–1720), UGGI. *Assoc. Int. Hydrol. Sci. Assoc. Gen. Helsinki, 1960* **54**, 388–91.

1961   Solar variations, climatic change and related geophysical problems. *Nature, Lond.* **190**, 967–8.

1961   *Geography: our planet, its peoples and resources*, G. Manley, Sir Gerald Barry, J. Bronowski, J. Fisher and Sir Julian Huxley (eds). London: Macdonald.

1961   Geographers at work. In *Geography, our planet, its people and resources*, G. Manley *et al.* (eds), 317–18. London: Macdonald.

1962   Early meteorological observations and the study of climatic fluctuation. *Endeavour* **21**, 43–50.

1962   The Late-Glacial climate of the Lake District. *Weather, Lond.* **17**, 60–4.

1962   A venturesome Victorian. *Weather, Lond.* **17**, 340–1.

1962   A Victorian explorer of the upper air. *New Scientist* **15**, 607–9.

1963   Seventeenth-century London temperatures: some further experiments. *Weather, Lond.* **18**, 98–105.

1963   The climatologist's view. In *The long winter 1962–3*, 14–18. Manchester: Manchester Guardian.

1963   Glaciology in the last 25 years. *Geog. Mag.* **35**, 729–38.

1963   Snowfall in Britain. *New Scientist* **17**, 68–70.

1963   The weather in Britain. *Anglia* **8**, 34–43 (in Russian).

1964   The evolution of the climatic environment. In *The British Isles: a systematic geography*, J. W. Watson and J. B. Sissons (eds), 152–76. London: Nelson.

1964   Cold winters in Manchester. *Weather, Lond.* **19**, 96.

1965 Possible climatic agencies in the development of Post-Glacial habitats. *Proc. R. Soc. Ser. B* **161**, 363–75.

1966 Climate and landscape architecture. *J. Inst. Landscape Archit.* **74**, 4–7.

1966 The problem of the climatic optimum: the contribution of glaciology. In *World climate from 8000 to 0 BC*, J. S. Sawyer (ed.), 34–9. London: Royal Meteorological Society.

1967 *Weather comes of age. Weather, Lond.* **22**, 171–3.

1967 This North-western environment. In *Inaugural lectures 1965–1967, University of Lancaster*, 1–11. Lancaster: University.

1967 Climate in Britain. In *Congress Proceedings of the 20th International Geographical Congress*, J. W. Watson (ed.), 34–45. London: Nelson.

1968 Dalton's accomplishment in meteorology. In *John Dalton and the progress of science*, D. S. L. Cardwell (ed.), 140–58. Manchester: University Press.

1968 Climatic variation (climatic change). In *Encyclopaedic Dictionary of Physics* vol.3, 25–8. London: Pergamon.

1969a *Corrected temperature data 1771–1798 at Lyndon in Rutland, taken from Thomas Barker's observations.* National Meteorological Library, Bracknell (unpublished).

1969b *Daily observations 1782–1796 kept by William Godschell at Albury near Guildford and at Huxham near Plymouth 1725–1752.* National Meteorological Library, Bracknell (unpublished).

1969c *Monthly means 1786–1833, kept by 'Cary in the Strand'.* National Meteorological Library, Bracknell (unpublished).

1969 Snowfall in Britain over the past 300 years. *Weather, Lond.* 24, 428–37.

1970 Climate in Britain over 10 000 years. *Geog. Mag.* 43, 100–7.

1970 The climate of the British Isles. *World Survey of Climatology* **5**, 81–133.

1971 The mountain snows of Britain. *Weather, Lond.* **26**, 192–200.

1971 Interpreting the meteorology of the Late and Post-Glacial. *Palaeogeogr., Palaeoclimatol., Palaeoecol.* **10**, 163–75.

1971 Scotland's semi-permanent snows. *Weather, Lond.* **26**, 458–71.

1972 Manchester rainfall since 1765. *Mem. Proc. Manchr Lit. Phil Soc.* **114**, 70–89.

1972 The Society's headquarters. *Weather, Lond.* **27**, 88–9.

1972 Scotland's semi-permanent snows. *Scott. Mountaineering Club J.* **30**, 4–15.

1972 Foreword (introductory preface). In *Times of feast, times of famine: a history of climate since the year 1000*, E. Le Roy Ladurie. xiii–xvi. London: George Allen & Unwin.

1973 Climate. In *The Lake District*, W. H. Pearsall and W. Pennington, 106–20. London: Collins.

1973 *Snow in London 1669–1958.* National Meteorological Library. Bracknell (unpublished).

1973 Climate in Britain over 10 000 years. In *Man made the land*, A. R. H. Baker and J. B. Harley (eds), 9–21. Newton Abbot: David and Charles.

1974 *Enjoy Cumbria's climate.* Windermere: Cumbria Tourist Board.

1974 Central England temperatures; monthly means 1659–1973. *Q. J. R. Met. Soc.* **100**, 389–405.

1975 Reply to 'Central England temperatures; monthly means of the Radcliffe Meteorological Station, Oxford' C. G. Smith. *Q. J. R. Met. Soc.* **101**, 387–9.

1975 Fluctuations of snowfall and persistence of snow-cover in marginal oceanic climates. In *Proceedings of the WMO/IAMP Symposium on Long Term Climatic Fluctuations, Norwich, 18–23 August 1975*, 183–7. Geneva: WMO.

1975 Snowfalls in June. *Weather, Lond.* **30**, 308.

1975 1684: the coldest winter in the English instrumental record. *Weather, Lond.* **30**, 382–8.

1975 Weather and climate of the Lake Counties. In *Lake District. National Park guide* No. **6**. Countryside Commission (ed.), 129–38. London: HMSO.

1975 Manchester rainfall since 1765: further comments, meteorological and social. *Mem. Proc. Manchr Lit. Phil Soc.* **117**, 93–103.

1976 The summer of 1976. *Weather, Lond.* **31**, 395–6.

1977 Sir Daniel Fleming's meteorological observations at Rydal 1689–1693. *Trans Cumb. Westm. Antiq. Archaeol Soc.* **77**, 121–6.

1977 A note on the summer of 1976 in Lancashire. *Mem. Proc. Manchr Lit. Phil Soc.* **119**, 96–9.

1977 1684: the coldest winter in the English instrumental record. *Marine Observer* **47**, 77–81.

1977 Annual rainfall in England since 1725: some comments. *Q. J. R. Met. Soc.* **103**, 820–22.

1978 Reflections on climatological research. A climatologist's view. In *Climate change and variability, a southern perspective*, A. B. Pittock, L. A. Frakes, D. Jenssen, J. A. Peterson and J. Zillman (eds), 360–3. Cambridge: Cambridge University Press.

1978 Meteorological observations on Royal Deeside. *Weather, Lond.* **33**, 457–9.

1978 Variation in the frequency of snowfall in east-central Scotland, 1708–1975. *Met. Mag., Lond.* **107**, 1–16.

1979 The climatic environment of the Outer Hebrides. *Proc. R. Soc. Edinb.* **77** B, 47–59.

1979 Temperature records on Fountains Fell, with some Pennine comparisons. *Field Studies* **5**, 85–92.

1980 The northern Pennines revisited: Moor House, 1932–1978. *Met. Mag., Lond.* **109**, 281–92.

1980 Cold winters at Oxford. *Weather, Lond.* **35**, 182–5.

1981 The use of archives and written records in meteorological research. *Archives* **15**(65), 3–10.

List of articles by Gordon Manley in *The Manchester Guardian*

PETER McNIVEN *and* MICHAEL J. TOOLEY

1954	26 August	p. 4, cols 6 & 7	The summer of 1954: not quite the worst
	9 September	p. 6, cols 4 & 5	Summer in England (letters to the Editor)
	1 December	p. 12, cols 5 & 6	Depressions off their courses upset November weather: on the way to a year of record wetness?
	31 December	p. 4, cols 6 & 7 p. 5, col. 6	The weather of 1954: a meteorological retrospect
1955	11 March	p. 6, cols 6 & 7 p. 7, col. 5	Arctic air over England: the lion in March
	26 May	p. 6, cols 6 & 7	Prospects for summer: a meteorological survey
	30 July	p. 1, cols 6 & 7	The warmest July for 21 years?
	24 August	p. 4, col. 6	Near the maximum hours of sunshine: present and previous summers compared
	1 September	p. 1, cols 4 & 5	The summer in statistics: possibility of 200 year record in NW
	17 December	p. 6, cols 6 & 7	The weather ahead: some puzzling questions
	21 December	p. 10, cols 5 & 6	Snow deceived forecasters: little changes had big effect
1956	2 January	p. 10, cols 5 & 6	The sunniest year of the century in the north-west? But 1955 was far from the warmest
	4 February	p. 6, cols 6 & 7	An Arctic anticyclone – our coldest weather
	29 February	p. 6, cols 6 & 7	The weather in February: fifth coldest in 250 years
	1 May	p. 6, cols 6 & 7	This year's dry spring: prospects of drought
	30 July	p. 12, cols 5 & 6	All on a summer's day . . . retreat of an anticyclone brings remarkable July weather

	4 August	p. 4, cols 6 & 7	Chances of a hot day and August floods
	1 September	p. 4, cols 6 & 7	End of a dismal August: faint hopes for September
	31 December	p. 1, col. 7	Perverse and cool: the weather in 1956
1957	9 February	p. 5, cols 1 & 2	Assessing the worth of a January spring: milder weather from the Atlantic
	19 March	p. 5, cols 1 & 2	Cautious optimism flowers in spring warmth: but short cold 'snaps' likely
	22 April	p. 1, col. 4 p. 2, cols 4 & 5	Weather in a pleasant mood: will 1921 experience be repeated this year?
	6 June	p. 5, cols 1 & 2	Best-season for a chance of sunshine: Whitsuntide's good record
	2 August	p. 1, col. 3	More settled in the North: week-end prospects
	10 October	p. 10, cols 3 & 5	The ideal indoor climate
	30 November	p. 6, cols 6 & 7	This curious autumn: prospect of a cold winter
1958	4 January	p. 4, cols 6 & 7	The weather in 1957: two notable months
	15 February	p. 4, cols 4, 5 & 6	Winds blowing hot and cold: the erratic British winter
	3 April	p. 5, cols 1 & 2	Meteorologist makes March journey: no escaping the weather
	26 May	p. 3, cols 1 & 2	Quest for good weather at Whitsuntide: England's favoured north-west
	13 September	p. 4, cols 6 & 7	Not so bad after all? A summer of ill repute
	13 December	p. 4, cols 6 & 7	Onward from the fog: cold winter in prospect?
	31 December	p. 4, cols 4 & 5	Wet summer, muddy autumn: mild end to a disappointing year
1959	21 March	p. 3, cols 1 & 2	More cold due before the summer: chance of a drier June
	25 April	p. 6, cols 6 & 7	An umbrella in reserve: guesses about the summer
	1 August	p. 4, cols 6 & 7	Bank Holiday weather: another vintage summer?
	29 September	p. 5, cols 1 & 2	Second, sixth or 18th best summer: measuring months in the sun
	10 October	p. 6, cols 4 & 5	The best summer ever known: 1959's place in weather history (Letters to the Editor)

	5 December	p. 4, cols 6 & 7	The prospects of fog: and steps to avoid it
1960	1 January	p. 16, cols 3 & 4	11 months warmer than average: stormy December ended splendid year
	2 April	p. 4, cols 6 & 7	The English spring: is it getting warmer?
	11 June	p. 6, cols 6 & 7	After the warm spring: an alpha-minus summer?
	1 August	p. 3, cols 3, 4 & 5	This English summer
	31 December	p. 1, col. 7 p. 2, cols 4 & 5	Rainy but warmer: 1960 maintained the trend
1961	3 March	p. 3, cols 1 & 2	Mildest February since 1869? Record-seekers count the points
	8 April	p. 3, cols 5 & 6	Catching the sun on the Celtic fringe: chance of brightness
	15 July	p. 1, col. 5	Under the sway of Atlantic depressions
	22 July	p. 6, cols 3, 4 & 5	St Swithin and the summer outlook
	26 August	p. 4, cols 3, 4 & 5	Well dressed for a windy August
	7 September	p. 8, col. 4	Warm and gusty winds (Letters to the Editor)
	17 December	p. 6, cols 3, 4 & 5	Will it be a cold winter?

Author index

Subject and place index

Italic numerals refer to a page on which a figure, figure caption or table is indexed.